The Most Successful Way of JEE Preparation

DPP

Daily Practice Problems

JEE Main & Advanced

Modern Physics & Semiconducting Devices

with Communication System

PHYSICS Vol-9

Deepak Paliwal

arihant

ARIHANT PRAKASHAN (SERIES), MEERUT

DPP
Daily Practice Problems

arihant

ARIHANT PRAKASHAN (SERIES), MEERUT

Administrative & Production Offices

Corporate Office 'Ramchhaya' 4577/15, Agarwal Road, Darya Ganj, New Delhi -110002
Tele: 011- 47630600, 43518550; Fax: 011- 23280316

Head Office Kalindi, TP Nagar, Meerut (UP) - 250002
Tel: 0121-2401479, 2512970, 4004199; Fax: 0121-2401648

Sales & Support Offices

Agra, Ahmedabad, Bengaluru, Bareilly, Chennai, Delhi, Guwahati, Hyderabad, Jaipur, Jhansi, Kolkata, Lucknow, Meerut, Nagpur & Pune

ISBN : 978-93-13193-39-5

Published by Arihant Publications (I) Ltd.

Production Team

Publishing Managers	Mahendra Singh Rawat & Keshav Mohan	*Inner Designer*	Mazher Chaudhary
Project Head	Sorabh Chaudhary	*Page Layouting*	Sunil Sharma, Hemant
Cover Designer	Shanu Mahad	*Proof Reading*	Ashish

For further information about the products from Arihant, log on to www.arihantbooks.com or email to info@arihantbooks.com

/arihantpub   /@arihantpub   Arihant Publications   /arihantpub

ABOUT DPP

Daily Practice Problems

Indian Institutes of Technology better known as IITs have always been the dream destination of each science group students who want to pursue their career as Engineers. These are the institutes which have got very high value not even in India but in the whole world. To get the admission in such institutes one need to face entrance Examinations; JEE Main & JEE Advanced, out of which JEE Main is considered as the preliminary exam & JEE Advanced is the final exam to get into the IITs. Approximately per year 12 lakh students enroll themselves for JEE Main and only the top 2.25 lakh in JEE Main merit can attempt for JEE Advanced, the fact clearly express the degree of toughness of JEE Main & JEE Advanced.

To have success in such an exam one needs to be very planned & systematic in its preparation. A very popular wise saying 'Rome was not built in a day' suggests that to achieve a big target like IITs we need to work continuously. For continuous work, one need to have proper planning that how much work he/she has to do on regular basis, i.e., on daily basis. Each student preparing for JEE has many books but unfortunately none of them gives the students a plan for daily studies, which is the most effective success key for JEE.

Keeping the daily practice needs of the students in mind we have come up with DPP Series (Daily Practice Problems) having approximately 26 books. The primary focus of this series is on achieving success through practice & proper planning. For that we have divided the whole syllabus of JEE Main & JEE Advanced; Physics, Chemistry & Mathematics, into different volumes (around 8 volumes of Mathematics & 9 volumes each of Physics & Chemistry), each volume has Daily Practice Problems Sheets having specific questions on various Topics of the individual chapter of the volume ensuring the complete Practice & Assessment of the topic.

DPP
Daily Practice Problems

In this way, we have tried to give you an effective daily planning for JEE keeping in mind the daily hours that a student can devote to his practice, you are just required to practice one sheet of each subject on daily basis. This habit of daily practice will keep you away from exam stress and thoughts like; how much I have prepared/am I preparing properly or not/ do I need to study more on daily basis or what I am doing is ok, etc., which regularly tease your mind and can give you panic at any stage of your preparation.

Some special features that we have tried to incorporate in our Daily Practice Problems (DPP) series are discussed below

1. Micro Level coverage of all the topics of a Chapter/ Unit given in JEE Main & Advanced Syllabus.
2. Each DPP has various types of questions; Subjective Questions, Single/ Multiple Correct Options, Statements type, Comprehension based, Integer Value Answer type, Matching, etc., making you practice & ready for all the question formats of JEE.
3. New & innovative questions in each DPP challenging & improving your problem solving skills.
4. Along with Topical Coverage, Revisal DPPs for JEE Main & JEE Advanced with each Chapter, covering multidisciplinary questions proving comprehensive practice.
5. JEE Main & JEE Advanced Archive (collection of last 13 years' exams questions) with each chapter.
6. Complete Solutions for each DPP.

We hope that DPP series will prove to be very beneficial for the students preparing for JEE, by giving them daily practice in a systematic & planned manner. The suggestions/criticism from the students, teachers & parents about DPP series will be welcomed whole heartedly and we promise to update this series accordingly from time to time.

DPP
Daily Practice Problems

DAILY TASKS...

1 DUAL NATURE OF MATTER AND RADIATION

2 ATOMS

DPP
Daily Practice Problems

3 NUCLEI

4 ELECTRONIC DEVICES

5 COMMUNICATION SYSTEM

1. Dual Nature of Matter and Radiation

DPP

DAILY PRACTICE PROBLEMS

DPP-1 Dual Nature of Electromagnetic Waves

Subjective Questions

Working Space

Direction *(Q. Nos. 1-3) These questions are subjective in nature, need to be solved completely on notebook.*

1 Let electric field of an electromagnetic wave in vacuum is

$$\mathbf{E} = \left(3.1\,\frac{\text{N}}{\text{C}}\right)\cdot\cos\left[\left(1.8\,\frac{\text{rad}}{\text{m}}\right)y + \left(5.4\times10^{8}\,\frac{\text{rad}}{\text{s}}\right)t\right]\cdot\hat{\mathbf{i}}$$

i. What is the direction of travel of wave?

ii. What is wavelength of wave when it passes through a medium of refractive index $\frac{3}{2}$?

iii. What is the frequency of wave?

iv. What change in frequency occurs, when wave travelling through vacuum passes into a medium of refractive index 3/2?

v. What is the amplitude of magnetic field of wave?

vi. Give an expression for magnetic part of wave.

vii. What is the energy density of wave?

viii. What is the energy imparted by the wave per square metre in case of normal incidence?

ix. What radiation pressure this wave causes over each square metre of a surface in case of the

a. total reflection?

b. total absorption?

c. 50% absorption and 50% reflection, by the surface?

2 A 30 pF air gap capacitor has circular plates of area $A = 100\ \text{cm}^2$. A 70 V battery with a 2Ω series resistance charges this capacitor. At $t = 0$ s, switch S is closed.

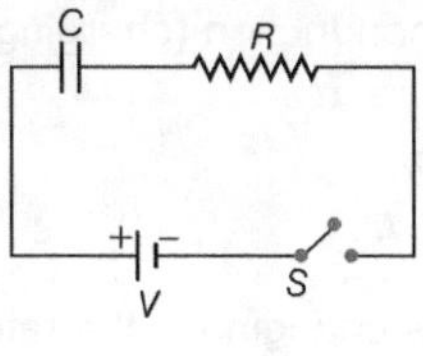

Calculate :

i. Current flowing in the circuit at $t = 0$ s.

ii. Value of displacement current at $t = 0$ s.

iii. Rate of change of electric field between the plates.

iv. Determine the magnetic field between the plates (neglect fringing).

Working Space

3 General theory of relativity given in 1916 by Einstein, gives principle of equivalence and it follows from equivalence principle that light should be subjected to gravity. This prediction is confirmed by taking photographs of distant stars.

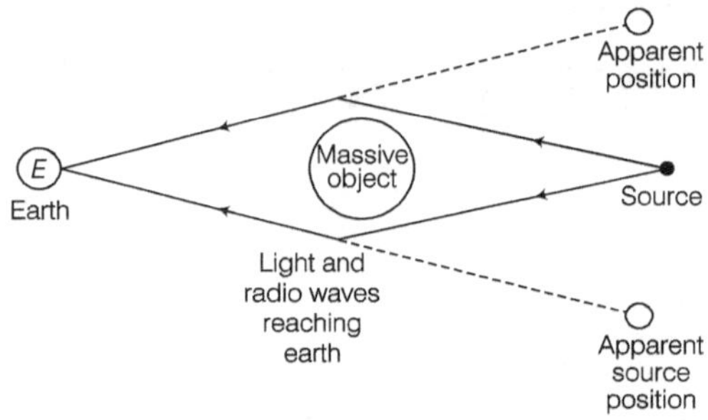

A massive object like a galaxy or a black hole deviates light from a distant pulsar. Light received on earth seems to come (identically) from two or more identical source.

If the mass of Sun is 2×10^{30} kg and radius is 7×10^{8} m, find the approximate red shift in light of wavelength 500 nm emitted from the Sun. For this purpose, assume particle nature of light and take moving mass of a photon.

Only One Option Correct Type

Direction (Q. Nos. 4-11) *This section contains 8 multiple choice questions. Each question has four choices (a), (b), (c) and (d), out of which ONLY ONE is correct.*

4 Radiation from the Sun reaches Earth at rate of 1350 $Js^{-1}m^{-2}$. The magnitudes of electric and magnetic fields are

a. 1.01×10^{6} Vm^{-1}, 3.37×10^{-3} T

b. 3.37×10^{3} Vm^{-1}, 1.01×10^{-6} T

c. 1.01×10^{3} Vm^{-1}, 3.37×10^{-6} T

d. 3.37×10^{6} Vm^{-1}, 1.01×10^{-3} T

5 Assume light as particles, then force required to bring a light's particle to rest will be

a. $\frac{\Delta p}{U}$

b. $\frac{\Delta U}{p}$

c. $\frac{\Delta U}{c}$

d. $\frac{\Delta p}{c}$

6 For a parallel plate capacitor with square plates of side 5.8 cm, the electric field is changing at a rate of 2×10^{6} $Vm^{-1}s^{-1}$. The value of conduction (charging) current at that instant is

a. 1.2×10^{15} A

b. 6×10^{-8} A

c. 1.1×10^{-5} A

d. 1.2×10^{-8} A

7 Potential difference across plates of a capacitor 6μF is changing at the rate of 12 Vs^{-1}. Displacement current at that instant will be

a. 1.2 A

b. 12×10^{-6} A

c. 7.2×10^{-6} A

d. 72×10^{-6} A

Working Space

8 A point source emits light energy uniformly in all directions at an average rate of P_0 with a single frequency f. The magnitude of peak electric field vector at a distance r from the source will be

a. $\sqrt{\dfrac{2\mu_0 c P_0}{\pi r^2}}$
b. $\sqrt{\dfrac{\mu_0 c P_0}{3\pi r^2}}$
c. $2\sqrt{\dfrac{\mu_0 c P_0}{\pi r^2}}$
d. $\sqrt{\dfrac{\mu_0 c P_0}{2\pi r^2}}$

9 What must be the maximum power level for a 4G relay tower to avoid breakdown of air (air breaks down at $3 \times 10^6\ \text{Vm}^{-1}$) at a distance of 0.5 m from the transmitting tower?

a. 40×10^{10} W
b. 400×10^{10} W
c. 4×10^{10} W
d. 0.4×10^{10} W

10 Estimate the force exerted by a 75 W bulb on your finger-tip, when tip is at a distance of 8 cm from the bulb. (Assume 100% absorption)

a. 1 N
b. 3×10^{-10} N
c. 4×10^{-6} N
d. 2×10^{-4} N

11 Let there are two bulbs in your house, one is red and other one is blue. Both bulbs are of same power. If n_r and n_b are the number of photons per second reaching towards you in a certain time, then

a. $n_r = n_b$
b. $n_r > n_b$
c. $n_r < n_b$
d. $n_r \cdot n_b = c^2$

One or More than One Options Correct Type

Direction (Q. Nos. 12-19) *This section contains 8 multiple choice questions. Each question has four choices (a), (b), (c) and (d), out of which ONE or MORE THAN ONE are correct.*

12 An electromagnetic wave travels along z-direction. Electric vector of wave is given by $\mathbf{E} = (E_1\hat{\mathbf{i}} + E_2\hat{\mathbf{j}})\cos(kz - \omega t)$. Mark the correct options.

a. Associated magnetic field is $\mathbf{B} = \dfrac{1}{c}(E_1\hat{\mathbf{i}} + E_2\hat{\mathbf{j}})\cos(kz - \omega t)$
b. Associated magnetic field is $\mathbf{B} = \dfrac{1}{c}(E_1\hat{\mathbf{i}} + E_2\hat{\mathbf{j}})\cos(kz + \omega t)$
c. Associated magnetic field is $\mathbf{B} = \dfrac{1}{c}(E_1\hat{\mathbf{i}} - E_2\hat{\mathbf{j}})\cos(kz - \omega t)$
d. Given electromagnetic wave is plane polarised

13 A charged particle oscillates about its mean position with a frequency of 10^9 Hz. The electromagnetic waves produced

a. will have a frequency of 10^9 Hz
b. will have a frequency of 2×10^9 Hz
c. will have a wavelength of 0.3 m
d. fall in the region of radio waves

14 The source of electromagnetic waves can be

a. charge moving with a constant speed
b. charge moving with a constant velocity
c. charge at rest
d. charge falling in an electric field

Working Space

15 An electromagnetic wave of intensity I falls on a surface kept in vacuum and exerts a radiation pressure p on it. Then, mark the correct options.

a. Radiation pressure is $\frac{I}{c}$ if wave is totally absorbed

b. Radiation pressure is $\frac{I}{c}$ if wave is totally reflected

c. For a reflecting surface, radiation pressure is more than an absorbing surface

d. For any real surface, radiation pressure is more than $\frac{I}{c}$

16 A microwave oven heats up a food item. Then, mark the correct options.

a. Frequency of electromagnetic waves used is 2450 MHz

b. Quantum energy of a microwave photon is about 1×10^{-5} eV

c. Radiation interaction produces rotation and torsion of water molceules

d. Frequency of microwave matches the natural oscillating frequency of water molecules

17 A variable frequency of AC source is connected to a capacitor. On decreasing the frequency of supply,

a. both conduction and displacement current increase

b. displacement current decreases

c. conduction current decreases

d. circuit reactance decreases

18 A plane electromagnetic wave travelling along z-direction is described by

$$\mathbf{E} = E_0 \sin(kz - \omega t) \cdot \hat{\mathbf{j}}$$

and

$$\mathbf{B} = B_0 \sin(kz - \omega t) \cdot \hat{\mathbf{i}}$$

Then, mark the correct options.

a. Average energy density of wave is

$$x_{av} = \frac{1}{2} \varepsilon_0 E_0^2 + \frac{1}{2} \frac{B_0^2}{\mu_0}$$

b. Average energy density of wave is

$$u_{av} = \frac{1}{4} \varepsilon_0 E_0^2 + \frac{1}{4} \frac{B_0^2}{\mu_0}$$

c. Average intensity of wave is

$$I_{av} = \frac{1}{2} c \, \varepsilon_0 E_0^2$$

d. Average intensity of wave is

$$I_{av} = \frac{1}{4} c \, \varepsilon_0 E_0^2$$

19 A 100 mW laser beam falls over a surface of $1 m^2$. Then, choose the correct options.

a. If surface is a perfect reflector, then force on the surface is 6.67×10^{-10} N

b. If surface is a perfect absorber, then force on the surface is 3.33×10^{-10} N

c. If surface is a perfect reflector, then force on the surface is 3.33×10^{-10} N

d. If surface is a perfect absorber, then force on the surface is 6.67×10^{-10} N

Working Space

Matching List Type

Direction *(Q. Nos. 20-22) Choices for the correct combination of elements from Column I and Column II are given as options (a), (b), (c) and (d), out of which one is correct.*

20 Match the Column I with Column II and mark the correct option from the codes given below.

	Column I		Column II
i.	Wave nature of light	p.	Young's double slit experiment
ii.	Particle nature of light	q.	Photoelectric effect
iii.	Wave nature of electrons	r.	Electron microscope
iv.	Particle nature of electrons	s.	Cathode rays

Codes

	i	ii	iii	iv		i	ii	iii	iv
a.	p	q	s	r	b.	q	p	s	r
c.	q	p	r	s	d.	p	q	r	s

21 Match the frequency values given in Column I with the region of electromagnetic spectrum given in Column II and mark the correct option from the codes given below.

	Column I		Column II
i.	2 MHz	p.	Gamma ray
ii.	2 GHz	q.	X-ray
iii.	2 THz	r.	Microwave
iv.	2 EHz	s.	Infrared ray
v.	2 ZHz	t.	Radio wave

Codes

	i	ii	iii	iv	v
a.	q	r	s	p	q
b.	p	q	r	s	t
c.	t	r	s	q	p
d.	t	r	s	p	q

22 A body at temperature T produces a continuous spectrum of wavelengths. Obtain the temperature values in Column II corresponding to different part of electromagnetic spectrum given in Column I. Match the Column I with Column II and mark the correct option from the codes given below.

	Column I Wavlength of radiation (m)		Column II Temperature (K)
i.	1 m	p.	About 6000 K
ii.	1 cm	q.	About 3000 K
iii.	1 μm	r.	About 0 K
iv.	500 nm		

Codes

	i	ii	iii	iv
a.	r	r	p	q
b.	p	p	q	r
c.	p	q	q	r
d.	p	r	q	p

Comprehension Type

Direction (Q. Nos. 23-26) *This section contains 2 paragraphs, each describing theory, experiments, data, etc. Four questions related to the paragraphs have been given. Each question has only one correct answer among the four given options (a), (b), (c) and (d).*

Working Space

Passage I

A plane electromagnetic wave has an intensity of 750 Wm^{-2} is falling perpendicularly over a surface of dimensions 50 cm $\times$ 100 cm. The surface reflects half of energy and absorbs half of energy.

23 Total energy absorbed by surface in 1 min is

a. 75 kJ
b. 45 kJ
c. 22.5 kJ
d. 11.3 kJ

24 Total momentum transferred to the surface in 1 min is

a. 7.53×10^{-5} kg-ms^{-1}
b. 3.76×10^{-5} kg-ms^{-1}
c. 1.13×10^{-5} kg-ms^{-1}
d. 1.13×10^{-4} kg-ms^{-1}

Passage II

An astronaut tries to reach back to space station at a distance of 100 km from himself by using his flashlight of 1000 W as a propellent in space. Reflector of his flashlight is of 0.5 m^2 and total mass of astronaut and his gear is around 150 kg.

25 What is the radiation pressure obtained from flashlight?

a. 0.67×10^{-5} Nm^{-2}
b. 1.33×10^{-5} Nm^{-2}
c. 2.66×10^{-5} Nm^{-2}
d. 0

26 How long will it take to reach space station? (Assume astronaut has enough oxygen supply in his gear)

a. In less than 1 month
b. In more than 1 month
c. In more than 1 day
d. Never reaches the spaceship

Statement Type

Direction (Q. Nos. 27 and 28) *This section is based on Statement I and Statement II. Select the correct answer from the codes given below.*

a. Both Statement I and Statement II are correct and Statement II is the correct explanation of Statement I
b. Both Statement I and Statement II are correct but Statement II is not the correct explanation of Statement I
c. Statement I is correct but Statement II is incorrect
d. Statement II is correct but Statement I is incorrect

27 **Statement I** An accelerating charge radiates energy in form of electromagnetic wave and frequency of emitted radiation increases with increase of speed of accelerating charge.

Statement II As charge moves with certain acceleration, the field behind and ahead of charge is different and symmetry of field in lost. So, energy must be transferred to space to adjust the field.

28 **Statement I** An incandescent bulb is less efficient than a tubelight of similar filament. (The tubelight has two filaments whereas a bulb has only one filament)

Statement II In tubelight most of ultraviolet radiation is converted into visible region by phosphor coating.

DPP-2 Matter Waves

Working Space

Subjective Question

Direction (Q. No. 1) *This question is subjective in nature, need to be solved completely on notebook.*

1 Find

i. The de-Broglie wavelength associated with the charged particles, when they are accelerated through a potential difference of V volts.

ii. The de-Broglie wavelength associated with thermal neutrons and that associated with a gas molecule at temperature of T kelvins.

Only One Option Correct Type

Direction (Q. Nos. 2-11) *This section contains 10 multiple choice questions. Each question has four choices (a), (b), (c) and (d), out of which ONLY ONE is correct.*

2 The de-Broglie wavelength associated with protons changes by 0.25%, if its momentum is changed by p_0. Initial momentum of proton is

a. $100\,p_0$ b. $\frac{p_0}{400}$

c. $401\,p_0$ d. $\frac{p_0}{100}$

3 The de-Broglie wavelength for electron in an orbit of hydrogen atom is 10^{-9} m. Principal quantum number for this electron is

a. 1 b. 2

c. 3 d. 4

4 What minimum potential must be applied on an electron microscope to view two point distinctively, which are about 1.22 Å apart ?

a. 1000 V b. 100 V

c. 150 V d. 10^4 V

5 The de-Broglie wavelength associated with a particle moving with speed of light will be

a. zero b. infinity

c. $\frac{hv}{mc}$ d. $\frac{mc}{h}$

6 The de-Broglie wavelength associated with electron of hydrogen atom in its ground state is

a. 0.3 Å b. 3.3 Å

c. 6.26 Å d. 10 Å

Working Space

7 In Davisson-Germer experiment, observed value of de-Broglie's wavelength is 0.165 nm whereas theoretical value is 0.166 nm. Choose the correct reason for a lower experimental value.

a. Error in observation of angle at which maxima of intensity of diffracted electrons occurs
b. Temperature of nickel crystal is more than the ambient temperature
c. Electron speed inside the crystal is more than $\sqrt{\left(\frac{2qV}{m}\right)}$ and no account is made to calculate this increase of speed
d. Electrons behave like waves in this experiment

8 In 1927, Clinton Davisson and Lester Germer confirmed the de-Broglie's hypothesis by observing scattering pattern of electrons in which maxima and minima were observed whose position depends on the electron energy. Maxima and minima are observed more prominently, when

a. target nickel block is maintained at low temperature
b. nickel target is maintained at high temperature
c. observation is not temperature dependent
d. accelerating voltage of electrons is increased

9 Which of the following is not a possible de-Broglie's wavelength of a particle, which moves inside a cubical box of side length L, without losing any energy (elastically colliding with walls of cube)?

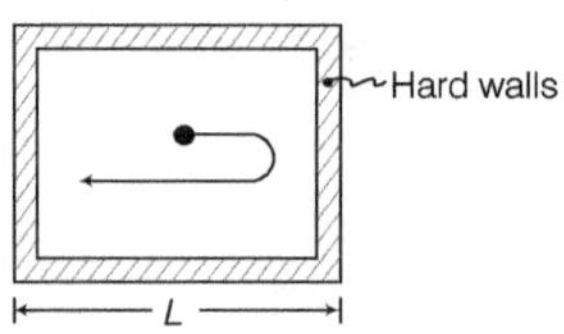

a. 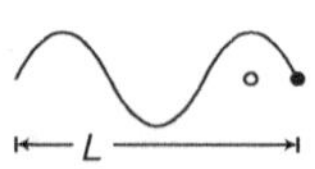

b. 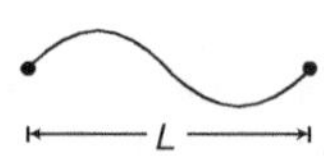

c. 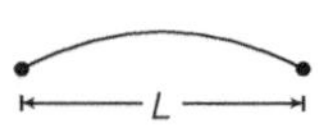

d. All of these

10 An electron with initial velocity $v_0\hat{\mathbf{i}}$ is in the region of electric and magnetic fields, $-E_0\hat{\mathbf{i}}$ and $B_0\hat{\mathbf{j}}$ respectively. The de-Broglie wavelength of the electron at time t is given by

a. $\lambda = \dfrac{h}{mv_0}$
b. $\lambda = \dfrac{h}{q\,t(E_0)}$
c. $\lambda = \dfrac{h}{qt\,(E_0^2 + B_0^2)^{1/2}}$
d. $\lambda = \dfrac{h}{qt\,(E_0^2 + v_0^2 B_0^2)^{1/2}}$

11 An α-particle moving horizontally makes a head on collision elastically with a proton (at rest). What are the ratio of de-Broglie's wavelength's associated with α-particle and proton just after collision ?

a. 2 : 1
b. 4 : 3
c. 1 : 2
d. 2 : 3

Working Space

One or More than One Options Correct Type

Direction (Q. Nos. 12-16) *This section contains 5 multiple choice questions. Each question has four choices (a), (b), (c) and (d), out of which ONE or MORE THAN ONE are correct.*

12 An α-particle of mass m_α and charge q_α moves with a velocity $v_0 = (2\hat{i} + 3\hat{j} - \hat{k})\,\text{ms}^{-1}$ through a region, where both a uniform magnetic field and a uniform electric field exist. Which of the following statements is/are correct? (Given, $\mathbf{B} = (2\hat{i} + 4\hat{j} + \hat{k})\,\text{T}$ and, $\mathbf{E} = (4\hat{i} - \hat{j} - 2\hat{k})\,\text{Vm}^{-1}$)

a. Acceleration of α-particle is perpendicular to Z-axis
b. Acceleration of α-particle is parallel to Z-axis
c. de-Broglie's wavelength associated with α-particle is increasing with time
d. de-Broglie's wavelength of α-particle keeps on decreasing with time

13 A neutron beam of energy E scatters from atoms of a metal surface with spacing $d = 0.1\,\text{nm}$. First maximum intensity of reflected beam occurs at $30°$. Then, mark the correct options.

a. Wavelength associated with neutron beam is 0.1 nm
b. Wavelength associated with neutron beam is 0.05 nm
c. $E = 0.21\,\text{eV}$
d. $E = 0.34\,\text{eV}$

14 At what de-Broglie wavelength will relativistic corrections are necessary for a moving electron?

a. $\lambda = 10^{-4}\,\text{nm}$ b. $\lambda = 10\,\text{nm}$ c. $\lambda = 10^{-1}\text{nm}$ d. $\lambda = 10^{-6}\,\text{nm}$

15 A particle moves such that, force acting over it is always directed towards a fixed point. Range of de-Broglie wavelength for the particle is (λ_1, λ_2) with $(\lambda_2 > \lambda_1)$. Then, mark the correct options.

a. Path of particle is circular
b. Path of particle is elliptical
c. Distance of fixed point and particle is least, when $\lambda = \lambda_1$
d. Distance of fixed point and particle is least, when $\lambda = \lambda_2$

16 Two particles A and B of de-Broglie wavelengths λ_1 and λ_2 respectively are combined to form a particle C. The process conserves momentum. If $\lambda_2 = 2\lambda_1$, then mark the correct options.

a. de-Broglie wavelength associated with particle C may be $\frac{2}{3}\lambda_1$
b. de-Broglie wavelength associated with the particle C may be $\frac{2}{3}\lambda_2$
c. de-Broglie wavelength associated with the particle C may be $2\lambda_1$
d. de-Broglie wavelength associated with the particle C may be $2\lambda_2$

Comprehension Type

Direction (Q. Nos. 17-21) *This section contains 2 paragraphs, each describing theory, experiments, data, etc. Five questions related to the paragraphs have been given. Each question has only one correct answer among the four given options (a), (b), (c) and (d).*

Passage I

The amplitude of de-Broglie waves that corresponds to a moving body reflects the probability that it will be found at a particular place at a particular time. So, a formula $y = A\cos(kx - \omega t)$ cannot be used to represent a de-Broglie's wave. Instead wave representation of a moving body corresponds to a wave packet or a wave group like one shown below.

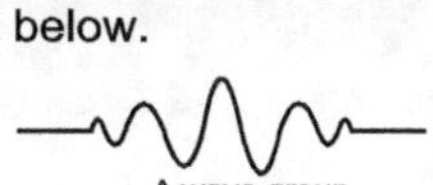

A wave group

Working Space

A familiar example of formation of a wave group is phenomena of beats.
A wave group travels with a velocity called group velocity.

i.e. $$v_g = \frac{\Delta\omega}{\Delta k} \text{ or } \frac{d\omega}{dk}$$

whereas, phase velocity associated with a de-Broglie wave is given by

$$v_p = \frac{\omega}{k}, \omega = 2\pi f \text{ and } k = \frac{2\pi}{\lambda}$$

where, ω = angular frequency
and k = wave number of de-Broglie's wave.

17 Let a body is moving with speed v ($v < c$). The group velocity v_g of de-Broglie waves associated with the body is

a. v
b. $2v$
c. $\frac{1}{2}v$
d. $\frac{v}{c}$

18 Which statements is incorrect for phase velocity v_p of the wave group?

a. $v_p > v_g$
b. $v_p > c$
c. $v_p > c$ does not violate special relatively theory
d. v_p has a physical significance

Passage II

In their pioneer experiment, CJ Davisson and L Germer, elastically scattered a beam of low speed electrons, from a polycrystalline nickel target. Initially, for constant electron energies of about 100 eV, the scattered intensity rapidly decreases as ϕ increases, there is no maxima.

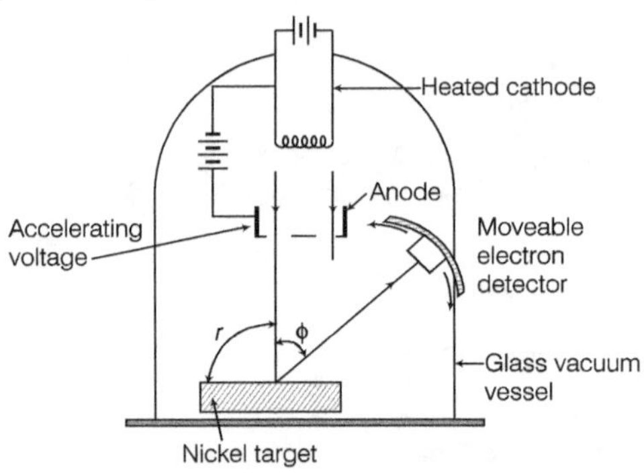

During experimentation, some air enters the vacuum enclosure. To remove the oxide layer formed, the sample is "backed".
When the apparatus was re-assembled, quite different results were found. There is strong variation in the intensity of scattered electrons with angle ϕ. With $\alpha = 90°$, $V = 54$ V and $\phi = 50°$, there is a intensity maxima of scattered electrons.

19 If an accelerating voltage of 100 V instead of 54 V in Davisson-Germer experiment, then the first peak in the distribution of scattered electrons is observed (Take, $d = 0.091$ m for target metal)

a. at an angle of 50°
b. at an angle less than 50°
c. at an angle of around 43°
d. No peak (maxima) is observed

Working Space

20 In Davisson-Germer experiment, path difference between two adjacent scattered electrons is

a. $\frac{nh}{p}$

b. $\frac{nh}{2p}$

c. $\frac{2nh}{p}$

d. zero

21 In Davisson-Germer experiment, electrons are elastically scattered from nickel target forming a diffraction pattern. Scattering of electrons is from

a. multiple crystals in sample
b. single crystal in sample
c. from oxide layer on top surface of sample
d. All of the above

Statement Type

Direction (*Q. Nos.* 22 *and* 23) *This section is based on Statement I and Statement II. Select the correct answer from the codes given below.*

a. Both Statement I and Statement II are correct and Statement II is the correct explanation of Statement I
b. Both Statement I and Statement II are correct but Statement II is not the correct explanation of Statement I
c. Statement I is correct but Statement II is incorrect
d. Statement II is correct but Statement I is incorrect

22 **Statement I** There is a strong practical evidence that a less massive particle becomes a wave at high speeds.

Statement II When electrons or thermal neutrons are used instead of light in Young's double slit experiment, similar pattern of maxima and minima are observed.

23 **Statement I** In case of Davisson-Germer experiment, $2d\sin\theta = n\lambda$.

Statement II When electrons are scattered from two different crystal planes, path difference is $2d\sin\theta$.

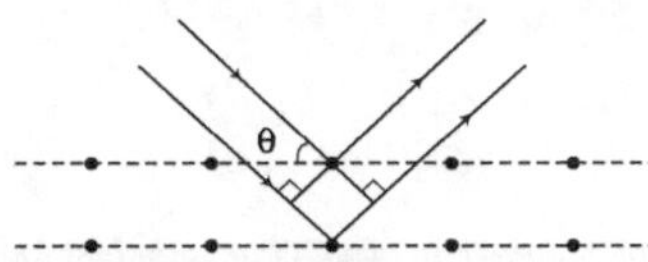

One Integer Value Correct Type

Direction (*Q. No.* 24) *This section contains a question. When worked out will result in an integer from* 0 *to* 9 (*both inclusive*).

24 When a beam of electron passes through two slits separated by 3μm, an interference pattern is obtained in which a second order maxima is observed at an angle of 55°.
If speed of electrons targetted at slits is 118 K ms^{-1}, then find the value of K. (Take, $h = 6.63 \times 10^{-34}$ Js, $m_e = 9.11 \times 10^{-31}$ kg and $\sin 55° = 0.82$) You can round off your answer to nearest integer.

DPP-3 Emission of Electrons and Photoelectric Effect

Subjective Questions

Working Space

Direction (Q. Nos. 1 and 2) *These questions are subjective in nature, need to be solved completely on notebook.*

1 A uniform monochromatic beam of light, of wavelength 4×10^{-7} m falls on a photosensitive material having a work function of 2 eV. The beam has an intensity of 3×10^{-9} Wm^{-2}. Assume that one proton ejects one electron then calculate

i. the number of electrons emitted per square metre per second.
ii. the energy absorbed per square metre, per second by metal sample.
iii. the maximum kinetic energy of emitted photoelectrons.
iv. the minimum kinetic energy of emitted electrons.

2 A photocell is operating in saturation mode with a photocurrent of 4.8 μA, when monochromatic radiation of wavelength 3000 Å and power 1 mW is incident over cathode of photocell which is coated with a photosensitive material.

When monochromatic radiation of wavelength 1650Å and power 5 mW is used over same photocell, it is observed that maximum speed of emitted photoelectrons increases by a factor of 2?

Assuming photoelectron generation efficiency remains constant. Calculate

i. threshold wavelength for the cathode.
ii. efficiency of photoelectron generation.
iii. saturation current in second case.

Only One Option Correct Type

Direction (Q. Nos. 3-12) *This section contains* 10 *multiple choice questions. Each question has four choices (a), (b), (c) and (d), out of which ONLY ONE is correct.*

3 In an experiment in photoelectric effect, an student found the slope maximum kinetic energy (of emitted electron) *versus* frequency of incident light around to be 4.12×10^{-15} (SI Units).

a. Observations taken by student are correct and graph plotted by the student is also correct
b. Observations taken by student are correct but the graph plotted is between maximum speed *versus* frequency
c. Observations taken are correct but the graph plotted is between stopping potential and frequency
d. Observations taken by student are faulty

Working Space

4 As energy associated with light changes with its wavelength, often the reciprocal of the wavelength $\left(\frac{1}{\lambda}\right)$, is used to describe energy associated with that wavelength. Then, mark the correct equivalence.

a. $1 \text{ eV} = 5092.6 \text{ cm}^{-1}$ and $1 \text{cm}^{-1} = 1.2398 \times 10^{-4} \text{ eV}$
b. $1 \text{ eV} = 1239.8 \text{cm}^{-1}$ and $1 \text{cm}^{-1} = 8068 \times 10^{-7} \text{ eV}$
c. $1 \text{ eV} = 1.6 \times 10^{-19} \text{ cm}^{-1}$ and $1 \text{cm}^{-1} = 6.65 \times 10^{19} \text{ eV}$
d. $1 \text{ eV} = 8065.8 \text{ cm}^{-1}$ and $1 \text{cm}^{-1} = 1.2398 \times 10^{-4} \text{ eV}$

5 In a photoemissive cell, with exciting wavelength λ, the maximum kinetic energy of the electron is K. If the exciting wavelength is changed to $\frac{3\lambda}{4}$, the kinetic energy of the fastest emitted electron will be

a. $\frac{3K}{4}$ b. $\frac{4K}{3}$
c. less than $\frac{4K}{3}$ d. more than $\frac{4K}{3}$

6 Light coming from a discharge tube filled with pure hydrogen falls on the cathode of the photoelectric cell. The work function of the surface of the cathode is 4 eV. Which of the following value of anode voltage with respect to the cathode will likely to make the photocurrent zero?

a. -4 V b. -6 V
c. -8 V d. -10 V

7 In a photoelectric experiment, the potential difference V that must be maintained between the illuminated surface and the collector 80 as just to prevent any electron from reaching the collector is determined at different frequencies f of the incident illumination. Graph obtained is shown in the figure. The maximum KE of the electrons emitted at frequency f_1 is

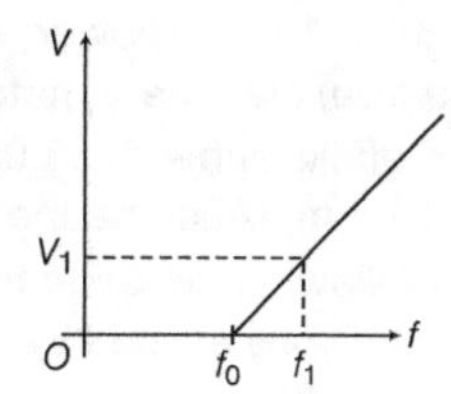

a. hf_1 b. $\frac{V_1}{f_1 - f_0}$
c. $h(f_1 - f_0)$ d. $eV_1(f_1 - f_0)$

8 Cut off potentials for a metal surface for light of wavelengths λ_1, λ_2 and λ_3 are V_1, V_2 and V_3 respectively. If V_1, V_2 and V_3 are in arithmatic progression, then λ_1, λ_2 and λ_3 are in

a. arithmatic progression b. geometric progression
c. harmonic progression d. None of these

9 A small potassium foil is placed perpendicular to the incident light at a distance of 0.5 m from a light source of 1.0 W. Work function for the potassium foil is 1.8 eV. The radius of a potassium atom is 1.3×10^{-10} m. Assuming wave nature of light, time elapsed before the first electron is emitted is

a. 17 h b. 17 ns
c. 17 min d. 17 s

10 A beam of light consists of five wavelengths 4000 Å, 4800Å, 6000Å, 7000Å and 7800Å. The light beam is falling normally over a metal surface of area 10^{-4} m^2 with a work function of 1.9 eV. Intensity of light beam is 7.5×10^{-3} Wm^{-2}, which is equally divided among the constituent wavelengths. If there is no loss of light energy, number of photoelectrons emitted per second is

a. 1.12×10^{12} b. 3.15×10^{12}
c. 1.77×10^{12} d. 4.06×10^{12}

11 A light beam, $E = 100[\sin(\omega_1 t) + \sin(\omega_2 t)]$ Vm^{-1} with $\omega_1 = 5 \times 10^{15}$ $s^{-1}$and $\omega_2 = 8 \times 10^{15} s^{-1}$, falls on a metal surface of work function 2.0 eV. Maximum KE of emitted photoelectrons is

a. 3.20 eV b. 1.5 eV

c. 3.27 eV d. 2.1 eV

12 Let a light beam of total intensity $1 \mu Wcm^{-2}$ falls on a clean iron sample piece with work function 4.5 eV and of 1.0 cm^2 area. Assume that iron sample reflects 96% of light and that only 3% of the absorbed energy lies in the violet region which can cause photoemission (of wavelength 250 nm). Number of electrons emitted will be

a. 1.5×10^{10} electrons per second

b. 3.0×10^9 electrons per second

c. 3.0×10^{10} electrons per second

d. 1.5×10^9 electrons per second

One or More than One Options Correct Type

Direction (Q. Nos. 13-18) *This section contains 6 multiple choice questions. Each question has four choices (a), (b), (c) and (d), out of which ONE or MORE THAN ONE are correct.*

13 Consider the following photoelectric experiment with following observations.

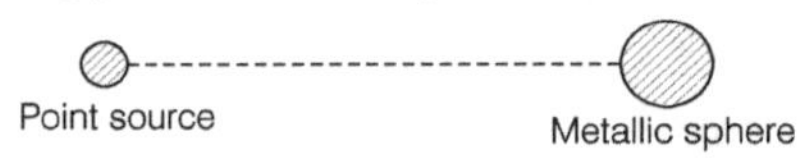

Point source is of 3.2×10^{-3} W power, emitting photons of 5.0 eV energy.
Distance of source from stationary metallic sphere is 0.8 m.
Work function for metallic sphere is 3.0 eV also assume 100% absorption. Radius of sphere is 8.0×10^{-3} m. (Assume the experiment starts at $t = 0$ s)
Then, which of the following statements is/are correct?

a. Total number of electrons emitted by sphere in first 3 min is around 10^7

b. Total number of electrons emitted by sphere in first 3 min is around 10^8

c. There is no emission of electron at $t = 1$s

d. There is no emission of electron after $t = 180$ s

14 A point source S is placed at a distance L from a cardsheet with a hole of diameter 0.1 m. A photosensitive metal plate P is placed at a distance D from this hole at other side of card-sheet as shown in the figure.

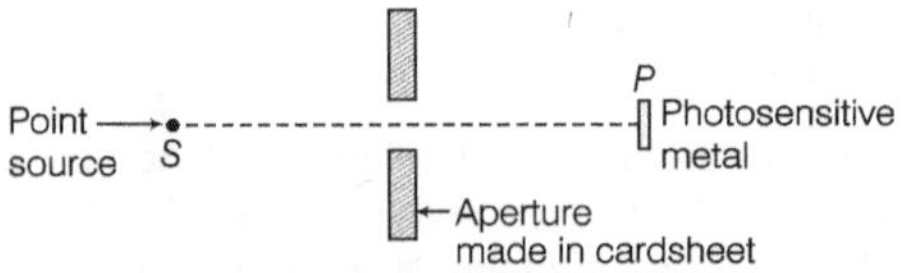

If source emits light of frequency greater than threshold frequency of the metal plate P then, mark the correct options.

a. When a concave lens of focal length L is placed in aperture, photoemission rate of P increases

b. When a convex lens of focal length L is placed in aperture, photoemission rate of P increases

c. When a glass slab is placed in aperture, photoemission rate of P decreases

d. When a prism is placed in the aperture photoemission from P may be stop

Assume that lens, slab or prism is of same size as that of aperture and P is much smaller than aperture.

Working Space

15 In the given set-up of photoelectric effect, two different cathode materials C_1 and C_2 are used.

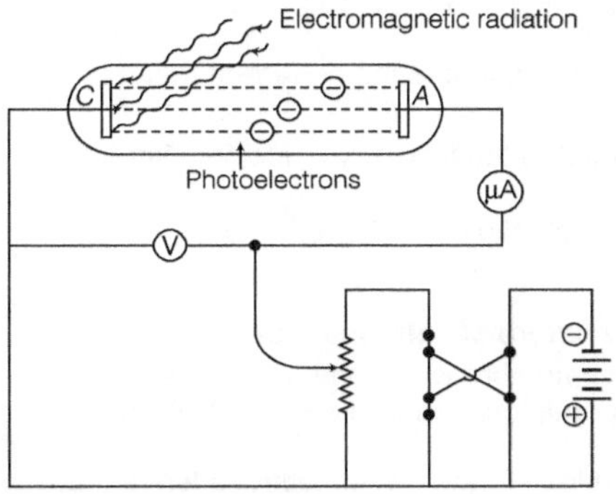

Let I_e = photoelectron current,
V = impressed electrical potential,
I_ε = intensity of electromagnetic radiation
f = frequency of electromagnetic radiation and
C = cathode metal material.

Which of the given graphs are correct ?

a. A plot of I_e *versus* V for constant f

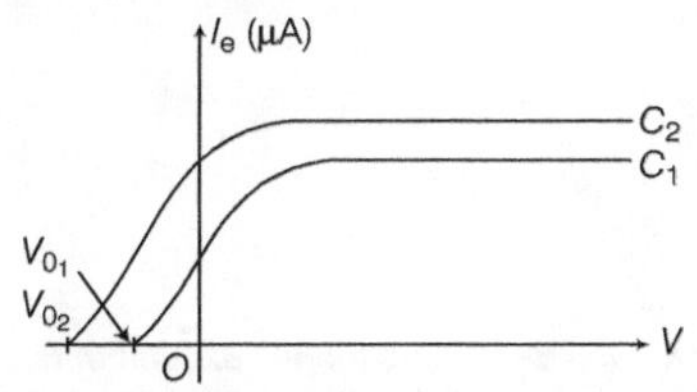

b. A plot of I_e *versus* f for constant V

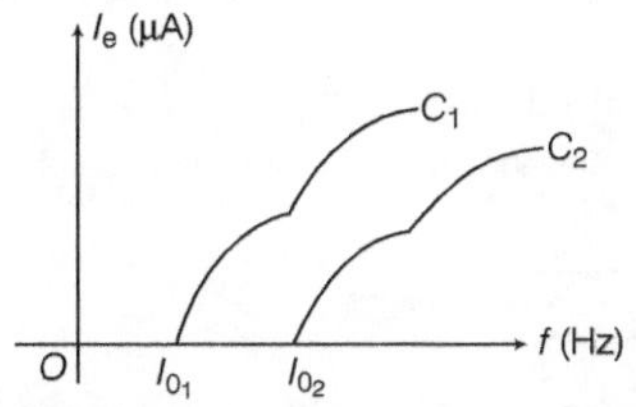

c. A plot of V_s (stopping potential) *versus* f for constant I_ε

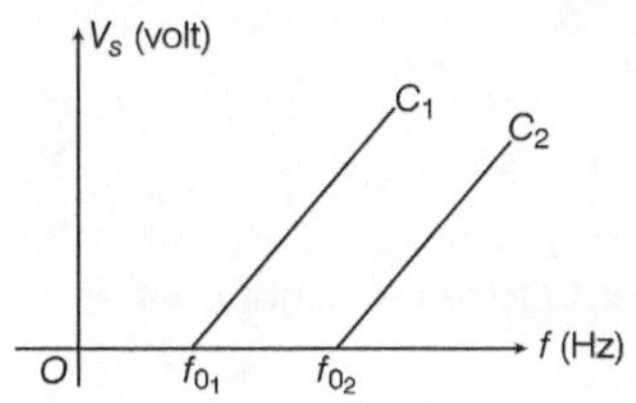

d. A plot of V_s (stopping potential) *versus* I_ε for constant f

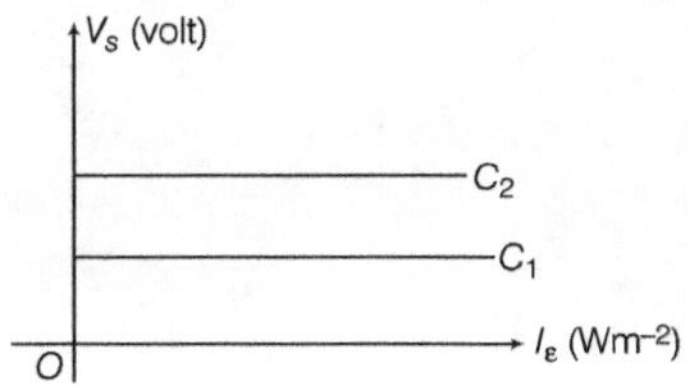

16 Which of the following are correct according to wave theory of light regarding photoelectric effect?

a. Amplitude of vibration of an electron is directly proportional to the maximum amplitude E_m of incident wave, $A_e \propto E_m$
b. Average KE of photoelectrons is directly proportional to the square of their vibrational amplitude, $KE_{av} \propto A_e^2$
c. Average KE of photoelectrons is proportional to the intensity of incident light, $KE_{av} \propto I_\varepsilon$
d. Stopping potential for the most energetic photoelectrons is proportional to the intensity of incident light, $V_s \propto I_\varepsilon$

17 Photoelectric effect does not support wave nature of light because

a. there is a minimum frequency below, which no photoelectrons are emitted
b. the maximum KE of emitted electrons depends only on frequency of light but not on its intensity
c. even when the metal surface is faintly illuminated, the photoelectrons leave the surface immediately
d. electric charge of photoelectrons is quantised

18 When a monochromatic light source is at a distance of 0.2 m from a photoelectric cell, the cut off voltage and the saturation current are respectively 0.6 V and 18.0 mA. If same source is placed at a distance of 0.6 m from the photocell, then

a. the stopping potential will be 0.2 V
b. the stopping potential will be 0.6 V
c. the saturation current will be 6.0 mA
d. the saturation current will be 2.0 mA

Comprehension Type

Direction (Q. Nos. 19-23) *This section contains 2 paragraphs, each describing theory, experiments, data, etc. Five questions related to the paragraphs have been given. Each question has only one correct answer among the four given options (a), (b), (c) and (d).*

Passage I

Adhip, a class XIIth student is experimenting using a photocell.

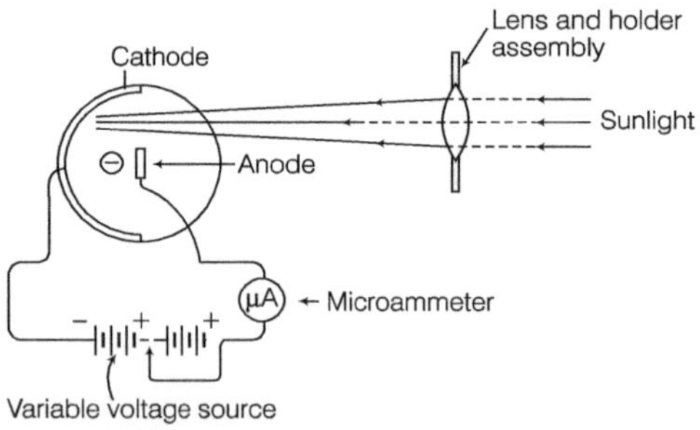

He uses a convex lens fitted in an opaque holder to focus sunlight on the photoemissive cathode of photocell.
Light falls on the cathode through the lens only.

Working Space

19 If for some position of lens, current through photocell is I, then current through the photocell when central part of lens (of diameter $d/2$) is covered with a black opaque paper as shown in the figure.

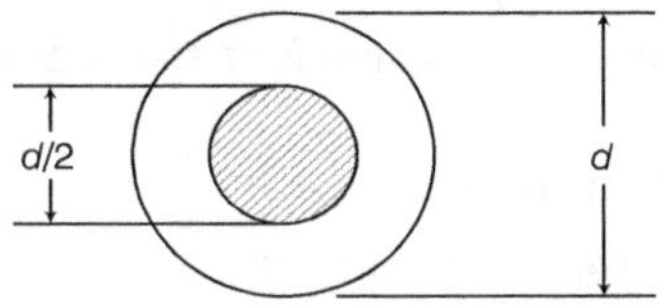

Keeping all other parameters same will be

a. I

b. $\frac{I}{2}$

c. $\frac{3I}{4}$

d. $\frac{2}{3}I$

20 If Sun subtends an angle of 30 min on the earth and focal length of lens is 100 cm, least diameter of image of Sun formed over the cathode will be

a. 3 mm

b. 6 mm

c. 9 mm

d. 17 mm

Passage II

In an experimental setup of photoelectric effect, a monochromatic point source is kept infront of an aperture such that light which passes through the aperture falls over a photosensitive surface as shown in the figure.

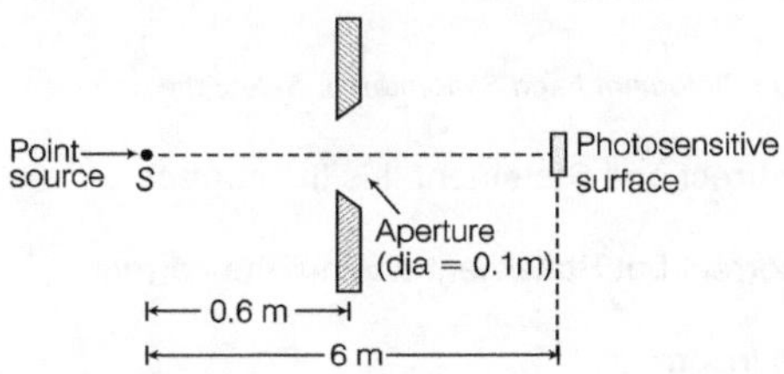

Distance of source and aperture is 0.6 m and distance of source and metal surface is 6 m. Dia of aperture is 0.1 m. Area of photosensitive metal surface is $0.5\ cm^2$ and efficiency of photoelectron generation is 90%. Source emits radiation of 6000Å and power of source is 2 W.

21 Photon flux reaching the metal surface is

a. $1.052 \times 10^{16}\ s^{-1}$

b. $6.06 \times 10^{18}\ s^{-1}$

c. $1.435 \times 10^{9}\ s^{-1}$

d. $2.87 \times 10^{13}\ s^{-1}$

22 A concave lens of focal length 0.6 m is placed at the opening of the aperture. If lens transmits 80% of the light falling over it, the new photon flux reaching the metal surface is

a. $1.435 \times 10^{9}\ s^{-1}$

b. $2.87 \times 10^{13}\ s^{-1}$

c. $0.841 \times 10^{16}\ s^{-1}$

d. $2.06 \times 10^{13}\ s^{-1}$

23 Ratio of stopping potentials when the light goes through aperture and when the light goes through lens will be

a. 1 : 2

b. 2 : 1

c. 1 : 3.2

d. 1 : 1

Working Space

Matching List Type

Direction (Q. No. 24) Choices for the correct combination of elements from Column I and Column II are given as options (a), (b), (c) and (d), out of which one is correct.

24 Match the Column I with Column II and mark the correct option from the codes given below.

	Column I		Column II
i.	Photocell	p.	Cell supplies current without using an external battery
ii.	Photoemissive cell	q.	Electrons emitted are attracted by an anode
iii.	Photoconductive cell	r.	Used in exposure metres of camera and in solar batteries
iv.	Photovoltaic cell	s.	Electrical conductivity increases when exposed to light
		t.	Converts light energy into electrical energy

Codes

	i	ii	iii	iv
a.	t	q	s	t,r,p
b.	t	p,q	r	s
c.	r	p,q	r,s	t
d.	s	r	q	p

Statement Type

Direction (Q. Nos. 25 and 26) This section is based on Statement I and Statement II. Select the correct answer from the codes given below.

a. Both Statement I and Statement II are correct and Statement II is the correct explanation of Statement I
b. Both Statement I and Statement II are correct but Statement II is not the correct explanation of Statement I
c. Statement I is correct but Statement II is incorrect
d. Statement II is correct but Statement I is incorrect

25 **Statement I** Work function for every emitted electron is same in case of photoemission.

Statement II Work function is minimum energy required by an electron to come out of metal surface with zero speed.

26 **Statement I** When photoemission occurs in a gaseous target, photoemission can occurs at all frequencies of incident photons.

Statement II In a gas sample, speed of an atom may be different from other atoms.

SET - 1

Revisal Problems
for JEE Main

Only One Option Correct Type

Working Space

Direction (Q. Nos. 1-20) *This section contains 20 multiple choice questions. Each question has four choices (a), (b), (c) and (d), out of which ONLY ONE option is correct.*

1 An electromagnetic radiation has photons of energy of 14.4 keV. Part of electromagnetic spectrum it belongs is called

a. Infrared region b. Visible region
c. X-ray region d. Gamma ray region

2 A point source is radiating at rate of 800 W. The maximum value of electric field at a distance of 3.5 m from the source will be

a. 56.7 Vm^{-1} b. 62.6 Vm^{-1} c. 39.3 Vm^{-1} d. 44.2 Vm^{-1}

3 Energy of a light photon is equal to the kinetic energy of photon. Energy of photon is E, let λ_1 be the de-Broglie wavelength of photon and λ_2 be the wavelength of the photon. The ratio $\lambda_1 : \lambda_2$ is proportional to

a. E^0 b. $E^{1/2}$ c. E^{-1} d. E^{-2}

4 In the given graph, let I_a, I_b and I_c be intensities and f_a, f_b and f_c be the frequencies for the curves a, b and c, respectively.

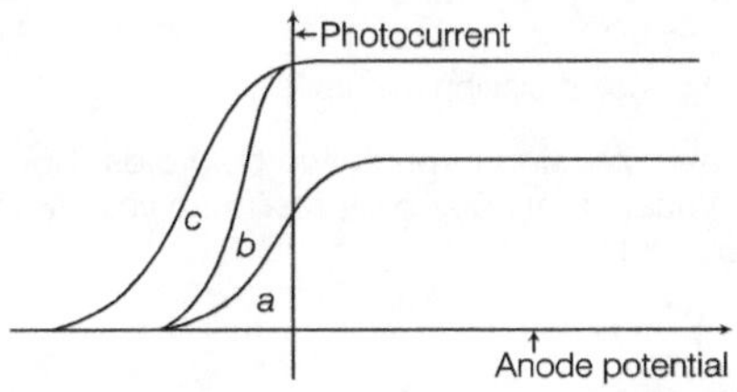

Then, which of these are correct ?

a. $f_a = f_b$ and $I_a \neq I_b$ b. $f_a = f_c$ and $I_a = I_c$
c. $f_a = f_b$ and $I_a = I_b$ d. $f_b = f_c$ and $I_b = I_c$

5 For thermionic emission, a cathode or emitter is heated in evacuated space to emit electrons. Choose the incorrect option regarding thermionic emission.

a. Tungsten is preferred as a suitable material as it has low work function
b. Metal alloys (a mixture of two metals) is preferred as they have low work function
c. Oxide coated cathodes are preferred as they have low operating temperatures and low work function
d. Oxide coated cathode are used for low voltage values

6 Light of wavelength 2475 Å is incident on Barium surface. Photoelectrons emitted describe a circle of radius 100 cm in a perpendicular magnetic field of intensity $\frac{1}{\sqrt{17}} \times 10^{-5}$ tesla.

Work function of the Barium is

$\left[\text{Take, } \frac{e}{m} = 1.7 \times 10^{11} \frac{\text{C}}{\text{kg}}\right]$

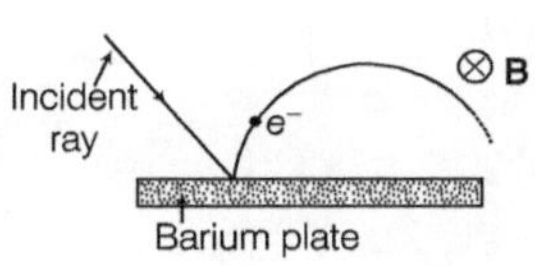

a. 1.8 eV b. 2.1 eV
c. 4.5 eV d. 3.3 eV

7 If momentum of an electron is changed by Δp, then the de-Broglie wavelength associated with it changes by 0.50%. The initial momentum of the electron will be

a. $\frac{\Delta p}{200}$ b. $\frac{\Delta p}{199}$
c. $199\Delta p$ d. $400\Delta p$

8 A particle of mass $3m$ at rest, decays into two particles of masses m and $2m$ having non-zero speeds. What is the ratio of de-Broglie wavelengths of the particles ?

a. 1 : 1 b. 1 : 2
c. 2 : 1 d. 2 : 3

9 A surface of size 10 cm × 10 cm is placed in the path of violet light ($\lambda = 4000$ Å) of intensity $4 Wm^{-2}$. Initially surface is perpendicular to the path of beam, then surface is tilted at an angle of 30° with the light beam. The ratio of number of photons falling per second over the surface in two cases will be

a. $\frac{1}{2}$ b. $\frac{2}{3}$
c. $\frac{3}{4}$ d. 1

10 Light from a discharge tube containing hydrogen falls over sodium surface. Work function of sodium metal is 1.82 eV and kinetic energy of fastest photoelectron emitted from sodium surface is 0.73 eV. Recoil speed of photon emitting hydrogen atom (assuming initially at rest) will be

a. $5.2\ ms^{-1}$ b. $0.5\ ms^{-1}$
c. $0.814\ ms^{-1}$ d. $8.14\ ms^{-1}$

11 An electromagnetic radiation represented by $E = a(1 + \cos\omega t)\cos\omega_0 t$ falls over a lithium surface with work function of 2.39 eV. Maximum kinetic energy of emitted photoelectrons will be (Given, $a = 4\ N/C$, $\omega = 6 \times 10^{14}\ rads^{-1}$ and $\omega_0 = 3.6 \times 10^{15}\ rads^{-1}$)

a. 0 b. 0.37 eV
c. 0.70 eV d. No photoemission occurs

12 A beam of light falls over cathode of a photocell after passing through two poloroids. Now, one the poloroids is rotated keeping other fixed. Variation of square of maximum velocity of emitted electrons with angle θ of rotation of polaroid is

a. 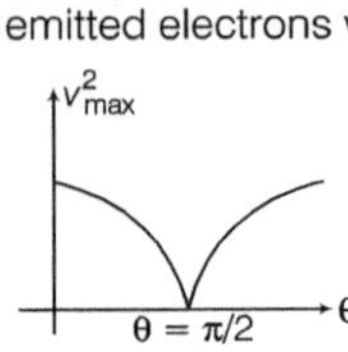

b. 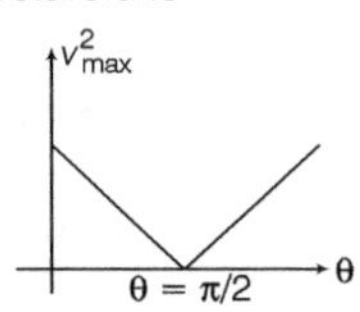

c. 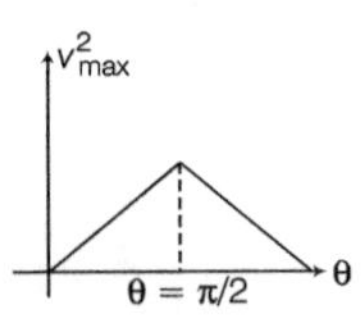

d. 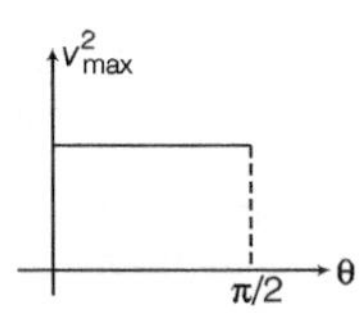

13 Estimate the value of work function of cesium, from the experimental results plotted on the graph given below.

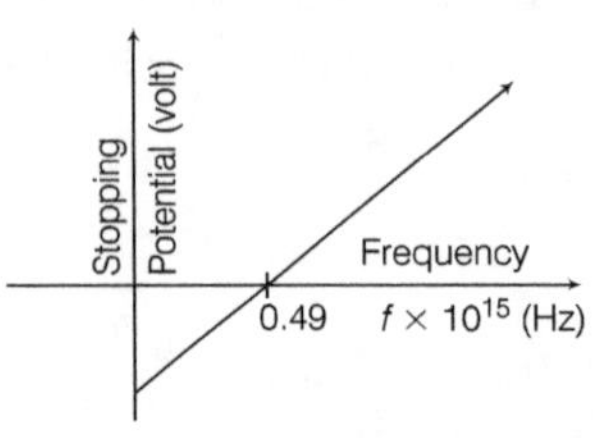

a. 3.2 eV b. 1.75 eV
c. 2 eV d. 2.5 eV

Working Space

Working Space

14 For a particle having kinetic energy K, an accurate value of de-Broglie wavelength associated with the particle is

a. $\lambda = \dfrac{hc}{\sqrt{K^2}}$ b. $\lambda = \dfrac{hc}{\sqrt{2mc^2K}}$

c. $\lambda = \dfrac{hc}{\sqrt{K^2 + 2mc^2K}}$ d. $\lambda = \dfrac{h}{p^2}$

15 Taking rest mass of an electron as 0.511 MeV/c^2 ($E = mc^2$, $m = \dfrac{E}{c^2}$), for 1% accuracy, an electron need to be treated relativistically, when its kinetic energy exceeds

a. 10 eV b. 10^3 eV

c. 10^5 eV d. None of these

16 An electon in a hydrogen-like atom is in an excited state. It has a total energy of −3.4 eV. The de-Broglie wavelength associated with electron is

a. 6.63 Å b. 10.25 Å

c. 7.96 Å d. 2.8 Å

17 Assume that the de-Broglie wave associated with an electron form an standing wave between the atoms, arranged in a one-dimensional array with nodes on each of the atomic sites. If a standing wave is formed, then distance d between atoms in array is 2 Å and again at 2.5 Å, but not at any intermediate value of d. Kinetic energy of this electron is around

a. 15 eV b. 150 eV

c. 1500 eV d. 1.5×10^4 eV

18 Consider the following photoelectric emission experiment.

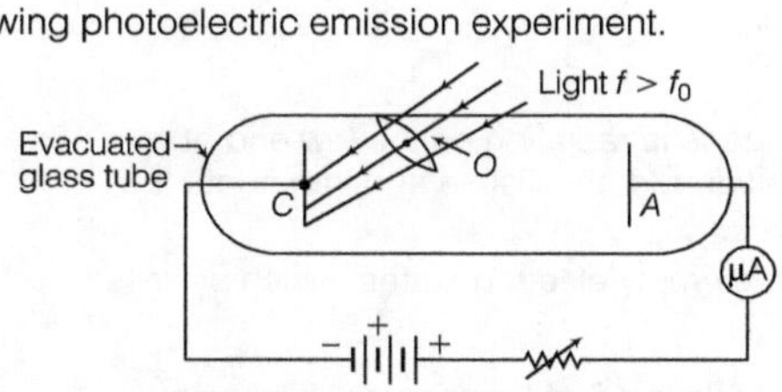

Distance of lens and centre of cathode (OC) is initially less than focal length of lens. Diameter of photosensitive cathode is more than diameter of lens. Lens is moved away from cathode slowly at a constant speed. Which of the following graphs is correct ? Assume that light falls over cathode only after passing through lens.

a. 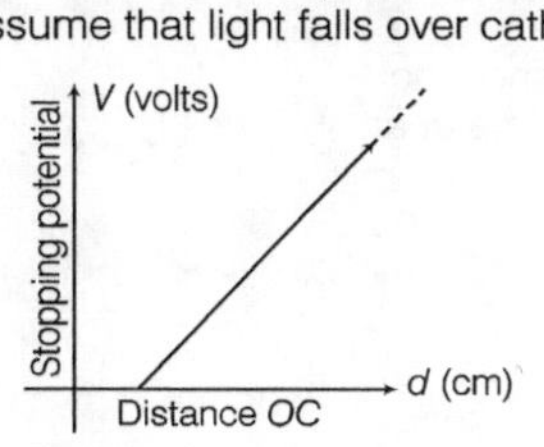

b. 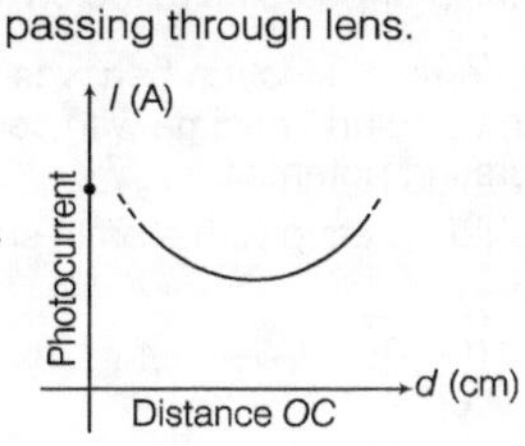

c. 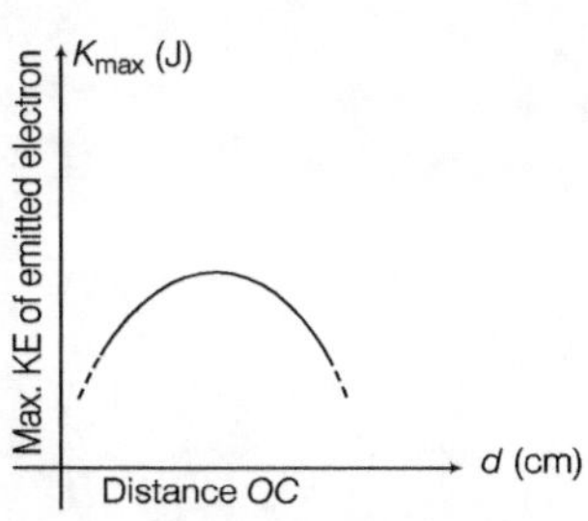

d. 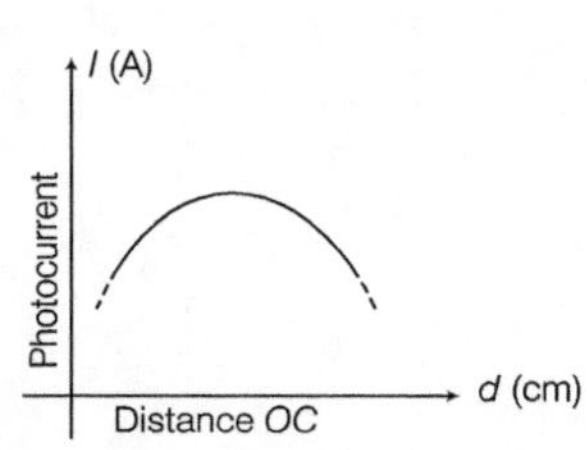

19 A student calculated de-Broglie wavelength associated with electron in ground state of hydrogen atoms as λ_1. He takes zero of potential energy at infinity. Another student take zero of potential energy at 1.06 Å distance and his calculated value of de-Broglie wavelength is λ_2. Then, λ_1/λ_2 is

a. 1 : 2
b. 2 : 1
c. 1 : 1
d. 2 : 3

Working Space

20 Which of these are incorrect regarding work function of a photosensitive surface?

I. Work function of a metal is not a constant quantity.
II. Work function of a metal surface increases with photoemission.
III. Work function of a metal surface decreases, when the surface is heated or coated with a suitable material.
IV. Work function of a metal surface is same as the energy of photon, when an electron is emitted with a non-zero speed.

a. Both I and III
b. Both II and III
c. II, III and IV
d. Only IV

Statement Type

Direction (Q. Nos. 21-23) ***This section is based on Statement I and Statement II. Select the correct answer from the codes given below.***

a. Both Statement I and Statement II are correct and Statement II is the correct explanation of Statement I
b. Both Statement I and Statement II are correct but Statement II is not the correct explanation of Statement I
c. Statement I is correct but Statement II is incorrect
d. Statement II is correct but Statement I is incorrect

21 **Statement I** Metals are opaque to all electromagnetic radiation on the low end of frequency spectrum from radio waves, through infrared, the visible and into about the middle of the ultraviolet radiation.

Statement II Metals have continuously available empty electron states, which permits electron transitions.

22 **Statement I** Aluminium and silver metals simply deflect incident photon following conservation of momentum like a bat deflects a ball and so, they appear shining.

Statement II Most of the absorbed radiation is re-emitted from the surface in form of visible light of the same wavelength,which appears as reflected light.

23 **Statement I** In Davission-Germer experiment, electron behaves as wave and hence their scattering pattern shows a series of maxima and minima with second order maxima at an angle greater than 50° for 54 V accelerating potential.

Statement II Crystal acts like a diffraction grating with atomic spacing d and for a maxima,

$$d \sin \theta = n\lambda, \; n \; 0, 1, 2, \ldots$$

Revisal Problems for JEE Advanced

Comprehension Type

Working Space

Direction *(Q. Nos. 1-10) This section contains 4 paragraphs, each describing theory, experiments, data, etc. Ten questions related to the paragraphs have been given. Each question has only one correct answer among the four given options (a), (b), (c) and (d).*

Passage I

For a typical case of photoemission, take work function of sodium 2.28 eV and energy absorbed per unit area per second is 1.00×10^{-7} mWcm^{-2}. Area of cross-section of a sodium atom is $\pi \times 10^{-16}$ cm^2. Assume

i. Classical wave theory
ii. Work function for every electron remains constant.

1 Maximum possible kinetic anergy of an emitted photoelectron is

a. more for an electron which is emitted earlier
b. less for an electron which is emitted later
c. equal for all electrons
d. more for an electron which is emitted later

2 Time lag in given case is

a. 130 s b. 130 min c. 130 h d. 130 days

Passage II

Let total intensity of light falling over a clean iron surface of 1.0 cm^2 be 1.0 μWcm^{-2}. Nearly 96% of the light is reflected and only 3% of the light lies in the violet region of the spectrum. Experimentation shows that only violet region causes photoemission. Assume that all photons of violet region have an effective wavelength of 250 nm and photoemission efficiency of iron surface is 100%.

3 Actual intensity available for the photoemission is nearly

a. 1 μWcm^{-2} b. 1 pWcm^{-2}
c. 1 nWcm^{-2} d. 1 fWcm^{-2}

4 Number of electrons emitted per second is

a. 1.5×10^9 b. 2.4×10^{10}
c. 1.1×10^{15} d. 1.1×10^8

5 If cut off frequency is 1.1×10^{15} Hz, then stopping voltage for emitted electrons is

a. 0.5 V b. 1.1 V c. 2.4 V d. 0.46 V

Passage III

For any source of light we construct a SPD graph (spectral power distribution graph). Most of the light sources are 'Heterochromatic' (hetero : different and chromatic : colour), i.e. they produces a beam of many wavelengths.

SPD over the surface of earth at noon looks as shown in the figure.

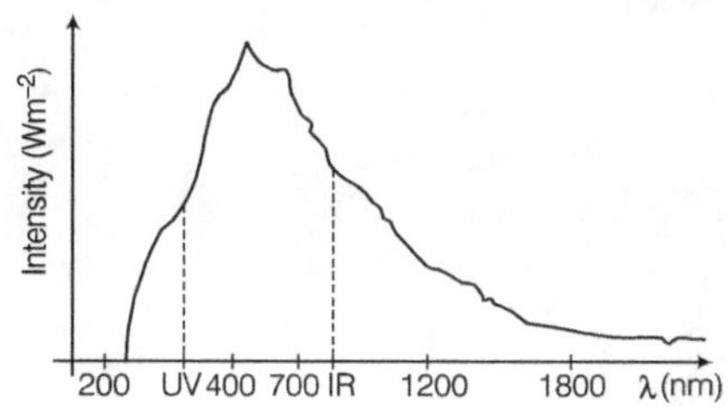

Working Space

6 Photosynthesis in plants is mainly due to absorption of photons of

a. red colour
b. yellow colour
c. violet colour
d. green colour

7 Sun's spectral power is mainly distributed in

a. UV and higher frequencies
b. IR and lower frequencies
c. visible and IR frequencies
d. equally distributed among different wavelengths

8 In the SPD graph given, mark the correct options.

a. No IR photon reaches earth's surface
b. No UV photon reaches earth's surface
c. No X-ray photon reaches earth's surface
d. No radio waves from sun reaches earth's surface

Passage IV

A bound electron can absorb a photon (as it does in case of photoelectric effect) but a free electron cannot absorb a photon. According to quantum theory of light, photons behave like particles except for their lack of rest mass.

We treat electrons and photons both as particles (if both are billiard balls). Let an X-ray photon strikes an electron.

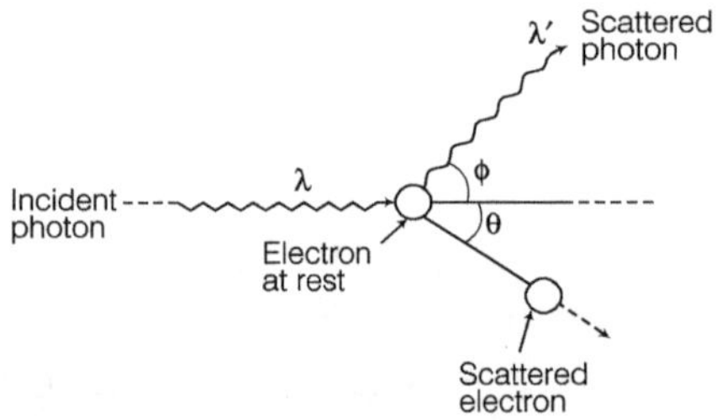

The calculations of momentum and energy show that in such collisions, a photon looses a certain amount of energy and an electron gains that part of energy. This is called Compton effect. If λ is wavelength of incident photon and λ' is wavelength of scattered photon, then Compton shows that,

$$\lambda' - \lambda = \frac{h}{mc}(1 - \cos\phi)$$

where, h = Planck's constant,

m = mass of electron,

c = speed of light in vacuum,

ϕ = angle between directions of incident and scattered photon.

The Compton effect is the chief means by which the X-rays lose energy, when they pass through matter.

9 X-rays of wavelength 10 pm are scattered after colliding with an electron. Maximum wavelength present in scattered X-ray spectrum will be

a. 10 pm
b. 10.7 pm
c. 14.9 pm
d. 5.1 pm

10 Maximum kinetic energy of recoiling electron is

a. 40.8 keV
b. 40.8 MeV
c. 40.8 eV
d. 40.8 J

Working Space

Only One Option Correct Type

Direction *(Q. Nos. 11-17) This section contains 7 multiple choice questions. Each question has four choices (a), (b), (c) and (d), out of which ONLY ONE is correct.*

11 A student puts a small metallic sphere of work function 4.76 eV in path of a beam of electromagnetic wave of uniform intensity.

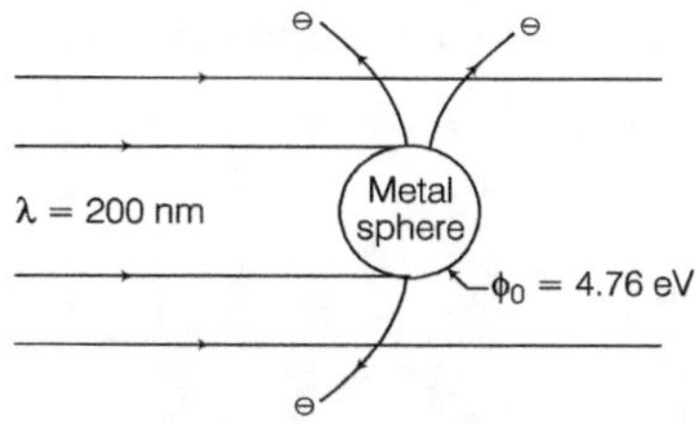

If radius of sphere is 1 cm and it is left in the radiation for a long time, then its mass is reduced by an amount of (Take, $hc = 1240$ eV-nm and m_e is mass of an electron)

a. $10^{19} m_e$
b. $10^{10} m_e$
c. $10^{7} m_e$
d. $10^{6} m_e$

12 One of your classmates comes with following graph.

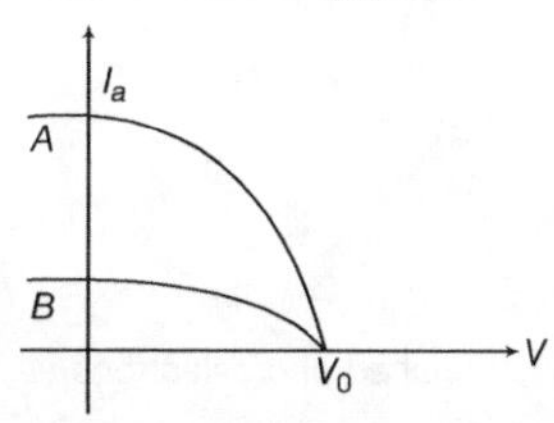

He informed you that he had taken photocurrent values on *Y*-axis and potential difference between cathode and anode on *X*-axis. What will be your opinion ?

a. Graph is incorrect
b. Intensity for metal *A* is more than intensity for metal *B*
c. Intensity for metal *B* is more than intensity for metal *A*
d. Intensity *A* = Intensity *B*

13 The photon interpretation of the photoelectric effect predicts that there is no measurable time lag between photon striking the metal surface and the start of photoemission. What would be its order ?

a. 10^{-10}s
b. 10^{-22}s
c. 10^{-15}s
d. 10^{-34}s

14 Let I is intensity of light which falls over a metal surface with work function ϕ_0. If metal's mass density is ρ and A is its atomic mass, then photoelectric time delay assuming classical wave model is

a. $\frac{\phi_0 \rho}{IA}$
b. $\frac{2\phi_0}{I}\left(\frac{\rho}{A}\right)^{2/3}$
c. $\frac{\phi_0}{I}\left(\frac{\rho}{A}\right)^{2/3}$
d. $\frac{\phi_0}{I} \cdot \frac{\rho}{(A)^{1/3}}$

15 According to classical theory, which of the following is correct for Davisson-Germer experiment ?

a. All the electrons must scattered in same direction
b. Scattered electrons have few favourable directions
c. Scattered electrons have no favourable directions
d. Scattering of incoming electrons have a strong dependence on their kinetic energies

Working Space

16 A moving electron has energy and a momentum. Which of these physical quantities does not have any physical significance ?

a. de-Broglie wavelength
b. Group speed of de-Broglie wave
c. Phase speed of the de-Broglie wave
d. Mass of moving electron

17 The coordinates of few points on stopping potential *versus* frequency graph are shown in the figure, (drawn for two different metals).

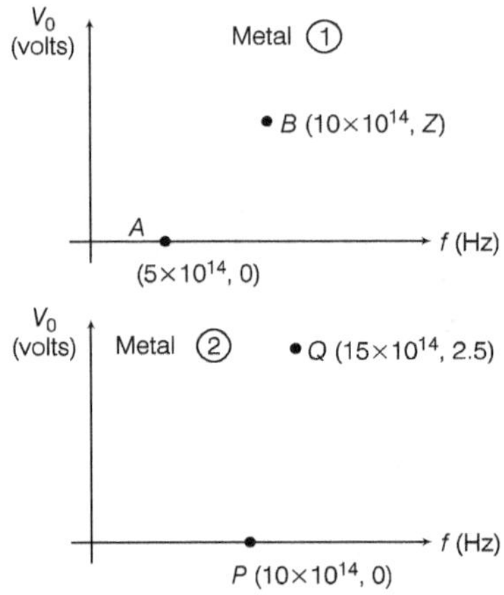

What do you conclude from these observations?

a. Both metals have same work function
b. Both metals have same value of maximum kinetic energy of emitted photoelectrons
c. $h = 6.04 \times 10^{-34}$ J-s
d. Experiment is not done properly

One or More than One Options Correct Type

Direction *(Q. Nos. 18-20) This section contains 3 multiple choice questions. Each question has four choices (a), (b), (c) and (d), out of which ONE or MORE THAN ONE are correct.*

18 A photon and a particle have same de-Broglie wavelength. Which of these are correct options ?

a. They have same linear momentum
b. They have same total energy
c. Energy of photon is less than total energy of particle
d. Kinetic energy of the particle is smaller than the photon's energy

19 Which of the following statements are correct?

a. Wavelength associated with a moving particle is
$$\lambda = \frac{hc}{\sqrt{K(K + 2m_0c^2)}}$$
b. Wavelength associated with a moving particle is $\lambda = \dfrac{h}{p}$
c. If total energy of particle exceeds its rest energy, then its de-Broglie wavelength is nearly the same as the wavelength of a photon with the same total energy
d. Frequency of de-Broglie wave is $f = \dfrac{pc}{h}$

20 In Davisson-Germer experiment, which of the following statements are correct?

a. Scattering of electrons is from many individual crystals in the target
b. Scattering of electrons is from only one crystal of the target
c. Kinetic energy of a missile electron decreases as it enters the crystal
d. Kinetic energy of a missile electron increases as it enters the crystal

Working Space

Matching List Type

Direction (Q. No. 21) ***Choices for the correct combination of elements from Column I and Column II are given as options (a), (b), (c) and (d), out of which one is correct.***

21 Let a graph is prepared between quantities given in Column I and value of Planck's constant is calculated in terms of x-intercept (n) and y-intercept (l) of the graph. Match values of Planck's constant in Column II with the entries in Column I and mark the correct option from codes given below.

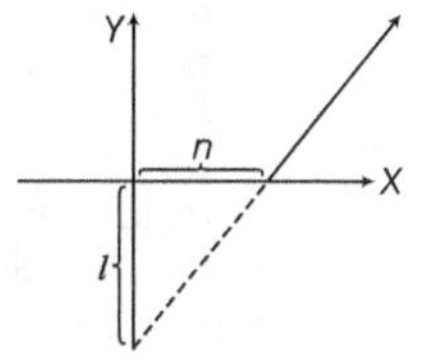

	Column I		Column II
i.	V_0, f	p.	$\frac{le}{n}$, e = charge of electron
ii.	K_{max}, f	q.	$\frac{l}{n}$
iii.	v^2_{max}, f	r.	$\frac{ml}{2n}$, m = mass of electron

Codes

	i	ii	iii		i	ii	iii
a.	p	q	r	b.	q	p	r
c.	r	q	p	d.	r	p	q

Statement Type

Direction (Q. Nos. 22 and 23) ***This section is based on Statement I and Statement II. Select the correct answer from the codes given below.***

a. Both Statement I and Statement II are correct and Statement II is the correct explanation of Statement I
b. Both Statement I and Statement II are correct but Statement II is not the correct explanation of Statement I
c. Statement I is correct but Statement II is incorrect
d. Statement II is correct but Statement I is incorrect

22 **Statement I** When an isolated moving electron absorbs a photon, its speed increases.
Statement II When an electron absorbs a photon, its energy increases by an amount hf.

23 **Statement I** When a mirror reflects 80% of light, energy of each reflected photon decreases by 20%.
Statement II In any real mirror 100% reflection of light is not possible.

One Integer Value Correct Type

Direction (Q. No. 24) ***This section contains a question. When worked out will result in an integer from 0 to 9 (both inclusive).***

24 Two metallic plates A and B each of area $5 \times 10^{-4} m^2$ are placed parallel to each other at separation of 1 cm. Plate B carries a positive charge of 33.7×10^{-12} C. A monochromatic beam of light with photons of energy 5 eV starts falling on plate A at $t = 0$ s.

Number of photons falling per square metre in 1 s is 10^{16}. If increase in KE of emitted photoelectrons, when they reach at plate B, is $10k$ eV. Find the value of k.

(Assume that 1 photoelectron is emitted for every 10^6 incident photons. Also, assume that all of emitted photoelectrons are collected by plate B and the work function of plate A remains constant at the value of 2 eV)

JEE Main & AIEEE Archive

(Compilation of Last 13 Years' Questions)

1 The radiation corresponding to $3 \rightarrow 2$ transition of hydrogen atom falls on a metal surface to produce photoelectrons. These electrons are made to enter a magnetic field of 3×10^{-4} T. If the radius of the largest circular path followed by these electrons is 10.0 mm, the work function of the metal is close to **(JEE Main 2014)**

a. 1.8 eV b. 1.1 eV
c. 0.8 eV d. 1.6 eV

2 The anode voltage of a photocells kept fixed. The wavelength λ of the light falling on the cathode is gradually changed. The plate current I of photocell varies as follows : **(JEE Main 2013)**

a. 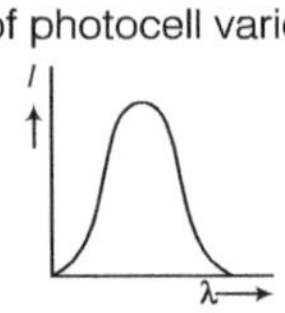

b. 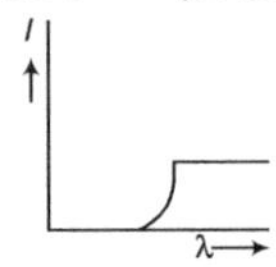

c. 

d. 

Direction (*Q. Nos. 3 and 4*) *This section is based on Statement I and Statement II. Select the correct answer from the codes given below.*

a. Both Statement I and Statement II are correct and Statement II is the correct explanation of Statement I
b. Both Statement I and Statement II are correct but Statement II is not the correct explanation of Statement I
c. Statement I is correct but Statement II is incorrect
d. Statement II is correct but Statement I is incorrect

3 **Statement I** Davisson-Germer experiment established the wave nature of electrons.

Statement II If electrons have wave nature, they can interfere and show diffraction. **(AIEEE 2012)**

4 **Statement I** A metallic surface is irradiated by a monochromatic light of frequency $v > v_0$ (the threshold frequency). The maximum kinetic energy and the stopping potential are K_{max} and V_0, respectively. If the frequency incident on the surface is doubled, both the K_{max} and V_0 are also doubled.

Statement II The maximum kinetic energy and the stopping potential of photoelectrons emitted from a surface are linearly dependent on the frequency of incident light. **(AIEEE 2011)**

5 After absorbing a slowly moving neutron of mass m_N (momentum ~ 0), a nucleus of mass M breaks into two nuclei of masses m_1 and $5m_1$, $(6m_1 = M + m_N)$, respectively. If the de-Broglie wavelength of the nucleus with mass m_1 is λ, then de-Broglie wavelength of the other nucleus will be **(AIEEE 2011)**

a. 25λ b. 5λ
c. $\lambda/5$ d. λ

6 If a source of power 4 kW produces 10^{20} photons/second, the radiation belong to a part of the spectrum called **(AIEEE 2010)**

a. X-rays
b. ultraviolet rays
c. microwaves
d. γ-rays

7 The surface a metal is illuminated with the light of 400 nm. The kinetic energy of the ejected photoelectrons was found to be 1.68 eV. The work function of the metal is (hc = 1240 eV-nm) **(AIEEE 2009)**

a. 3.09 eV b. 1.42 eV
c. 151 eV d. 1.68 eV

Passage for Q. Nos. (8 and 9)

Wave property of electrons implies that they will show diffraction effects. Davisson and Germer demonstrated this by diffracting electrons from crystals. The law governing the diffraction from a crystal is obtained by requiring that electron waves reflected from the planes of atoms in a crystal interfere constructively (see figure). **(AIEEE 2008)**

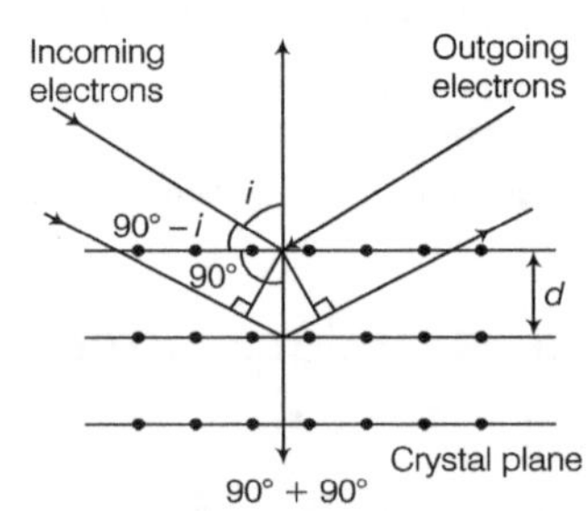

8 Electrons accelerated by potential V are diffracted from a crystal. If $d = 1$ Å and $i = 30°$, V should be about ($h = 6.6 \times 10^{-34}$ J-s, $m_e = 9.1 \times 10^{-31}$ kg, $e = 1.6 \times 10^{-19}$ C)

a. 2000 V
b. 50 V
c. 500 V
d. 1000 V

9 If a strong diffraction peak is observed when electrons are incident at an angle i from the normal to the crystal planes with distance d between them (see figure), de-Broglie wavelength λ_{dB} of electrons can be calculated by the relationship (n is an integer)

a. $d \sin i = n\lambda_{dB}$
b. $2d \cos i = n\lambda_{dB}$
c. $2d \sin i = n\lambda_{dB}$
d. $d \cos i = n\lambda_{dB}$

10 In an experiment, electrons are made to pass through a narrow slit of width d comparable to their de-Broglie wavelength. They are detected on a screen at a distance D from the slit (see figure).

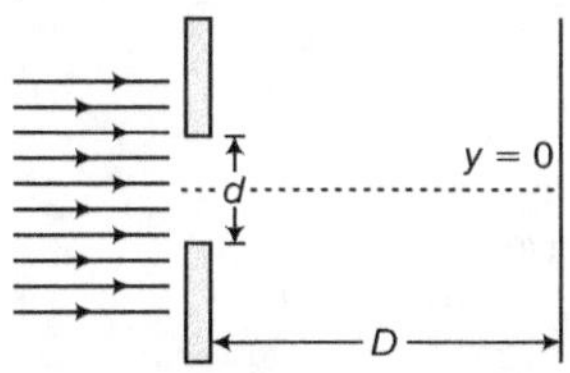

Which of the following graphs can be expected to represent the number of electrons N detected as a function of the detector position y ($y = 0$ corresponds to the middle of the slit)? **(AIEEE 2008)**

a. 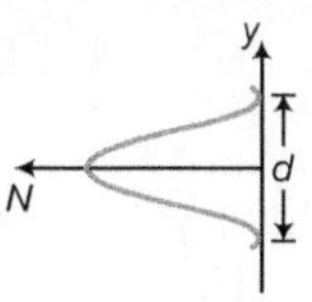

b. 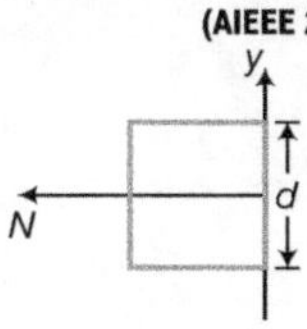

c. 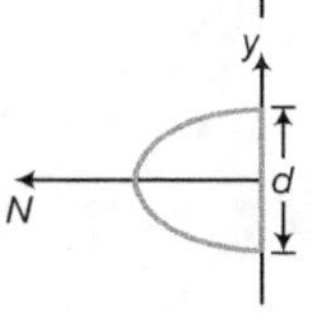

d. 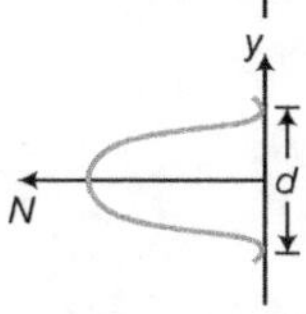

11 Photon of frequency ν has a momentum associated with it. If c is the velocity of light, then the momentum is **(AIEEE 2007)**

a. $\frac{\nu}{c}$ b. $h\nu c$
c. $\frac{h\nu}{c^2}$ d. $\frac{h\nu}{c}$

12 The threshold frequency for a metallic surface corresponds to an energy of 6.2 eV and the stopping potential for a radiation incident on this surface is 5 V. The incident radiation lies in **(AIEEE 2006)**

a. ultraviolet region
b. infrared region
c. visible region
d. X-ray region

13 The time taken by a photoelectron to come out after the photon strikes is approximately **(AIEEE 2006)**

a. 10^{-4} s b. 10^{-10} s
c. 10^{-16} s d. 10^{-1} s

14 The anode voltage of a photocell is kept fixed. The wavelength λ of the light falling on the cathode is gradually changed. The plate current I of the photocell varies as follows : **(AIEEE 2006)**

a. 

b. 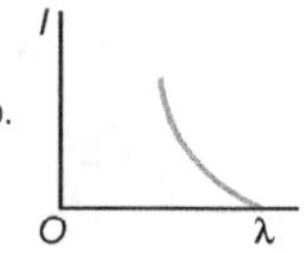

c. 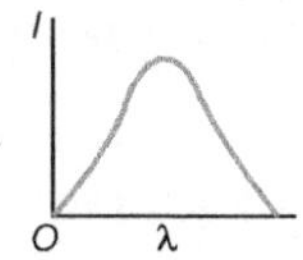

d. None of these

15 A uniform electric field and a uniform magnetic field are acting along the same direction in a certain region. If an electron is projected along the direction of the fields with a certain velocity, then **(AIEEE 2005)**

a. its velocity will decrease
b. its velocity will increase
c. it will turn towards right of direction of motion
d. it will turn towards left of direction of motion

16 If the kinetic energy of a free electron doubles, its de-Broglie wavelength changes by the factor **(AIEEE 2005)**

a. $\frac{1}{2}$ b. 2
c. $\frac{1}{\sqrt{2}}$ d. $\sqrt{2}$

17 A photocell is illuminated by a small bright source placed 1 m away. When the same source of light is placed 1/2 m away, the number of electrons emitted by photocathode would **(AIEEE 2004)**

a. decrease by a factor of 4
b. increase by a factor of 4
c. decrease by a factor of 2
d. increase by a factor of 2

18 According to Einstein's photoelectric equation, the plot of the kinetic energy of the emitted photoelectrons from a metal *versus* the frequency of the incident radiation gives a straight line whose slope **(AIEEE 2004)**

a. depends on the nature of the metal used
b. depends on the intensity of the radiation
c. depend both on the intensity of the radiation and the metal used
d. is the same for all metals and independent of the intensity of the radiation

19 The work function of a substance is 4.0 eV. The longest wavelength of light that can cause photoelectron emission from this substance is approximately **(AIEEE 2004)**

a. 540 nm
b. 400 nm
c. 310 nm
d. 220 nm

20 Two identical, photocathodes receive light of frequencies f_1 and f_2. If the velocities of the photoelectrons (of mass m) coming out are respectively v_1 and v_2, then **(AIEEE 2003)**

a. $v_1^2 - v_2^2 = \frac{2h}{m}(f_1 - f_2)$
b. $v_1 + v_2 = \left[\frac{2h}{m}(f_1 + f_2)\right]^{1/2}$
c. $v_1^2 + v_2^2 = \frac{2h}{m}(f_1 + f_2)$
d. $v_1 - v_2 = \left[\frac{2h}{m}(f_1 - f_2)\right]^{1/2}$

21 Formation of covalent bonds in compounds exhibits **(AIEEE 2002)**

a. wave nature of electron
b. particle nature of electron
c. both wave and particle nature of electron
d. None of the above

22 Sodium and copper have work functions 2.3 eV and 4.5 eV, respectively. Then, the ratio of the wavelengths is nearest to **(AIEEE 2002)**

a. 1 : 2
b. 4 : 1
c. 2 : 1
d. 1 : 4

JEE Advanced & IIT JEE Archive

(Compilation of Last 10 Years' Questions)

1 For photoelectric effect with incident photon wavelength λ, the stopping potential is V_0. Identify the correct variation(s) of V_0 with λ and $1/\lambda$.
(2015, Only One Option Correct Type)

a. 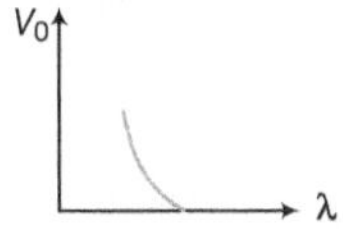

b. 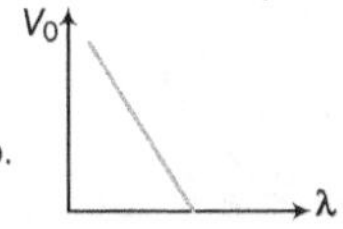

c. 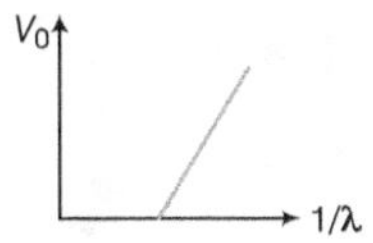

d. 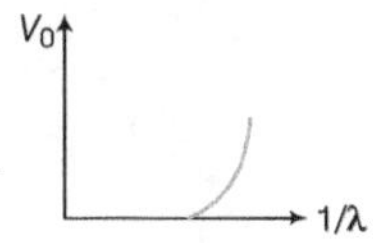

2 An electron in an excited state of Li^{2+} ion has angular momentum $3h/2\pi$. The de-Broglie wavelength of the electron in this state is $P\pi a_0$ (where, a_0 is the Bohr radius). The value of P is **(2015 Adv., Integer Type)**

3 The work functions of silver and sodium are 4.6 and 2.3 eV, respectively. The ratio of the slope of the stopping potential *versus* frequency plot for silver to that of sodium is **(2013 Adv., Integer Type)**

4 A silver sphere of radius 1 cm and work function 4.7 eV is suspended from an insulating thread in free-space. It is under continuous illumination of 200 nm wavelength light. As photoelectrons are emitted, the sphere gets charged and acquires a potential. The maximum number of photoelectrons emitted from the sphere is $A \times 10^Z$ (where, $1 < A < 10$). The value of Z is **(2011, Integer Type)**

5 An α-particle and a proton are accelerated from rest by a potential difference of 100 V. After this, their de-Broglie wavelengths are λ_α and λ_p respectively. The ratio $\frac{\lambda_p}{\lambda_\alpha}$, to the nearest integer, is
(2010, Integer Type)

Passage for Q. Nos. (6-8)

When a particle is restricted to move along X-axis between $x = 0$ and $x = a$, where a is of nanometre dimension, its energy can take only certain specific values. The allowed energies of the particle moving in such a restricted region, correspond to the formation of standing waves with nodes at its ends $x = 0$ and $x = a$. The wavelength of this standing wave is related to the linear momentum p of the particle according to the de-Broglie relation. The energy of the particle of mass m is related to its linear momentum as $E = \frac{p^2}{2m}$.

Thus, the energy of the particle can be denoted by a quantum number n taking values 1, 2, 3, ...($n = 1$, called the ground state) corresponding to the number of loops in the standing wave.

Use the model described above to answer the following three questions for a particle moving in the line $x = 0$ to $x = a$. [Take, $h = 6.6 \times 10^{-34}$ Js and $e = 1.6 \times 10^{-19}$ C] **(2009, Comprehension Type)**

6 The allowed energy for the particle for a particular value of n is proportional to

a. a^{-2} b. $a^{-3/2}$ c. a^{-1} d. a^2

7 If the mass of the particle is $m = 1.0 \times 10^{-30}$ kg and $a = 6.6$ nm, the energy of the particle in its ground state is closest to

a. 0.8 MeV b. 8 MeV c. 80 MeV d. 800 MeV

8 The speed of the particle that can take discrete values is proportional to

a. $n^{-3/2}$ b. n^{-1}

c. $n^{1/2}$ d. n

9 Photoelectric effect experiments are performed using three different metal plates p, q and r having work functions $\phi_p = 2.0$ eV, $\phi_q = 2.5$ eV and $\phi_r = 3.0$ eV, respectively. A light beam containing wavelengths of 550 nm, 450 nm and 350 nm with equal intensities illuminates each of the plates. The correct I-V graph for the experiment is **(2009, Only One Option Correct Type)**

a. 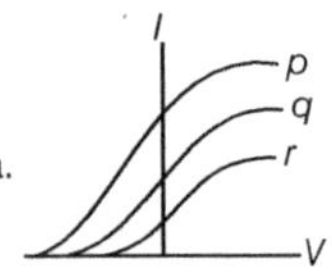

b. 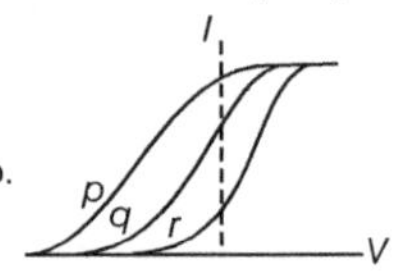

c. 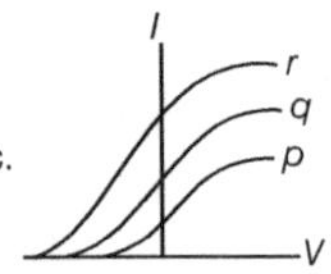

d. 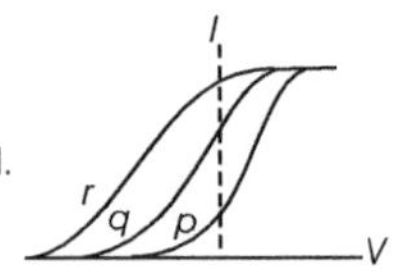

10 Electrons with de-Broglie wavelength λ fall on the target in an X-ray tube. The cut-off wavelength of the emitted X-rays is **(2007, Only One Option Correct Type)**

a. $\lambda_0 = \dfrac{2mc\lambda^2}{h}$ b. $\lambda_0 = \dfrac{2h}{mc}$

c. $\lambda_0 = \dfrac{2m^2c^2\lambda^3}{h^2}$ d. $\lambda_0 = \lambda$

11 The graph between $1/\lambda$ and stopping potential (V) of three metals having work functions ϕ_1, ϕ_2 and ϕ_3 in an experiment of photoelectric effect is plotted as shown in the figure. Which of the following statement(s) is/are correct? (Here, λ is the wavelength of the incident ray). **(2006, One or More than One Options Correct Type)**

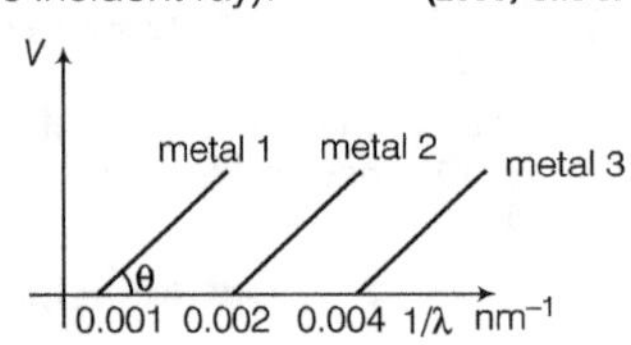

a. Ratio of work functions $\phi_1 : \phi_2 : \phi_3 = 1 : 2 : 4$
b. Ratio of work functions $\phi_1 : \phi_2 : \phi_3 = 4 : 2 : 1$
c. $\tan\theta$ is directly proportional to hc/e, where h is Planck's constant and c is the speed of light
d. The violet colour light can eject photoelectrons from metals 1 and 2

12 A photon collides with a stationary hydrogen atom in ground state inelastically. Energy of the colliding photon is 10.2 eV. After a time interval of the order of microsecond, another photon collides with same hydrogen atom inelastically with an energy of 15 eV. What will be observed by the detector? **(2005, Only One Option Correct Type)**

a. 2 photons of energy 10.2 eV
b. 2 photons of energy 1.4 eV
c. One photon of energy 10.2 eV and an electron of energy 1.4 eV
d. One photon of energy 10.2 eV and another photon of energy 1.4 eV

13 If the wavelength of the nth line of Lyman series is equal to the de-Broglie wavelength of electron in initial orbit of a hydrogen like element ($Z = 11$). Find the value of n. **(2005, Integer Type)**

2. Atoms

DPP

DAILY PRACTICE PROBLEMS

DPP-1 Early Atomic Models

Subjective Question

Working Space

Direction *(Q. No. 1) This question is subjective in nature, need to be solved completely on notebook.*

1 For scattering by an inverse square field, like produced by a charged nucleus in Rutherford's model, the relation between impact parameter b and scattering angle θ is given by $b = \dfrac{Ze^2 \cot(\theta/2)}{4\pi\varepsilon_0(K)}$

where, K = Kinetic energy of α-particle, Ze = Charge of nucleus.

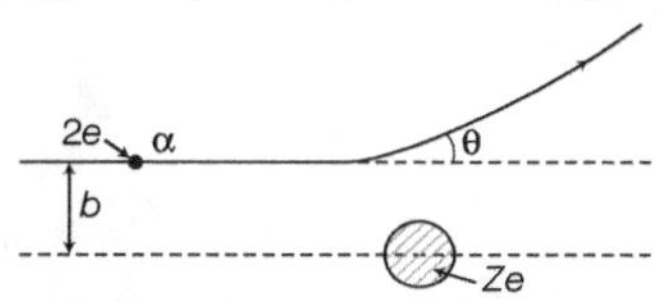

i. What is the scattering angle for $b = 0$?

ii. For a given value of b, does the angle of deflection increase or decrease with increasing energy of α-particle?

iii. What is the impact parameter at which the scattering is 90° for $Z = 79$ and $K = 10\,\text{MeV}$?

iv. Why does the mass of a nucleus not enter the formula given above but the charge does so?

v. For a given energy of the projectile, does the scattering angle increase or decrease with decrease in impact parameter?

vi. Draw a graph of number of α-particles deflected at some given angle θ and also given the formula governing relation of $N(\theta)$ and θ.

Only One Option Correct Type

Direction *(Q. Nos. 2-11) This section contains 10 multiple choice questions. Each question has four choices (a), (b), (c) and (d), out of which ONLY ONE is correct.*

2 Dalton's atomic theory describes an atom as

a. a sphere with only electrons and protons
b. a sphere which is divisible in chemical reactions only
c. a sphere with electrons, photons and neutrons
d. an indivisible particle

3 Which of these are correct regarding atom of same element?

a. Size of atom in Thomson's model is larger than the Rutherford's model
b. Size of atom in Rutherford's model is larger than the Thomson's model
c. Size of atom is same in both Rutherford's and Thomson's model
d. No comparison can be made for both models

4 In α-particle scattering experiment, probability of backward scattering is more in

a. Dalton's atom
b. Thomson's atom
c. Rutherford's atom
d. Same in all of the above

Working Space

5 In α-particle scattering experiment, for a small thickness of gold foil and keeping other parameters constant, the number of α-particles scattered at moderate angles is proportional to thickness ($N \propto t$). This is due to the fact that

a. increase in thickness is proportional to number of collisions of particle
b. increase in thickness is inversely proportional to number of collisions of particle
c. scattering is predominantly due to single collision and chance of a single collision increases with the number of target atoms
d. scattering in thick sheets is due to multiple collision of α-particles

6 What would be the trajectory of an electron in JJ Thomson's experiment without any magnetic field?

a. 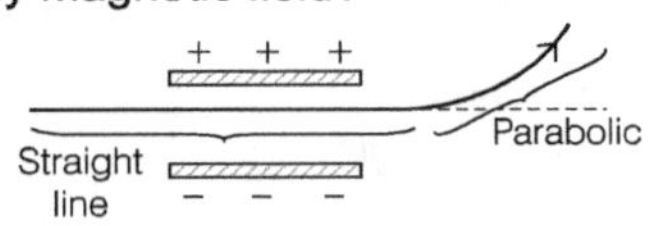

b. 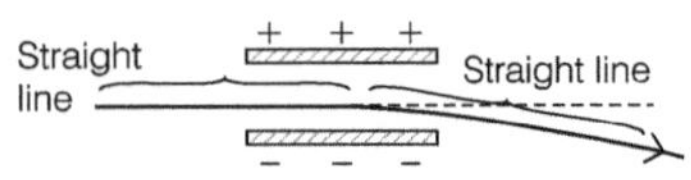

c. 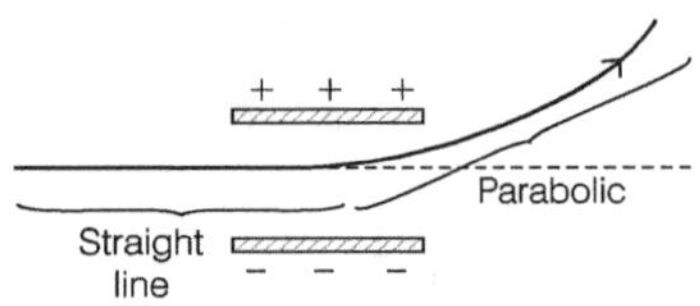

d. 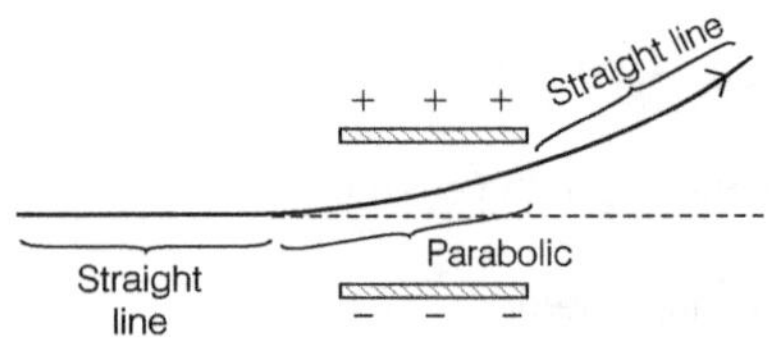

7 An α-particle is scattered by a nucleus as shown in the figure.

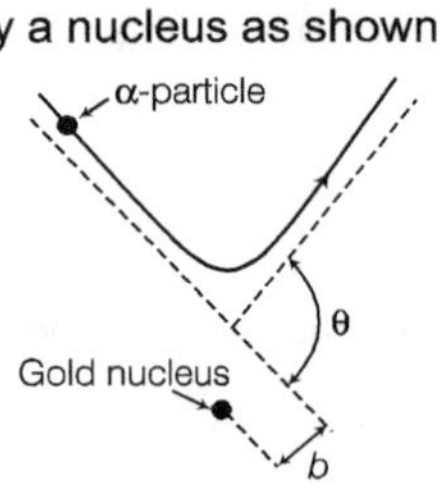

Let mass of α-particle is m and its speed is v. Assuming nucleus remains stationary, change in momentum of α-particle is

a. $2mv\sin\theta$ b. $2mv\cos\theta$
c. $2mv\cos\frac{\theta}{2}$ d. $2mv\sin\frac{\theta}{2}$

8 In Geiger-Marsden experiment, relation between impact parameter b and scattering angle θ is

a. $b \propto \sin\frac{\theta}{2}$ b. $b \propto \cos\frac{\theta}{2}$
c. $b \propto \tan\frac{\theta}{2}$ d. $b \propto \cot\frac{\theta}{2}$

Working Space

9 In α-particle scattering experiment for a given angle θ, fraction of α-particles scattered by angle θ or more by a foil of thickness *t* and containing *n* atoms per unit volume will be

a. $nt\pi b^2$
b. $nt^2\pi b$
c. $n^2t^3b^3$
d. $n\pi$

10 In Thomson's experiment of finding specific charge of electrons, if electrons are replaced by muons (m = 208 × mass of an electron, q = same as electron), then the no deflection is observed when

a. B is increased 208 times
b. B is increased 14.4 times
c. E is increased 208 times
d. E is increased 14.4 times

11 In Thomson's experiment the potential difference between cathode and anode in the accelerating column is same as the potential difference between the deflecting plates (in the region of crossed **E** and **B**). If potential difference is doubled, by what factor should the magnetic field be increased to ensure that the electron beam remains undeflected?

a. $\sqrt{2}$
b. $2\sqrt{2}$
c. 2
d. 4

Comprehension Type

Direction *(Q. Nos. 12-15) This section contains 2 paragraphs, each describing theory, experiments, data, etc. Four questions related to the paragraphs have been given. Each question has only one correct answer among the four given options (a), (b), (c) and (d).*

Passage I

A Thomson's mass spectrograph is used to measure the masses of various isotopes in a gas. This is done by measuring q/m ratio of the gas ions.
Gas ions are produced by ionising gas sample at low pressure.
These ions are accelerated and are made to pass through a region of parallel electric and magnetic fields.

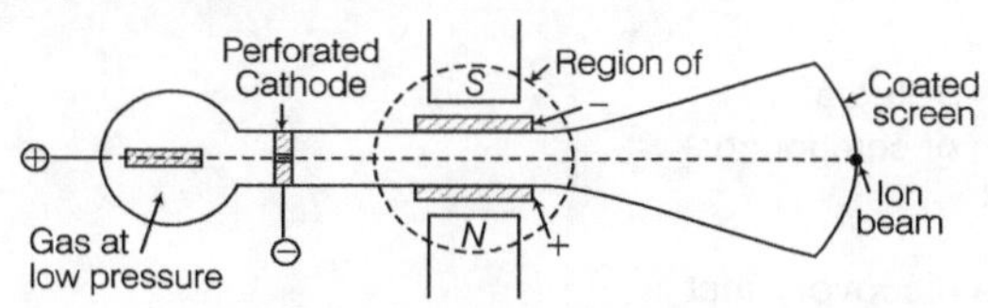

Let initially ions are moving along positive X-axis and electric and magnetic fields are applied along $+y$-direction.

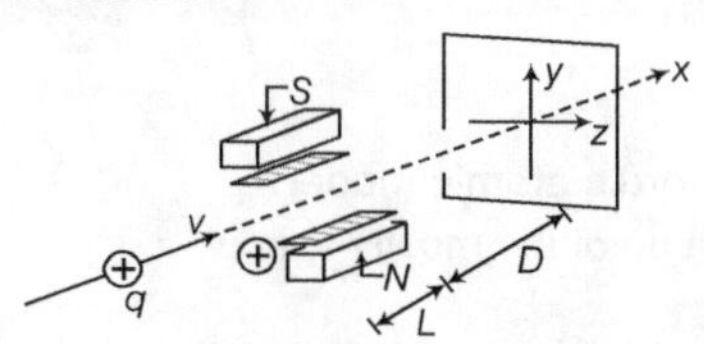

Let L = length of region in which electric and magnetic fields exists and D = distance of region of electric and magnetic fields from screen.

12 Equation of curve traced on screen, when gas contains only one isotope will be (Take, C = a constant)

a. $z^2 = C \cdot y$
b. $y^2 = C \cdot z$
c. $z^2 + Cy^2 = 1$
d. $y^2 + Cz^2 = 1$

Working Space

13 When gas contains more than one isotopes, then there are more them one parabola on screen. Then, mark the correct option.

a. Latusrectum for heavy isotope > Latusrectum for lighter isotope
b. Latusrectum for heavy isotope < Latusrectum for lighter isotope
c. Latusrectum for heavy isotope = Latusrectum for lighter isotope
d. Data insufficient to reach any conclusion

Passage II

In some of Rutherford's experiments, a silver foil is used. Let kinetic energy of α-particles is 4.8 MeV, mass of α-particles is 6.64×10^{-27} kg, charge on silver nucleus is $47e$.

14 For a head-on collision impact parameter for an α-particle, which is scattered at an angle of 45° is

a. 2.8×10^{-14} m
b. 3.36×10^{-14} m
c. 1.41×10^{-14} m
d. 5.60×10^{-14} m

15 Change in momentum of the α-particle when it is scattered by an angle of 45° is

a. 3.38×10^{14} kg - ms^{-1}
b. 5.6×10^{14} kg - ms^{-1}
c. 1.41×10^{14} kg - ms^{-1}
d. 2.3×10^{14} kg - ms^{-1}

One or More than One Options Correct Type

Direction (Q. Nos. 16-19) *This section contains 4 multiple choice questions. Each question has four choices (a), (b), (c) and (d), out of which ONE or MORE THAN ONE are correct.*

16 Which of the following statements of Dalton's atomic model are still known to be correct?

a. Each element has its own type of atom
b. All matter is made up of tiny particles called atom
c. Atoms can be joined together to form of compounds
d. Atoms of same element are identical

17 Gold foil experiment contributed to the discovery that

a. atoms contains electrons
b. atoms are mostly empty space
c. atoms contains a nucleus
d. nucleus is positively charged

18 What are the key reasons for failure of Rutherford's atomic model?

a. Rutherford's model has the atom that is made up of the mostly space
b. Rutherford's model includes neutrons
c. Energy of Rutherford's model keeps on the decreasing
d. Rutherford's model does not explain line emission spectrum

19 JJ Thomson's experimental observations revealed followings options.

a. Atoms are divisible
b. Atoms contains charges
c. Atoms have nucleus
d. Atoms contains few uncharged particles

Matching List Type

Working Space

Direction (Q. No. 20) *Choices for the correct combination of elements from Column I and Column II are given as options (a), (b), (c) and (d), out of which one is correct.*

20 Considering the outcomes of Geiger-Marsden experiment, match the Column I with Column II and mark the correct option from the codes given below.

	Column I		Column II
i.	Most pass undeflected	p.	There is a positively charged centre inside atom
ii.	Few deflected at small angles	q.	Nucleus is very small and massive
iii.	Very few are deflected at large angles	r.	Matter is mostly empty

Codes

	i	ii	iii		i	ii	iii
a.	r	p	q	b.	r	q	p
c.	p	r	q	d.	p	q	r

Statement Type

Direction (Q. Nos. 21 and 22) *This section is based on Statement I and Statement II. Select the correct answer from the codes given below.*

a. Both Statement I and Statement II are correct and Statement II is the correct explanation of Statement I
b. Both Statement I and Statement II are correct but Statement II is not the correct explanation of Statement I
c. Statement I is correct but Statement II is incorrect
d. Statement II is correct but Statement I is incorrect

21 **Statement I** As total energy of an electron in Rutherford's model is negative, this model proposses a stable orbit.

Statement II A negative total energy is necessary for a stationary stable orbit.

22 **Statement I** In cathode ray tube a low pressure is applied to keep number of collisions of electrons (with gas atoms) as low as possible.

Statement II Number of collisions per unit time is proportional to mean free path length of a gas molecule.

DPP-2 Bohr's Theory for Hydrogen and Hydrogen-like Atoms

Subjective Questions

Working Space

Direction (Q. Nos. 1 and 2) *These questions are subjective in nature, need to be solved completely on notebook.*

1 Apply Bohr's atomic model to

i. Estimate the radius of first Bohr orbit of a hydrogen atom, if electron and proton are bound by gravitational attraction.

ii. Estimate the value of $n\left(\text{the integer in } \frac{nh}{2\pi}\right)$ for planetary motion of the Earth around the Sun.

iii. Why do we never speak of quantisation of orbits of planets around the Sun?

2 Bohr's atomic model does not account for screening effect produced due to presence of other electrons.
For a helium atom, ground state ionisation potential measured experimentally is 24.46 V. Estimate the effective charge of nucleus as seen by each electron in its ground state.

Only One Option Correct Type

Direction (Q. Nos. 3-11) *This section contains 9 multiple choice questions. Each question has four choices (a), (b), (c) and (d), out of which ONLY ONE is correct.*

3 In Bohr's atom, energy is absorbed or radiated when an electron makes a transition between two stationary states. If f is the frequency of emitted radiation and E_1 and E_2 are the energies of electron in two states, then practically which is correct for a hydrogen like atom for the transition as shown in the figure.

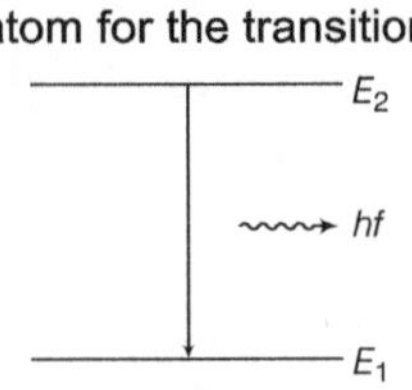

a. $hf = |E_1 - E_2|$
b. $hf < |E_1 - E_2|$
c. $hf > |E_1 - E_2|$
d. $|E_1 - E_2| \le hf \le |E_1 + E_2|$

Working Space

4 Which of the following conservation laws must be fulfilled in a transition in which an atom, a molecule or nucleus absorbs or emits electromagnetic radiation?

I. Energy conservation
II. Momentum conservation
III. Mass conservation

a. Only I
b. Both I and II
c. Both I and III
d. All of these

5 A beam of fast moving electrons accelerated through a potential difference of 10^4 V passed through a monatomic gas at room temperature. Then, mark the correct option.

a. Electrons undergo an elastic collision
b. Electrons undergo an inelastic collision
c. Elastic collision occurs when energy of electrons is greater than first excitation energy of gas atoms
d. Inelastic collision occurs when energy of electrons is less than first excitation energy of gas atoms

6 Photodissociation of carbon monoxide molecule occurs with a minimum energy of 11 eV. This can be made possible by incidenting following electromagnetic radiation.

a. Ultraviolet light
b. Red-Yellow visible light
c. Infrared light
d. Microwaves

7 Two H-atoms in the ground state collides inelastically and are excited to reach first excited state, the maximum amount by which their combined KE is reduced is

a. 10.20 eV
b. 20.40 eV
c. 13.6 eV
d. 27.20 eV

8 For the ground state the electron in H-atom has an angular momentum of $h/2\pi$, according to simple Bohr model angular momentum is a vector and hence there will be infinitely many orbits with the vector pointing in all possible directions.In reality, this is not correct because

a. Bohr model gives incorrect value of angular momentum
b. only one of these orientations corresponds to minimum energy
c. angular momentum is in the direction of spin of electron
d. electrons go around in horizontal orbits

9 Simple Bohr model cannot be applied to calculate the energy levels of a multi-electron atom because

a. electrons collide with each other
b. non-validity of the Coulomb's law
c. electrons not being subjected to a central force
d. None of the above

10 If M = mass of H_2 atom, m_e = mass of an electron and m_p = mass of a proton, then which of the following is correct? (Assume the model is Bohr's atomic model)

a. $M - m_e - m_p = 0$
b. $M - (m_e + m_p) > 0$
c. $m_e + m_p - M \approx 2.42 \times 10^{-35}$ kg
d. $m_e + m_p - M \approx 9.31$ MeV/c^2

11 Equivalent electric current created by electron of a H-atom in its ground state using Bohr's model is nearly

a. 2.42×10^{-35} A
b. 4.48×10^{-30} A
c. 24×10^{-30} A
d. 70×10^{-35} A

One or More than One Options Correct Type

Working Space

Direction (Q. Nos. 12-16) *This section contains 5 multiple choice questions. Each question has four choices (a), (b), (c) and (d), out of which ONE or MORE THAN ONE are correct.*

12 What are the shortcomings with the Bohr's atomic model?

a. It violates the Heisenberg uncertainty principle
b. It provides an incorrect value for angular momentum
c. It does not explain relative intensities of spectral lines
d. It does not predict spectrum of larger atoms

13 A single electron orbits around a stationary nucleus. If 47.2 eV is required to excite the electron from the second orbit to third orbit. (Assume the model is, Bohr's atomic model) Then, mark the correct options.

a. Number of protons in the nucleus is five
b. Total energy of electron is in ground state 340 eV
c. Kinetic energy of electron in ground state is 340 eV
d. Potential energy of electron in ground state is –680 eV

14 Let potential energy of electron in Bohr's first orbit of hydrogen atom is zero. Then, mark the correct options.

a. Kinetic energy of electron in second orbit is 3.40 eV
b. Potential energy of electron in second orbit is 20.4 eV
c. Total energy of electron in second orbit is – 23.80 eV
d. Kinetic energy of electron in second orbit is always 3.40 eV irrespective of choice of zero of potential energy

15 A neutron of kinetic energy K undergoes a head on collision with a hydrogen atom at rest in ground state and is free to move.

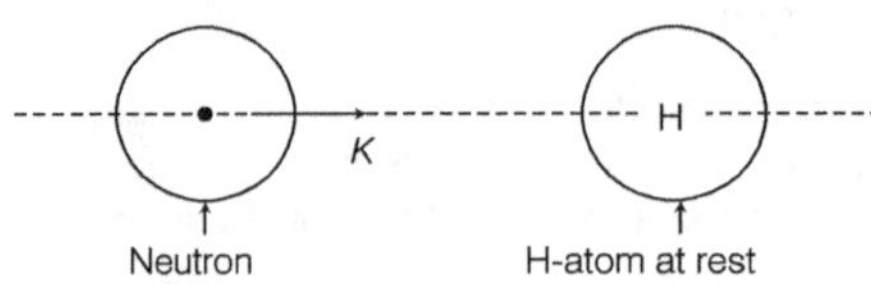

Then, mark the correct option(s).

a. For $K < 20.4$ eV collision is perfectly elastic
b. For $K > 20.4$ eV collision is perfectly inelastic
c. For $K \geq 20.4$ eV collision may be elastic or it may be perfectly inelastic
d. For $K = 24.18$ eV collision may be elastic, inelastic or perfectly inelastic

16 Consider a hypothetical hydrogen like atom having a nucleus of mass $2m$ with an electron of mass m revolving around the nucleus. Then, mark the correct option(s) is/are.

a. Radius of orbit of electron in nth stationary state is $0.795 \frac{n^2}{Z}$ (Å)

b. Energy of electron in nth state $-9.1\left(\frac{Z^2}{n^2}\right)$ (eV)

c. Radius of orbit of electron in nth stationary state is $0.53 \frac{n^2}{Z}$ (Å)

d. Energy of electron in nth state is $-\frac{13.6}{n^2}$ (eV)

Matching List Type

Direction (Q. No. 17) *Choices for the correct combination of elements from Column I and Column II are given as options (a), (b), (c) and (d), out of which one is correct.*

Working Space

17 For a hydrogen atom match the physical quantity in Column I with the dependence of it to principal quantum number in Column II and mark the correct option from the codes given below.

	Column I		Column II
i.	Magnetic moment	p.	$(n)^{-5}$
ii.	Magnetic field	q.	$(n)^{-3}$
iii.	Angular frequency	r.	$(n)^{0}$
iv.	Corresponding current	s.	$(n)^{1}$

Codes

	i	ii	iii	iv		i	ii	iii	iv
a.	s	q	p	r	b.	r	s	q	p
c.	p	q	r	s	d.	s	p	q	q

One Integer Value Correct Type

Direction (Q. Nos. 18 and 19) *This section contains 2 questions. When worked out will result in an integer from 0 to 9 (both inclusive).*

18 Radius of first orbit of a Bohr's atom is 0.265 Å. When this atom makes a transition from first excited state to ground state, it emits a wavelength of 300 Å. Find the ionisation energy for this atom in Rydbergs. (Take, hc = 12375 eV Å and radius of hydrogen atom in ground state, r_H = 0.53 Å)

19 In a mu-meson a particle with charge equal to charge of electron $(-e)$ and mass 208 times of mass of electron moves in a circular orbit around a nucleus of charge $+3e$. Assuming a static nucleus, if in its nth state the radius of orbit of electron is same as Bohr's first orbit for the hydrogen atom, then find the value of $n/5$.

Statement Type

Direction (Q. Nos. 20 and 21) *This section is based on Statement I and Statement II. Select the correct answer from the codes given below.*

a. Both Statement I and Statement II are correct and Statement II is the correct explanation of Statement I
b. Both Statement I and Statement II are correct but Statement II is not the correct explanation of Statement I
c. Statement I is correct but Statement II is incorrect
d. Statement II is correct but Statement I is incorrect

20 **Statement I** Considering Bohr's atom, graph of $\log_e\left(\frac{r_n}{r_1}\right)$ and $\log_e n$ has a slope of 2.

Statement II In case of Bohr's atom, $r_n = 0.53\left(\frac{n^2}{Z}\right)$ Å.

21 **Statement I** In a hydrogen sample at room temperature, there may be a fraction of atoms in excited state.

Statement II Every molecule of sample is moving randomly and undergoes collisions with other molecules.

Comprehension Type

Working Space

Direction (Q. Nos. 22 and 23) This section contains a paragraph, describing theory, experiments, data, etc. Two questions related to the paragraph have been given. Each question has only one correct answer among the four given options (a), (b), (c) and (d).

Passage

In Bohr's atomic model, it is assumed that the mass of electron is very less compared to mass of nucleus and nucleus remains stationary with electron revolving around it. Let's make a correction in Bohr's theory for finite mass and motion of nucleus.

An a atom with nucleus of mass M is in a hydrogen like atom with a single electron of mass m. Let r = distance of nucleus and only electron of atom.

An ω = angular rotational speed of electron.

$\left(\text{Take, angular momentum } L = \frac{nh}{2\pi}\right)$

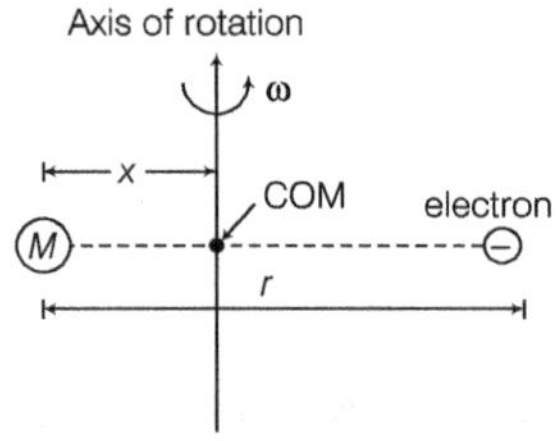

22 Angular momentum of the atom about axis through centre of mass will be

a. $mr^2\omega$

b. $m(r-x)^2\omega$

c. $\frac{r^2\omega(M+m)}{Mm}$

d. $\frac{Mm}{M+m} \cdot r^2\omega$

23 The new ground state energy of electron will be

a. same as ground state energy of Bohr atom

b. more than ground state energy of Bohr atom

c. less than ground state energy of Bohr atom

d. sometime it is less than and sometimes it is more ground state energy Bohr atom

DPP-3 Electronic Transitions in Hydrogen and Hydrogen-like Atoms

Subjective Question

Direction (Q. No. 1) This question is subjective in nature, need to be solved completely on notebook.

Working Space

1 A positronium atom is a system that consists of a positron and an electron that orbits each other. (Use Bohr's Quantisation).

i. Compare the radius of electron orbit of ground state in positronium atom with that of electron in hydrogen atom.

ii. Compare energy of nth state of electron of positronium atom with energy of nth state of electron in hydrogen atom.

iii. Compare the wavelengths of spectral lines of positronium with ordinary hydrogen.

iv. Compare the ionisation energy of positronium with ionisation energy of hydrogen.

v. An antiproton has a mass of a proton but a charge electron. If a proton and an antiproton orbited each other, how for apart would they be in ground state of such a system. Is the system is stable?

Only One Option Correct Type

Direction (Q. Nos. 2-12) This section contains 11 multiple choice questions. Each question has four choices (a), (b), (c) and (d), out of which ONLY ONE is correct.

2 In Hydrogen-like atoms, ratio of $E_{4n} - E_{2n}$ and $E_{2n} - E_n$ is proportional to

a. $\frac{Z^2}{n^2}$ b. $\frac{Z^4}{n^4}$

c. $\frac{Z}{n}$ d. $\frac{Z^0}{n^0}$

3 A hydrogen atom in $n = 5$ state, makes a transition to its ground state. Assuming hydrogen atom initially at rest, the recoil speed of hydrogen atom nearly is

a. 10^{-4} ms^{-1} b. 2×10^{-2} ms^{-1}

c. 4.2 ms^{-1} d. 3.8×10^{-2} ms^{-1}

4 The electron in a hydrogen atom makes transition from M-shell to L-shell. The ratio of magnitudes of initial to final centripetal acceleration of electron is

a. 9 : 4 b. 81 : 16

c. 4 : 9 d. 16 : 81

5 Monochromatic wavelength (λ) is made incident over a hydrogen sample at room temperature. Hydrogen atom subsequently emits radiations of 10 different wavelengths. The value of λ is

a. 95 nm b. 103 nm

c. 73 nm d. 88 nm

Working Space

6 An electron collides with a fixed hydrogen atom in its ground state. Hydrogen atom is excited and colliding electron loses of its kinetic energy. Consequently, the hydrogen atom may emit a photon corresponding to largest wavelength of the Balmer series. Minimum possible kinetic energy of colliding electron will be

a. 10.2 eV b. 1.9 eV
c. 12.1 eV d. 13.6 eV

7 In case of line absorption spectrum of hydrogen gas, colour of absorption line is

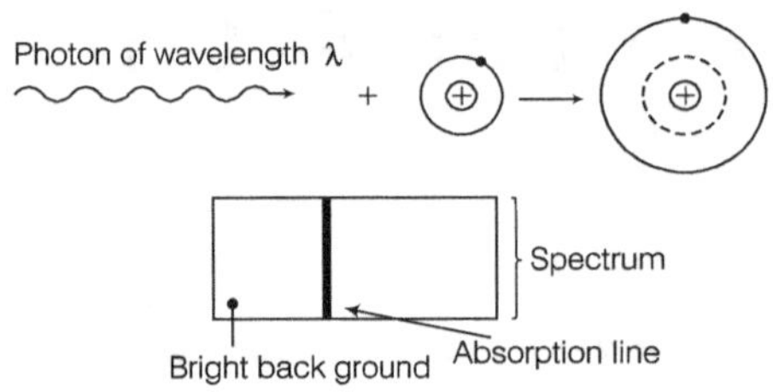

a. black-totally dark
b. bright but of complementary colour to that of emission line
c. dull complementary colour to that of emission line
d. dull and same colour as that of emission line

8 Consider the Bohr's hydrogen atom. An electron from $n = 2$ orbit drops to $n = 1$ orbit. The frequency of photon emitted is

a. equal to revolution frequency of revolution of electron in $n = 1$ orbit
b. equal to revolution frequency of revolution of electron in $n = 2$ orbit
c. equal to an intermediate frequency of revolution of electron in $n = 1$ and $n = 2$ orbit
d. not related to revolution frequency

9 A proton and an electron initially at rest, combined to form a hydrogen atom in ground state. A single photon is emitted in this process. The wavelength of the emitted photon is

a. R b. $2R$
c. $\frac{1}{R}$ d. $\frac{1}{2R}$

10 An excited atom with initial speed v emits a photon in direction of its motion. (Assume v is small comparison to c and relativistic considerations are not required). Then, the frequency of emitted photon (in comparison to frequency emitted when excited atom is initially at rest) is

a. higher than by an amount nearly $\frac{v}{c}$
b. lower than by an amount nearly $\frac{v}{c}$
c. equal to frequency emitted, when excited atom is at rest
d. zero (as a moving excited atom does not emit radiation)

11 Owing to motion of electron, predicted wavelengths of the spectral lines of hydrogen are

a. larger than observed wavelengths
b. smaller than observed wavelengths
c. equal to the observed wavelengths
d. arithmatic mean of just lower and higher values of wavelengths

Working Space

12 Ratio of series limit of Brackett and Pfund series of hydrogen atom spectrum is

a. $\frac{1}{4}$ b. $\frac{4}{9}$

c. $\frac{9}{16}$ d. $\frac{16}{25}$

One or More than One Options Correct Type

Direction (Q. Nos. 13-17) *This section contains 5 multiple choice questions. Each question has four choices (a), (b), (c) and (d), out of which ONE or MORE THAN ONE are correct.*

13 A beam of ultraviolet light of all wavelengths passes through the hydrogen gas at room temperature in *x*-direction. Assume that all photons emitted due to electronic transitions of the hydrogen gas are confined to move in *y*-direction.

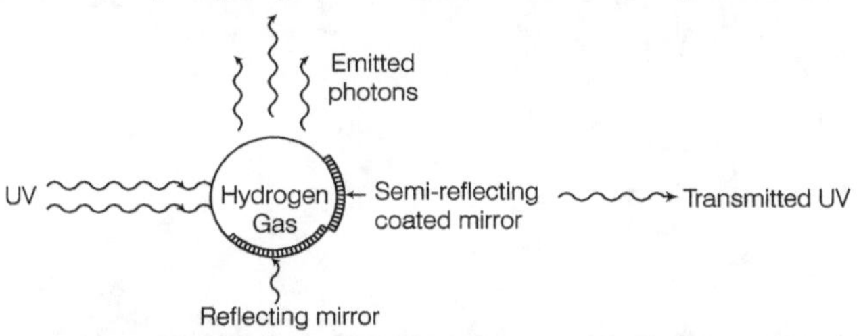

Mark the correct option(s).

a. Some of the incident wavelengths are absent in *x*-direction
b. Only wavelengths are present in *y*-direction which are absent in *x*-direction
c. In *y*-direction some visible light is present
d. In *y*-direction some infrared light is present

14 A particular hydrogen like atom has its ground state binding energy of 122.4 eV. Then, mark the correct option(s).

a. Its atomic number is 3
b. An electron of 90 V can excite it when atom is at room temperature
c. An electron of 91.8 eV kinetic energy can be brought almost to rest by this atom
d. An electron of KE of 2.6 eV may emerge when electron of KE of 125 eV collides with this atom

15 A sample of monatomic hydrogen gas is bombarded with a stream of electrons which are accelerated by a potential difference of 12.75 V. In the emission spectrum, one can observe.

a. Lyman series b. Balmer series
c. Paschen series d. All of these

16 Whenever a hydrogen atom emits a photon in Balmer series,

a. it may emit another photon in the Balmer series
b. it must emit another photon in the Lyman series
c. the second photon (if emitted) have a wavelength of around 122 nm
d. the second photon (if emitted) have a wavelength of around 50 nm

17 Wavelengths and frequencies of photons in transitions having serial numbers 1, 2 and 3 of a hydrogen like atom are $\lambda_1, \lambda_2, \lambda_3$ and f_1, f_2, f_3 respectively. Then, mark the correct option(s).

a. $\lambda_3 = \lambda_1 + \lambda_2$ b. $\lambda_3 = \frac{\lambda_1 \lambda_2}{\lambda_1 + \lambda_2}$

c. $f_3 = f_1 + f_2$ d. $f_3 = \frac{f_1 f_2}{f_1 + f_2}$

Working Space

Comprehension Type

Direction (Q. Nos. 18-22) *This section contains 2 paragraphs, each describing theory, experiments, data, etc. Five questions related to the paragraphs have been given. Each question has only one correct answer among the four given options (a), (b), (c) and (d).*

Passage I

An electron is rotating in a circular path under the influence of a magnetic field of constant magnitude and direction. Assume that Bohr's quantisation of angular momentum is valid for this electron. (Take, h = Planck's constant, q = Charge of electron, r = Radius of path and B = Magnetic field strength)

18 Radius of nth orbit of electron is

a. $\sqrt{\frac{2\pi h}{\pi n B q}}$ b. $\sqrt{\frac{nh}{2\pi B q}}$

c. $\sqrt{\frac{nqh}{2\pi B}}$ d. $\sqrt{\frac{2nh}{\pi B q}}$

19 Kinetic energy of the nth orbit will be

a. $\frac{nhBq}{4\pi m}$ b. $\frac{nhBq}{2\pi m}$

c. $\frac{nhB}{4\pi mq}$ d. $\frac{nhB}{2\pi mq}$

20 Let the electron makes a transition from second orbit to first orbit and let a photon is emitted in this process, then frequency of emitted photon will be

a. zero (no emission)
b. cannot be found
c. half of classical revolution frequency
d. classical revolution frequency

Passage II

Consider a diatomic gas molecule with moment of inertia.

$$I_{COM} = \frac{3}{16} \times 10^{-46}\ \text{kgm}^2$$

Now, assume that

i. A gas molecule in a sample can absorbs energy in form of rotational kinetic energy only.
ii. Angular momentum of rotating gas molecule is quantised $\Rightarrow L = \frac{nh}{2\pi}$, i.e. it follows Bohr's quantisation. (Take, $h = 2\pi \times 10^{-34}$ Js)

21 Ratio of kinetic energy in second possible state of rotation to that of first possible state of rotation is

a. 1 : 2 b. 2 : 1

c. 1 : 4 d. 4 : 1

22 A gas molecule in a sample of gas volume, is transferred to first excited state from ground state. Value of excitation frequency is

a. $\frac{8}{\pi} \times 10^{11}$ Hz b. $\frac{13}{\pi} \times 10^{11}$ Hz

c. $\frac{19}{\pi} \times 10^{11}$ Hz d. $\frac{4}{\pi} \times 10^{11}$ Hz

Statement Type

Direction (Q. Nos. 23-25) *This section is based on Statement I and Statement II. Select the correct answer from the codes given below.*

a. Both Statement I and Statement II are correct and Statement II is the correct explanation of Statement I
b. Both Statement I and Statement II are correct but Statement II is not the correct explanation of Statement I
c. Statement I is correct but Statement II is incorrect
d. Statement II is correct but Statement I is incorrect

Working Space

23 **Statement I** In a high resolution spectroscope, an emission line appears as a small width band diffused on edges.

Statement II When a light ray passes over a sharp edge, it produces a diffraction pattern.

24 **Statement I** A hypothetical one electron atom (hydrogen like) with energy levels defined by $E_n = \frac{-18.0}{n^2}$ eV, shows only three distinct wavelengths in its emission spectrum, when bombarded by electrons accelerated by a potential difference of 16.2 V.

Statement II Maximum possible number of emission lines produced by an electron, when it makes transition from nth stationary state is $\frac{n}{2}(n-1)$.

25 **Statement I** Between any two given energy levels, the number of absorption transitions is always less than the number of emission transition.

Statement II Absorption transition is a one step process it starts always from the ground level, whereas an emission transition may start from any higher energy level and end at any lower energy level.

Matching List Type

Direction (Q. Nos. 26 and 27) *Choices for the correct combination of elements from Column I and Column II are given as options (a), (b), (c) and (d), out of which one is correct.*

26 In Column I, maximum wavelengths of spectral serieses are given and in Column II minimum wavelengths are given. Match the Column I with Column II and mark the correct option from the codes given below.

	Column I (λ_{max})		Column II (λ_{min})
i.	$\frac{4}{3}R$	p.	$9/R$
ii.	$\frac{36}{5}R$	q.	$16/R$
iii.	$\frac{144}{7}R$	r.	$25/R$
iv.	$\frac{400}{9}R$	s.	$1/R$
v.	$\frac{900}{11}R$	t.	$4/R$

Codes

	i	ii	iii	iv	v
a.	s	t	p	q	r
b.	s	t	p	r	q
c.	s	p	t	q	r
d.	s	q	p	r	t

Working Space

27 On the basis of Bohr's atomic model match the Column I and Column II for a hydrogen like atom, which have atomic number Z and mark the correct option from the codes given below.

	Column I		Column II
i.	Radius of orbit	p.	Proportional to Z
ii.	Current due to motion of electron	q.	Inversely proportional to Z
iii.	Magnetic field at centre of atom	r.	Proportional to Z^2
iv.	Velocity of electron	s.	Proportional to Z^3

Codes

	i	ii	iii	iv
a.	q	r	s	p
b.	q	r	p	s
c.	q	s	r	p
d.	q	p	r	s

One Integer Value Correct Type

Direction (*Q. No.* 28) *This section contains a question. When worked out will result in an integer from* 0 *to* 9 (*both inclusive*).

28 A sample of monatomic hydrogen gas contains 100 atoms and all the atoms are in nth excited state.
As the atoms come down to the ground state following various transitions, they release a total energy of $\frac{4800}{49}$ Rch. (where, 1Rch = 13.6 eV)
Find the value of N, where maximum possible number of photons emitted in the process is $7N \times 100$.

DPP-4 X-Rays

Only One Option Correct Type

Working Space

Direction (Q. Nos. 1-10) *This section contains 10 multiple choice questions. Each question has four choices (a), (b), (c) and (d), out of which ONLY ONE option is correct.*

1 Generation of X-rays is

a. conversion of kinetic energy into radiant energy
b. conversion of momentum into energy
c. conversion of mass into energy
d. conversion of electric charge into energy

2 Ratio of energies of X-rays of wavelengths 0.01 Å and 0.5 Å will be

a. 1 : 1 b. 1 : 2
c. 1 : 50 d. 50 : 1

3 The coefficient of absorption of X-rays is related to the atomic number as

a. $\mu \propto \frac{1}{Z^4}$ b. $\mu \propto Z^3$
c. $\mu \propto \frac{1}{Z^3}$ d. $\mu \propto Z^4$

4 Intensity of X-rays from a Coolidge tube is plotted against wavelength λ as shown in the figure.

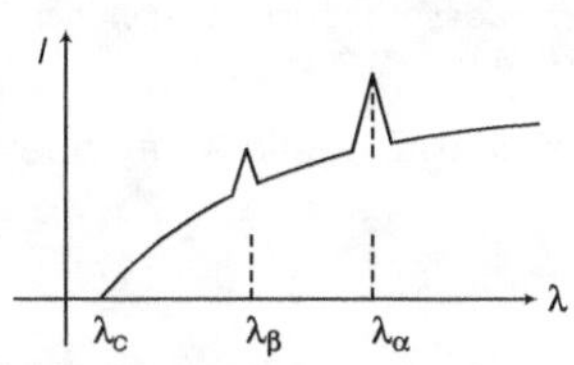

Minimum wavelength is λ_c and the wavelength of the K_α-line is λ_K. As, the accelerating voltage is increased.

a. $\lambda_K - \lambda_c$ increases
b. $\lambda_K - \lambda_c$ decreases
c. $\lambda_{K\alpha}$ increases
d. $\lambda_{K\beta}$ increases

5 Potential difference applied to a X-rays tube is 1.5×10^5 V, 1% energy of the electron is being converted into X-rays photons. Minimum wavelength of X-rays produced is

a. 0.082 Å b. 0.825 Å
c. 8.25 Å d. 82.5 Å

6 Ratio of K_α and K_β wavelengths of X-rays specturm is

a. same for all target materials
b. same for a particular target material
c. depends on filament current
d. depends on accelerating voltage of targetting electrons

Working Space

7 If K_α wavelength for molybdenum target is 0.17 Å. Then, wavelength of K_β-line of copper is nearly around ($Z_{Mo} = 42$, $Z_{Cu} = 29$)

a. 3 Å b. 1.52 Å
c. 1.28 Å d. 0.71 Å

8 For characteristics X-rays of minimum wavelength, an electron must transit.

a. $n = 2$ to $n = 1$ b. $n = 3$ to $n = 1$
c. $n = 4$ to $n = 1$ d. $n = \infty$ to $n = 1$

9 In a tungsten sample some impurity is mixed. If there is a high intensity $\lambda_{K_\alpha} = 0.21$ Å, and a low intensity $\lambda_{K_\alpha} = 1.537$ Å, then the impurity is

a. copper b. zinc
c. molybdenum d. nickel

10 A plot of $\sqrt{f}$ *versus* Z is given

f = frequency of K_α line

Z = Atomic number

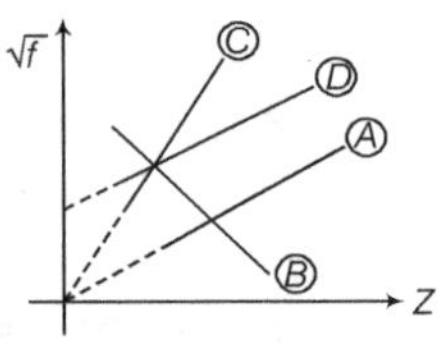

Which of the above lines shows Moseley law?

a. *A* b. *B*
c. *C* d. *D*

One or More than One Options Correct Type

Direction (Q. Nos. 11-15) *This section contains 5 multiple choice questions. Each question has four choices (a), (b), (c) and (d), out of which ONE or MORE THAN ONE are correct.*

11 For a given material the energy and wavelength of characteristic X-ray obey

a. $E(K_\alpha) > E(K_\beta) > E(K_\gamma)$
b. $E(M_\alpha) < E(L_\alpha) < E(K_\alpha)$
c. $\lambda(K_\alpha) > \lambda(K_\beta) > \lambda(K_\gamma)$
d. $\lambda(M_\alpha) > \lambda(L_\alpha) > \lambda(K_\alpha)$

12 An X-rays tube has 3 main controls.

I. Target material (atomic number Z)
II. Filament current
III. Accelerating voltage

The given figure shows a typical intensity distribution.

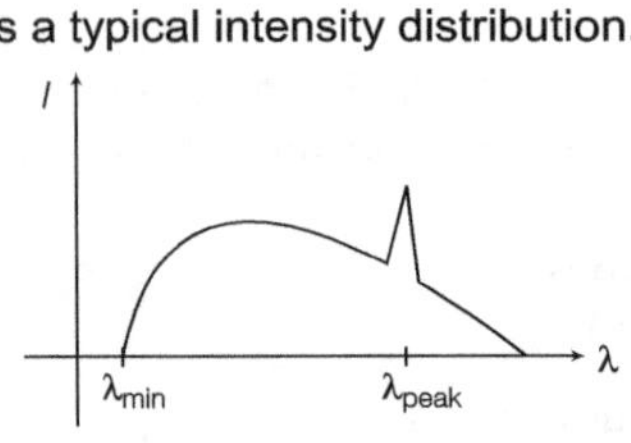

Then, mark the correct options.

a. $\lambda_{min} \propto V^{-1}$
b. λ_{peak} increases with increase in Z
c. Penetrating power of X-rays increases with accelerating voltage
d. Intensity for every wavelength increases for increase in filament current

Working Space

13 Wavelengths K_α for lead isotopes Pb^{208}, Pb^{206} and Pb^{204} are λ_1, λ_2 and λ_3 respectively. Then, mark the correct option(s).

a. $\lambda_1 = \lambda_2 = \lambda_3$
b. $\lambda_1 > \lambda_2 > \lambda_3$
c. $\lambda_1 < \lambda_2 < \lambda_3$
d. $\lambda_2 = \sqrt{\lambda_1 \lambda_3}$

14 An X-rays tube is operating at 50 kV and 20 mA. The target material of the tube has a mass of 1.0 kg and specific heat 495 J kg^{-1} °C^{-1}. One percent of supplied power is converted into X-rays and entire remaining energy goes into heating the target.
Then, mark the correct options.

a. Rate of rise of temperature of target is 2 °CS^{-1}
b. $\lambda_{min} = 0.25 \times 10^{-10}$ m
c. A suitable target material must have high melting temperature
d. A suitable target material must have low thermal conductivity

15 An X-ray tube is operating at 6.6 kV. In the continuous spectrum of the emitted X-rays which of the following frequencies will be missing?

a. 10^{18} Hz
b. 1.5×10^{18} Hz
c. 2×10^{18} Hz
d. 2.5×10^{18} Hz

Comprehension Type

Direction (Q. Nos. 16-20) *This section contains 2 paragraphs, each describing theory, experiments, data, etc. Five questions related to the paragraphs have been given. Each question has only one correct answer among the four given options (a), (b), (c) and (d).*

Passage I

A K_α photon is emitted when an $L(n = 2)$ electron undergoes a transition to a vacant $K(n = 1)$ state. The L-electron experiences a nuclear charge Ze that is reduced to an effective charge in the neighborhood of $(Z - 1)e$ due to the shielding effect of K-electron.

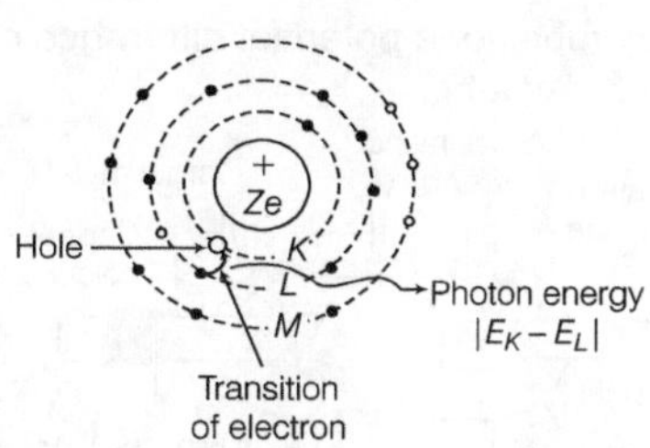

We can use expression for energy of electron of Bohr's model,

$$E_n = -\frac{mZ^2e^4}{8\varepsilon_0^2 h^2 n^2}$$

For a transition from some initial state i to a final state f,

$$\Delta E = hf = |E_f - E_i|$$

$\Rightarrow$ f = frequency of emitted photon of X-ray

$$= \frac{m(Z-1)^2 e^4}{8\varepsilon_0^2 h^3}\left(\frac{1}{n_f^2} - \frac{1}{n_i^2}\right) = cR(Z-1)^2\left(\frac{1}{1^2} - \frac{1}{2^2}\right)$$

$\Rightarrow K_\alpha$X-ray's frequency will be $f = \dfrac{3cR(Z-1)^2}{4}$,

where, $R = \dfrac{me^4}{8\varepsilon_0^2 ch^3} = 1.097 \times 10^7 \text{ m}^{-1}$ is the Rydberg constant.

So, energy of a K_α photon is $E(K_\alpha) = 10.2\,(Z - 1)^2$ eV.

In 1914, Moseley studied the characteristics X-ray spectrum of a number of heavy atoms and concluded that the chemical properties of an element must be a function of its atomic number (Z). As a result, the correct sequence of elements in the periodic table could be established.

Working Space

16 Wavelength of K_α line for an element of atomic number 43 is λ. Then, the wavelength for K_α line for an element of atomic number 29 is

a. $\frac{43}{29}\lambda$ b. $\frac{42}{28}\lambda$

c. $\frac{9}{4}\lambda$ d. $\frac{4}{9}\lambda$

17 Mark the correct graph between $\sqrt{f}$ and Z, (Root of frequency of K_α-line and atomic number of target element).

a.
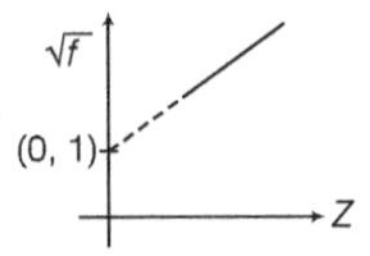

b.
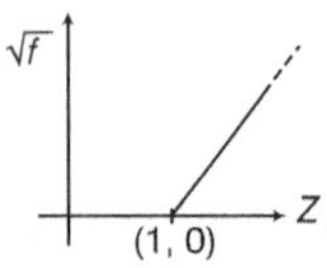

c.
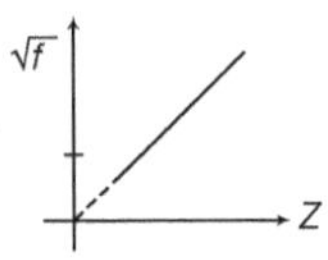

d.
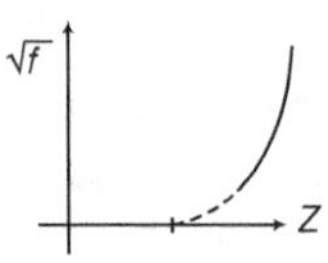

18 Determine the energy of K_β photon of characteristics X-rays (K_β) of tungsten ($Z = 74$) target when an electron from M-shell ($n = 3$) drops into a vacancy in the K-shell ($n = 1$).

a. 80530 eV b. 75690 eV

c. 63840 eV d. 66090 eV

Passage II

In a Coolidge X-ray tube, let a potential difference of 12.4 kV produce a current of 2 mA, in the given X-ray tube.

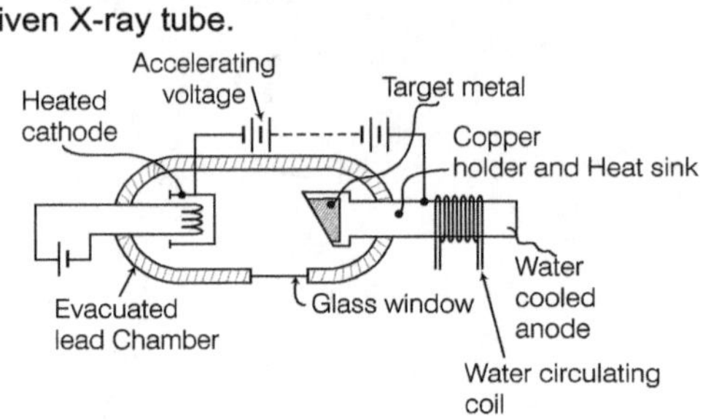

19 What is the minimum wavelength λ_{min} of X-rays produced?

a. 1 nm b. 1 mm

c. 1 Å d. 1 pm

20 For a given target, ratio of wavelengths K_α and K_β of characteristics X-rays is

a. $\frac{3}{4}$

b. $\frac{8}{9}$

c. $\frac{32}{27}$

d. Cannot be determined with given data

Working Space

Matching List Type

Direction (Q. No. 21) *Choices for the correct combination of elements from Column I and Column II are given as options (a), (b), (c) and (d), out of which one is correct.*

21 Match parameters given in Column I with the physical quantity in Column II on which they depends and mark the correct option from the codes given below.

	Column I		Column II
i.	$\lambda_{K\alpha}$ (Characteristics wavelength)	p.	Accelerating voltage
ii.	λ_{min} (cut off wavelength)	q.	Atomic number of target
iii.	Intensity of X-rays	r.	Less Absorption
		s.	More Absorption
		t.	Filament current
		u.	Maximum frequency for a given accelerating voltage
		v.	Not emitted from H-atom

Codes

	i	ii	iii
a.	p,t	q,s,u,v	r
b.	q,s,v	p,r,u,v	t
c.	p,r,v	q,s,u,v	t
d.	q,s,v	p,r,u,v	t

Statement Type

Direction (Q. Nos. 22 and 23) *This section is based on Statement I and Statement II. Select the correct answer from the codes given below.*

a. Both Statement I and Statement II are correct and Statement II is the correct explanation of Statement I
b. Both Statement I and Statement II are correct but Statement II is not the correct explanation of Statement I
c. Statement I is correct but Statement II is incorrect
d. Statement II is correct but Statement I is incorrect

22 **Statement I** X-rays are not used in radar systems.

Statement II Due to their high frequency, they are readily absorbed when they meet an obstacle in their path.

23 **Statement I** X-rays does not form interference pattern in Young's double slit experiment in a set up in which visible monochromatic coherent light forms an intererence pattern.

Statement I Coherence length of X-rays is much smaller than visible light.

SET - 1

Revisal Problems
for JEE Main

Only One Option Correct Type

Working Space

Direction *(Q. Nos. 1-20) This section contains 20 multiple choice questions. Each question has four choices (a), (b), (c) and (d), out of which ONLY ONE option is correct.*

1 In an electronic transition of an electron from $n = 2$ to $n = 1$ state, shortest wavelength is produced in case of
a. hydrogen atom
b. deuterium atom
c. singly ionised helum ion
d. doubly ionised lithium ion

2 From a Coolidge tube, X-rays emerges at
a. a single frequency
b. all wavelengths from 1Å to 100Å
c. a certain frequency band
d. wavelengths larger than a certain minimum wavelength

3 In Column I few physical quantities are given and in Column II values which they can take. Make the Column I with Column II and mark the correct option from the codes given below.

	Column I		Column II
i.	Kinetic energy of an air molecule at room temperature	p.	0.02 eV
ii.	X-ray photon energy	q.	0.06 eV
iii.	Work function of a photosensitive metal	r.	2.0 eV
iv.	Thermal energy of solid molecule at room temperature	s.	10 k eV

Codes

	i	ii	iii	iv		i	ii	iii	iv
a.	p	s	r	q	b.	q	r	s	p
c.	s	r	p	q	d.	r	s	p	q

4 For tungsten K_α wavelength is 0.021 nm. Energy required to excite on electron of tungsten to first excited state must be at least.
a. 0.59 eV
b. 0.59 MeV
c. 59 keV
d. 590 eV

5 In the given reaction $Li^{2+} \longrightarrow {}_{-1}^{0}e + Li^{3+}$ energy required as per Bohr's atomic model is
a. 1.51 eV
b. 13.6 eV
c. 40.8 eV
d. 122.4 eV

6 A heavy metal have electron of $n = 1$ state with energy 75 keV. To get an X-ray spectrum with $\lambda_{min} = 0.155$ Å, metal must be bombarded with electrons having a KE of (Take, $hc = 12400$ eV.Å)
a. at must 80 keV
b. at least 80 keV
c. 80 keV
d. 75 keV to 80 keV

7 The radius of last orbit of ${}_{92}^{132}U$ atom (assume Bohr's atomic model and the radius of last orbit of a multielectron atom is n times that of Bohr's radius) is
a. 0.53 Å
b. 3.047 Å
c. 0.092 Å
d. 48.05 Å

8 A beam of photons with wavelengths 100 nm and 80 nm respectively, is made to pass through a sample of hydrogen gas at room temperature.

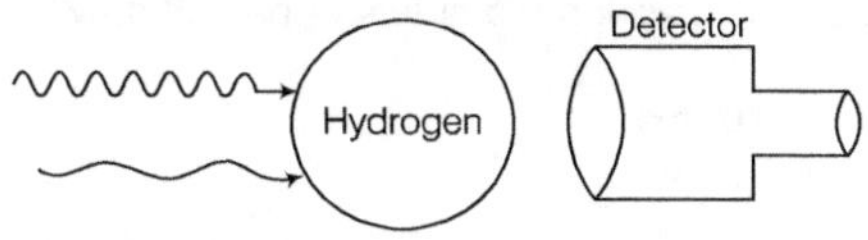

At the detector which of the followings is observed?

a. Light with 3-different wavelengths
b. Few photons and few fast moving electrons
c. White light
d. Light with 5-different wavelengths

9 A doubly ionised lithium ion at room temperature is excited by a photon of wavelength 10.2 eV. Number of emission lines observed in emission spectrum of this ion is

a. 1 b. 4
c. 6 d. 0

10 If a_0 = Bohr's radius, then the radius of the $n = 2$ electronic orbit in triply ionised beryllium is

a. $4a_0$ b. a_0
c. $\frac{a_0}{4}$ d. $\frac{a_0}{16}$

11 Light from a discharge tube containing hydrogen gas falls over a sodium piece (work function of sodium = 1.82 eV). As a result, electrons with KE upto 0.73 eV are emitted from the sodium. Change in angular momentum of electron of H-atom in the emission of light will be

a. $\frac{-h}{2\pi}$ b. $\frac{h}{2\pi}$
c. $\frac{h}{\pi}$ d. $\frac{-h}{\pi}$

12 A fast moving electron enters a region containing α-particles. Electron combines with an α-particle to form a helium ion (He^+), emitting a single photon in this process. He^+ ion formed is in its 4th excited state.
A detector detects photons of energy 3.4 eV, 3.84 eV, 7.56 eV, 48.36 eV, 40.80 eV, 1.2 eV, 51 eV, 52.2 eV, 11.4 eV, 2.64 eV and 10.2 eV in the light emitted from the region of experimentation. Kinetic energy of electron before entering the region of α-particles must be

a. 1.2 eV b. 3.4 eV
c. 10.2 eV d. 4 eV

13 For a hypothetical hydrogen like atom, wavelength λ in Å of photon emitted in a transition shown is $\lambda = \frac{1500\,n^2}{n^2 - 1}$, $n = 2, 3, 4, \cdots$

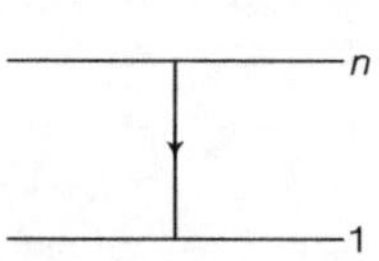

The ratio of momenta of least energetic and the most energetic photons will be

a. $\frac{3}{4}$ b. $\frac{4}{3}$
c. $\frac{2}{1}$ d. $\frac{1}{2}$

14 Ionisation energy of a hydrogen like Bohr atom is 4 Rydberg. Radius of ground state orbit of electron in this atom is

a. 0.53 Å b. 0.265 Å
c. 0.1325 Å d. 1.06 Å

Working Space

15 A neutron moving with a speed v strikes a hydrogen atom which is in ground state and moving towards neutron in same line and with same speed.
The minimum speed of the neutron for which inelastic collision (complete or partial) may takes place is

a. $4 \times 10^4 \text{ ms}^{-1}$
b. $3 \times 10^8 \text{ ms}^{-1}$
c. $4 \times 10^8 \text{ ms}^{-1}$
d. $3 \times 10^6 \text{ ms}^{-1}$

16 Consider the following statements.

I. Sun's spectrum contains few dark lines.
II. Sun's spectrum is line emission spectrum.

Choose the correct options.

a. Both I and II
b. Only I
c. Only II
d. None of the above

17 Let frequency of revolution of an e^- of H-atom in ground is f_1. Then, number of revolutions made by electron in an $n = 2$ Bohr orbit in 8s is

a. $8f_1$
b. $\frac{f_1}{64}$
c. $64f_1$
d. f_1

18 Bohr's atomic theory is first successful theory which explains line emission spectrum is put forward in 1913. Bohr's work is creditable because when in 1924 de-Broglie's hypothesis comes into light, the results are found exactly the same.
In H-atom in $n = 4$ state, the de-Broglie's wavelength λ associated with electron is

a. λ = circumference of Bohr's orbit
b. $\lambda = 4 \times$ circumference of Bohr's orbit
c. $\lambda = \frac{1}{4} \times$ circumference of Bohr's orbit
d. No relation exists

19 Ratio of maximum wavelengths of Brackett and Pfund series for hydrogen atom is

a. $\frac{5}{9}$
b. $\frac{9}{16}$
c. $\frac{16}{25}$
d. $\frac{44}{81}$

20 Consider the following statements.

I. Lymen series lies in ultraviolet region.
II. Balmer series lies in visible region.
III. Paschen, Brackett and Pfund serieses lies in infrared region.

Choose the correct options.

a. Both I and III
b. Both II and III
c. Both I and II
d. All of these

SET - 2

Revisal Problems
for JEE Advanced

Only One Option Correct Type

Working Space

Direction (Q. Nos. 1-5) *This section contains 5 multiple choice questions. Each question has four choices (a), (b), (c) and (d), out of which ONLY ONE is correct.*

1 In a Coolidge tube the potential difference of electron gun is increased from 12.4 kV to 24.8 kV. As a result value of $|\lambda_{K_\alpha} - \lambda_c|$ increases two fold. The wavelength of K_α-line is $\left(\frac{hc}{e} = 12.4\,\text{kV.Å}\right)$

a. 1 Å b. 0.5 Å
c. 1.5 Å d. 1.25 Å

2 Ionisation potentials for electron of a Pb atom are 48×10^{-17} J, 240×10^{-17} J, 1400×10^{-17} J ... etc. Energy of X-ray photon emitted due to transition $2 \rightarrow 1$ of Pb electron is

a. 1.92×10^{-15} J b. 2.40×10^{-15} J
c. 1.16×10^{-14} J d. 1.40×10^{-14} J

3 A 2.67 MeV photon produces an electron-positron pair. Which of these is correct?

a. 1.65 MeV of energy is produced as heat
b. 1.65 MeV of visible light is produced
c. 1.65 MeV of X-ray (photon) is produced
d. 1.65 MeV is the KE of electron positron pair

4 A metal piece when welded using an oxyacetylene flame gets hot and radiates most of the energy at 460 nm. Temperature of hot metal is

a. 2725 K b. 4800 K
c. 6300 K d. 7500 K

5 For three hydrogen transitions as given below.

S.No.	Initial state	Final state
Transition I	1	3
Transition II	6	2
Transition III	4	5

Then, mark the correct option.

a. Transition which involves largest energy cannot occur in steps
b. Transition which involves smallest energy can occurs in steps
c. Transition which involves intermediate energy (intermediate to other two) cannot occurs in steps
d. All of the transitions cannot occurs in steps

One or More than One Options Correct Type

Direction (Q. Nos. 6-9) *This section contains 4 multiple choice questions. Each question has four choices (a), (b), (c) and (d), out of which ONE or MORE THAN ONE are correct.*

6 Consider a doubly ionised Lithium atom and mark the correct options from below.

a. When an electron undergoes $1 \rightarrow 2$ transition, ΔKE $= 91.4$ eV
b. When an electron undergoes $1 \rightarrow 2$ transition Δ PE $= 182.8$ eV
c. When an electron undergoes $4 \rightarrow 3$ transition Δ KE $= 5.95$ eV
d. When an electron is in 4th stationary state, its total energy is -7.65 eV

Working Space

7 Ionisation energy for an electron is defined as $E_{ionisation} = E_{\infty} - E_n$. Then, choose the correct options.

a. An electron cannot absorbs a photon of frequency $f \geq \frac{E_{\infty} - E_n}{h}$

b. An electron can absorbs a photon of frequency $f \geq \frac{E_{\infty} - E_1}{h}$ when electron is in ground state

c. There is no energy level beyond E_{∞} state or $n = \infty$

d. Maximum PE of electron is zero

8 Which of these are correct for emission spectrum of H-like atom according to Bohr's theory ?

a. When an electron makes a transition from $n = 22$ state to $n = 12$ state just one emission line is obtained

b. When an electron makes a transition from $n = 22$ state to $n = 12$ state, there may be a maximum of 55 emission lines in emission spectrum

c. In transition $22 \rightarrow 12$, atom recoil with a momentum

$\Delta p = hRZ^2 \left(\frac{1}{22^2} - \frac{1}{12^2}\right)$

d. Frequency of emitted radiation in transition $n = 22 \rightarrow n = 21$ is

$f = RcZ^2 \left(\frac{1}{21^2} - \frac{1}{22^2}\right)$

9 A class XIIth student draws graph A, B, C and D to show relationships of principal quantum number n and different physical quantities related to H-atom.

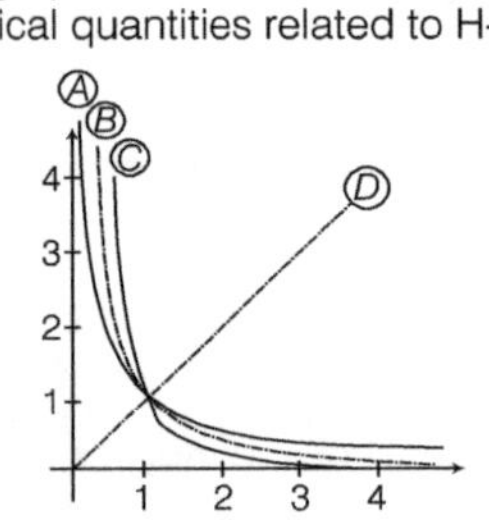

Then, mark the correct option(s).

a. Magnetic field produced due to Bohr's electron at centre of atom is given by graph marked (C)

b. Magnetic dipole moment of electron of H-atom is represented by graph marked (D)

c. Frequency of rotation of electron is represented by graph (B)

d. Energy of electron is represented by graph (A)

Comprehension Type

Direction (Q. Nos. 10-20) *This section contains 5 paragraphs, each describing theory, experiments, data, etc. Eleven questions related to the paragraphs have been given. Each question has only one correct answer among the four given options (a), (b), (c) and (d).*

Passage I

A hydrogen atom may undergoes an 'optical transition' when an electron transits from an higher state n_i to a lower state n_f. The surplus energy is carried away by the photon ($E = hf$). But not all energy conserving optical transitions may occur. As photons also carries momentum, to conserve total angular momentum (atom and photon), in optical transitions the angular momentum of the electron in the initial and final states must differ exactly by one unit i.e. $|l_f - l_i| = 1$ or $\Delta l = \pm 1$. This selection rule must be obeyed in all optical transitions.

10 Few states of H-atom for $n = 2$ are given

I. $n = 2, l = 1, m_l = -1$

II. $n = 2, l = 1, m_l = 0$

III. $n = 2, l = 1, m_l = \pm 1$

IV. $n = 2, l = 0, m_l = 0$

Which of the above states are possible?

a. Both I and II
b. Both II and III
c. Both III and IV
d. All of these

11 Which of the following transition is not frequently possible?

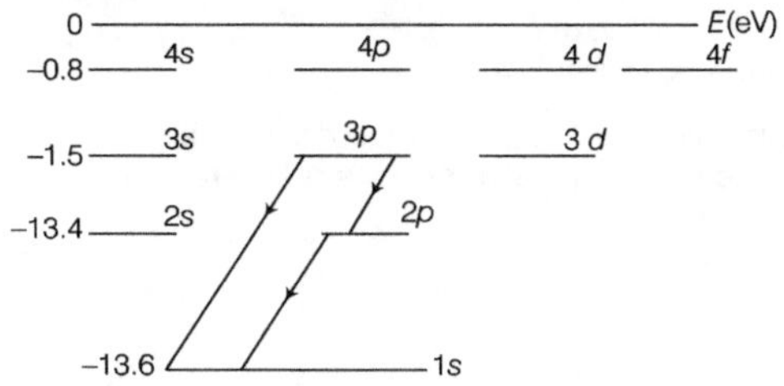

a. $3p \rightarrow 1s$
b. $2p \rightarrow 1s$
c. $3p \rightarrow 2p$
d. All of these

Passage II

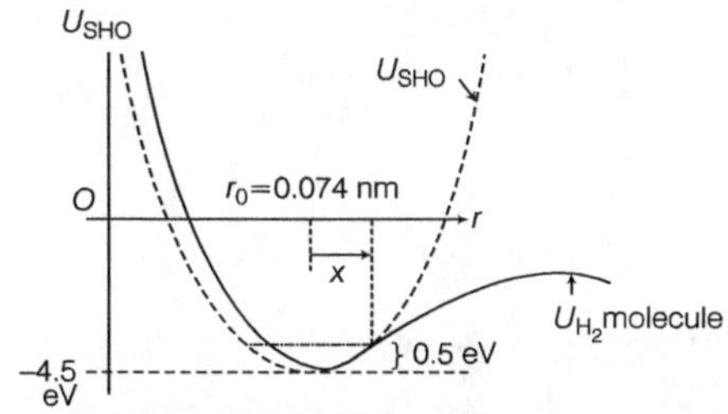

PE of H_2 molecule as a function of atomic separation r is shown above. This curve in the vicinity of equilibrium separation r_0 closely resembles the potential energy of a simple harmonic oscillator, $U = \frac{1}{2}kx^2$, which is shown superposed in dashed lines.

So, for small displacements from r_0, each atom experiences a restoring force approximately proportional to the displacement, and the molecule vibrates as a simple harmonic oscillator (SHO). The classical frequency of vibration is

$$f = \frac{1}{2\pi}\sqrt{\frac{k}{\mu}}$$

where, k = spring constant

and μ = reduced mass $= \frac{m_1 m_2}{m_1 + m_2}$

The "Schrodinger" equation for SHO potential energy yields solution for energy that are quantised according to $E_{vib} = \left(v + \frac{1}{2}\right)hf, v = 0, 1, 2, \ldots$

where, v = an integer called vibrational quantum number. So, allowed vibrational energies of a diatomic molecule are :

Transitions are subjected to selection rule $\Delta v = \pm 1$.

12 Stiffness constant k for H_2 molecule assuming $x^2 = 1.6 \times 10^{-22}\ m^2$ is (use the curve for estimation)

a. $550\ Nm^{-2}$
b. $2000\ Nm^{-2}$
c. $650\ Nm^{-2}$
d. $1000\ Nm^{-2}$

Working Space

Working Space

13 Fundamental wavelength corresponding to vibrational translation for H_2 is

a. $2700\,\pi$(nm)
b. $1000\,\pi$(nm)
c. $540\,\pi$ (nm)
d. $4000\,\pi$ (nm)

14 Hydrogen spectrum also contains wavelengths corresponding to vibrational transitions. Number of different wavelengths that arises due to vibrational transitions of H_2 are

a. ∞
b. 1
c. 4
d. 8

Passage III

If an high energy electron strikes a target atom and disloges an inner atomic electron, the excitation energy of atom may be lost without emission of an X-ray photon. This is called Auger effect.

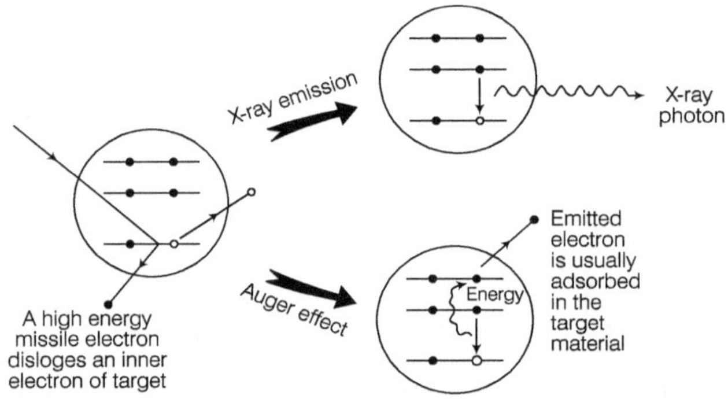

Auger effect is very useful in studying the surface properties in case of semiconductor devices. Auger effect is shown by surface or near surface atoms.
Let $n = 4$ electron is emitted from chromium atom ($Z = 24$) by absorbing the energy from $n = 2$ to $n = 1$ transition.

15 Energy released in transition $2 \rightarrow 1$ is

a. $\frac{5}{4}Z^2R$
b. $3Z^2R$
c. $\frac{3}{2}Z^2R$
d. $\frac{3}{4}Z^2R$

16 KE of emitted 'Auger' electron is

a. $\frac{3}{4}Z^2R$
b. $\frac{1}{16}Z^2R$
c. $\frac{11}{16}Z^2R$
d. $\frac{3}{16}Z^2R$

Passage IV

Aluminium has a molar mass of 26.89 g and atomic number 13. (Take, density of aluminium $= 3 \times 10^3\ \text{kg m}^{-3}$)

17 Atomic density for aluminium is

a. $6.6 \times 10^{28}\ \text{m}^{-3}$
b. $8.7 \times 10^{26}\ \text{m}^{-3}$
c. $7.5 \times 10^{28}\ \text{m}^{-3}$
d. $7.8 \times 10^{34}\ \text{m}^{-3}$

18 Electron density for aluminium is

a. $6.6 \times 10^{25}\ \text{m}^{-3}$
b. $8.7 \times 10^{26}\ \text{m}^{-3}$
c. $7.5 \times 10^{28}\ \text{m}^{-3}$
d. $7.8 \times 10^{29}\ \text{m}^{-3}$

Passage V

For production of X-rays, it is essential to keep high order vacuum in the tube. Value of λ_{min}, cut off wavelength is

$$\lambda_{min} = \frac{1.2375 \times 10^{-6}}{V}$$
$$= \frac{12375}{V} \cdot \text{Å}$$

Wavelength of K_α-line is given by $\frac{1}{\lambda_{K_\alpha}} = R(Z-1)^2\left(\frac{1}{1^2} - \frac{1}{2^2}\right)$.

Wavelength of L_α-line is given by $\frac{1}{\lambda_{L_\alpha}} = R(Z-7.4)^2\left(\frac{1}{2^2} - \frac{1}{3^2}\right)$.

19 If there is some gas left in the tube, then

a. there will be explosion
b. flames are produced in the tube
c. target metal of anode will be melt
d. filament of cathode will melt

20 Wavelength of K_α-line for an element of atomic number 43 is λ. For element with atomic number 29 it is

a. $\frac{43}{29}\lambda$
b. $\frac{42}{28}\lambda$
c. $\frac{9}{4}\lambda$
d. $\frac{4}{9}\lambda$

Statement Type

Direction (Q. Nos. 21 and 22) *This section is based on Statement I and Statement II. Select the correct answer from the codes given below.*

a. Both Statement I and Statement II are correct and Statement II is the correct explanation of Statement I
b. Both Statement I and Statement II are correct but Statement II is not the correct explanation of Statement I
c. Statement I is correct but Statement II is incorrect
d. Statement II is correct but Statement I is incorrect

21 Statement I In presence of a magnetic field , an emission line $2 \rightarrow 1$ splits into three spectral lines which are equally spaced.

Statement II In an external field energy of magnetic dipole depends on its orientation in the field.

22 Statement I Energy of a K_α photon is approximated as $E(K_\alpha) = 13.6 \times \frac{3}{4}(Z-1)^2$.

Statement II An electron of L-shell is partially screened by an electron of K-shell and so effective nuclear charge is $(Z-1)$ units.

Working Space

JEE Main & AIEEE Archive

(Compilation of Last 13 Years' Questions)

1 Hydrogen ($_1H^1$), deuterium ($_1H^2$), sngly ionised helium $(_2He^4)^+$ and doubly ionised lithium $(_3Li^8)^{++}$ all have one electron around the nucleus. Consider an electron transition form $n = 2$ to $n = 1$. If the wavelengths of emitted radiation are $\lambda_1, \lambda_2, \lambda_3$ and λ_4 respectively for four elements, then approximately which one of the following is correct? **(JEE Main 2014)**

a. $4\lambda_1 = 2\lambda_2 = 2\lambda_3 = \lambda_4$
b. $\lambda_1 = 2\lambda_2 = 2\lambda_3 = \lambda_4$
c. $\lambda_1 = \lambda_2 = 4\lambda_3 = 9\lambda_4$
d. $\lambda_1 = 2\lambda_2 = 3\lambda_3 = 4\lambda_4$

2 In a hydrogen like atom electron make transition from an energy level with quantum number n to another with quantum number $(n-1)$. If $n >> 1$, the frequency of radiation emitted is proportional to **(JEE Main 2013)**

a. $\frac{1}{n}$ b. $\frac{1}{n^2}$
c. $\frac{1}{n^3/2}$ d. $\frac{1}{n^3}$

3 Hydrogen atom is excited from ground state to another state with principal quantum number equal to 4. Then, the number of spectral lines in the emission spectra will be **(JEE Main 2013)**

a. 2 b. 3
c. 5 d. 6

4 A diatomic molecule is made of two masses m_1 and m_2 which are separated by a distance r. If we calculate its rotational energy by applying Bohr's rule of angular momentum quantisation, its energy will be given by (n is an integer) **(AIEEE 2012)**

a. $\frac{(m_1+m_2)^2 n^2h^2}{2m_1^2m_2^2r^2}$
b. $\frac{n^2h^2}{2(m_1+m_2)r^2}$
c. $\frac{2n^2h^2}{(m_1+m_2)r^2}$
d. $\frac{(m_1+m_2)n^2h^2}{2m_1m_2r^2}$

5 Energy required for the electron excitation in Li^{2+} from the first to the third Bohr orbit is **(AIEEE 2011)**

a. 36.3 eV b. 108.8 eV
c. 122.4 eV d. 12.1 eV

6 The transition from the state $n = 4$ to $n = 3$ in a hydrogen like atom results in ultraviolet radiation. Infrared radition will be obtained in the transition from **(AIEEE 2009)**

a. $2 \to 1$ b. $3 \to 2$
c. $4 \to 2$ d. $5 \to 4$

7 Suppose an electron is attracted towards the origin by a force $\frac{k}{r}$, where k is a constant and r is the distance of the electron from the origin. By applying Bohr model to this system, the radius of the nth orbital of the electron is found to be r_n and the kinetic energy of the electron to be T_n. Then, which of the following is true? **(AIEEE 2008)**

a. $T_n \propto \frac{1}{n^2}, r_n \propto n^2$
b. T_n is independent of n, $r_n \propto n$
c. $T_n \propto \frac{1}{n}, r_n \propto n$
d. $T_n \propto \frac{1}{n}, r_n \propto n^2$

8 Which of the following transitions in hydrogen atoms emit photons of highest frequency? **(AIEEE 2007)**

a. $n = 2$ to $n = 6$
b. $n = 6$ to $n = 2$
c. $n = 2$ to $n = 1$
d. $n = 1$ to $n = 2$

9 The diagram shows the energy levels for an electron in a certain atom. Which transition shown represents the emission of a photon with the most energy? **(AIEEE 2005)**

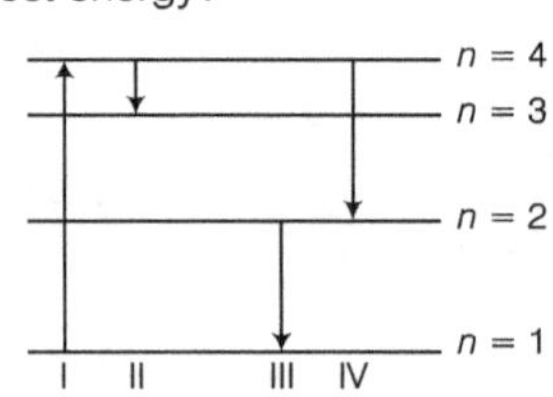

a. III b. IV
c. I d. II

10 If the binding energy of the electron in a hydrogen atom is 13.6 eV, the energy required to remove the electron from the first excited state of Li^{2+} is **(AIEEE 2003)**

a. 30.6 eV b. 13.6 eV
c. 3.4 eV d. 122.4 eV

JEE Advanced & IIT JEE Archive

(Compilation of Last 10 Years' Questions)

1 If λ_{Cu} is the wavelength of k_α, X-ray line of Copper (atomic number 29) and λ_{Mo} is the wavelength of the k_α X-ray line of Molybdenum (atomic number 42), then the ratio $\lambda_{Cu} : \lambda_{Mo}$ is close to **(2012, Only One Option Correct Type)**

a. 1.99 b. 2.14
c. 0.50 d. 0.48

2 The wavelength of the first spectral line in the Balmer series of hydrogen atom is 6561 Å. The wavelength of the second spectral line in the Balmer series of singly ionised helium atom is **(2011, Only One Option Correct Type)**

a. 1215 Å b. 1640 Å
c. 2430 Å d. 4687 Å

Passage (3-5)

The key feature of Bohr's theory of spectrum of hydrogen atom is the quantisation of angular momentum when an electron is revolving around a proton. We will extend this to a general rotational motion to find quantised rotational energy of a diatomic molecule assuming it to be rigid. The rule to be applied is Bohr's quantisation condition. **(2010, Comprehension Type)**

3 A diatomic molecule has moment of inertia I. By Bohr's quantisation condition its rotational energy in the nth level ($n = 0$ is not allowed) is

a. $\frac{1}{n^2}\left(\frac{h^2}{8\pi^2 I}\right)$ b. $\frac{1}{n}\left(\frac{h^2}{8\pi^2 I}\right)$
c. $n\left(\frac{h^2}{8\pi^2 I}\right)$ d. $n^2\left(\frac{h^2}{8\pi^2 I}\right)$

4 It is found that the excitation frequency from ground to the first excited state of rotation for the CO molecule is close to $\frac{4}{\pi} \times 10^{11}$ Hz. Then, the moment of inertia of CO molecule about its centre of mass is close to (take, $h = 2\pi \times 10^{-34}$ J-s)

a. 2.76×10^{-46} kg-m^2
b. 1.87×10^{-46} kg-m^2
c. 4.67×10^{-47} kg-m^2
d. 1.17×10^{-47} kg-m^2

5 In a CO molecule, the distance between C (mass = 12 amu) and O (mass = 16 amu), where 1 amu $= \frac{5}{3} \times 10^{-27}$ kg, is close to

a. 2.4×10^{-10} m b. 1.9×10^{-10} m
c. 1.3×10^{-10} m d. 4.4×10^{-11} m

6 The radius of the orbit of an electron in a Hydrogen-like atom is $4.5\,a_0$ where a_0 is the Bohr radius. Its orbital angular momentum is $\frac{3h}{2\pi}$. It is given that h is Planck constant and R is Rydberg constant. The possible wavelength (s), when the atom de-excites, is (are) **(2008, One or More than One Options Correct Type)**

a. $\frac{9}{32R}$ b. $\frac{9}{16R}$
c. $\frac{9}{5R}$ d. $\frac{4}{3R}$

7 Which one of the following statements is incorrect in the context of X-rays generated from an X-ray tube? **(2008, Only One Option Correct Type)**

a. Wavelength of characteristic X-rays decreases when the atomic number of the target increases
b. Cut-off wavelength of the continuous X-rays depends on the atomic number of the target
c. Intensity of the characteristic X-rays depends on the electrical power given to the X-ray tube
d. Cut-off wavelength of the continuous X-rays depends on the energy of the electrons in the X-ray tube

Passage (8-10)

In a mixture of H-He^+ gas (He^+ is singly ionized He atom), H atoms and He^+ ions are excited to their respective first excited states. Subsequently, H atoms transfer their total excitation energy to He^+ ions (by collisions). Assume that the Bohr model of atom is exactly valid. **(2008, Comprehension Type)**

8 The quantum number n of the state finally populated in He^+ ions is

a. 2 b. 3
c. 4 d. 5

9 The wavelength of light emitted in the visible region by He^+ ions after collisions with H atoms is

a. 6.5×10^{-7} m b. 5.6×10^{-7} m
c. 4.8×10^{-7} m d. 4.0×10^{-7} m

10 The ratio of the kinetic energy of the $n = 2$ electron for the H atom to that of He^+ ion is

a. 1/4 b. 1/2
c. 1 d. 2

11 Some laws/processes are given in Column I. Match these with the physical phenomena given in Column II and mark the correct option from the codes given below. **(2007, Matching Type)**

	Column I		Column II
i.	Transition between two atomic energy levels	p.	Characteristic X-rays
ii.	Electron emission from a material	q.	Photoelectric effect
iii.	Moseley's law	r.	Hydrogen spectrum
iv.	Change of photon energy into kinetic energy of electrons	s.	β-decay

Codes

	i	ii	iii	iv
a.	p	q	r,s	p,q
b.	p,r	p,q,s	p	q
c.	q	p,s	p,q	r,s
d.	q,s	p,s	q,r	p,q,s

12 The largest wavelength in the ultraviolet region of the hydrogen spectrum is 122 nm. The smallest wavelength in the infrared region of the hydrogen spectrum (to the nearest integer) is **(2007, Only One Option Correct Type)**

a. 802 nm b. 823 nm
c. 1882 nm d. 1648 nm

13 K_α wavelength emitted by an atom of atomic number $Z = 11$ is λ. Find the atomic number for an atom that emits K_α radiation with wavelength 4λ. **(2005, Only One Option Correct Type)**

a. $Z = 6$ b. $Z = 4$
c. $Z = 11$ d. $Z = 44$

14 X-rays are incident on a target metal atom having 30 neutrons. The ratio of atomic radius of the target atom and ${}^4_2He$ is $(14)^{1/3}$.

i. Find the mass number of target atom.
ii. Find the frequency of K_α line emitted by this metal. (Take, $R = 1.1 \times 10^7 m^{-1}$, $c = 3 \times 10^8 ms^{-1}$) **(2005, Subjective Type)**

JEE Archive

3. Nuclei

DPP

DAILY PRACTICE PROBLEMS

DPP-1 Atomic Nucleus (Composition, Size, Atomic Masses, Isotopes, Isobars, Isotones and General Stability Trends of Nucleus)

Subjective Question

Working Space

Direction *(Q. No. 1) This question is subjective in nature, need to be solved completely on notebook.*

1 i. Are the nucleons fundamental particles or do they consist of smaller parts? To find the answer of this question let a nucleon is probed just as Rutherford probed an atom. What should be the kinetic energy of an electron for it to probe a nucleon. (Assume that diameter of a nucleon is around at 10^{-15}m)

ii. Find the density of nuclear matter for $^{12}C_6$. Is it different for various elements? Radius of $^{12}C_6$ nucleus is around 2.7 fm.

iii. Find the repulsive electrostatic force on a proton, whose distance from another proton is 2.4 fm (centre to centre). (Assume proton as a uniformly positively charged sphere)

iv. The total Binding Energy of a nucleus is due to its volume energy (energy of nucleons deep inside the nucleus), surface energy (energy of nucleons lies on the surface), and coulomb energies.

$$E_b = E_v + E_s + E_c$$

$$= \left[a_1A - a_2A^{2/3} - a_3\frac{Z(Z-1)}{A^{1/3}} - a_4\frac{(A-2Z)^2}{A} \pm \frac{a^5}{A^{3/4}} \right]$$

This expression is found by CE von Weizsacker.

Where, $a_1 = 14.1\,\text{MeV}$, $a_2 = 13.0\,\text{MeV}$, $a_3 = 0.595\,\text{MeV}$,

$a_4 = 19.0\,\text{MeV}$, $a_5 = 33.5\,\text{MeV}$.

A = mass number and Z = atomic number.

Now using the above formula, derive a formula for atomic number (Z) of the most stable isobar of an element with given A. Also, use this for finding most stable isobar with $A = 25$.

Only One Option Correct Type

Direction *(Q. Nos. 2-12) This section contains 11 multiple choice questions. Each question has four choices (a), (b), (c) and (d), out of which ONLY ONE is correct.*

2 In $^{28}_{14}Si$ (silicon) and $^{85}_{37}Rb$ (rubidium) ratio of neutrons in their nuclei is

a. $\frac{28}{85}$

b. $\frac{14}{37}$

c. $\frac{7}{24}$

d. $\frac{17}{21}$

Working Space

3 Nucleus A and nucleus B have different number of photons and different number of neutrons. Then, mark the correct option.

a. They may have same number of electrons in their neutral atoms
b. They may have a rigid nucleus boundary
c. They may have same binding energy per nucleon
d. They may have equal radii

4 Chemical properties of isotopes of an element are same because

a. the energy required to remove a nucleon is about 10^2 eV
b. the energy required to remove a nucleon is about 10^4 kJ
c. the energy required to remove a nucleon is about few MeV
d. the outer wall of nucleus is made up of neutrons

5 A stable nucleus of an isobar is

a. more heavier than less stable nucleus
b. less heavier than less stable nucleus
c. of same mass as that of a less stable nucleus
d. sometimes more heavier and sometimes lighter than less stable nucleus

6 Which of the following is not correct?

a. Nuclear force is the strongest known force
b. Net nuclear force on a nucleon deep inside the nucleus is zero
c. The nuclear force between two protons is same as that between two neutrons
d. For any separation, the nuclear force between two protons is greater than the electrostatic repulsion between them

7 Consider the plot of N (neutron number) *versus* Z (photon number) for the different nuclei. Let three nuclides A, B and C are at the positions as shown in the figure. The order of their stability may be

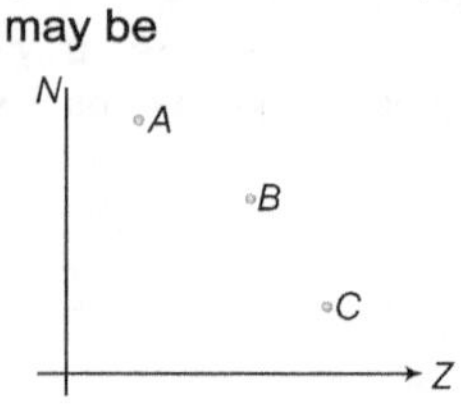

a. $A > B > C$
b. $A < B < C$
c. $B > A > C$
d. $C < A < B$

8 For larger value of proton number Z, the nuclear force is unable to hold the nucleus together against repulsion of protons. For stable nuclides, Z must be at most

a. 140
b. 72
c. 67
d. 83

9 Let V is volume of nucleus, A is mass number of nucleus and ρ is density of nucleus. Mark which of the following is a constant.

a. $A \cdot V$
b. ρV
c. $\frac{V^3}{A^2}$
d. $\frac{V}{A}$

10 Which of the following is correct ?

a. Nuclear forces are central force
b. Nuclear forces are conservative in nature
c. Nuclear forces are charge dependent
d. Nuclear forces are exchange forces

Working Space

11 The nuclei of ${}^7_3Li$ and ${}^7_4Be$ are

a. Isotopes
b. Isobars
c. Isotones
d. Mirror nuclei

12 Two nucleus are projected in a region of magnetic field such that their velocity vectors make same angle with the direction of magnetic field.

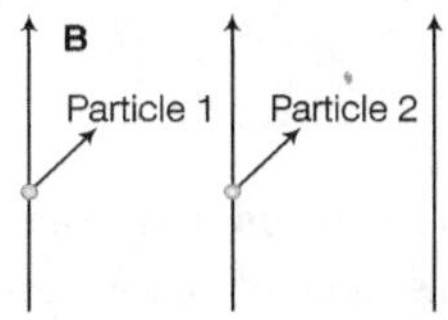

Then,

a. trajectories of isobars are similar helix
b. trajectories of isotones are similar helix
c. trajectories of isotopes are similar helix
d. None of the above

One or More than One Options Correct Type

Direction *(Q. Nos. 13-17) This section contains 5 multiple choice questions. Each question has four choices (a), (b), (c) and (d), out of which ONE or MORE THAN ONE are correct.*

13 Which of these are the characteristics of unstable nucleus?

a. $\frac{N}{Z}$ ratio is high
b. Low binding energy
c. High packing fraction
d. Smaller nuclear volume

14 If *A*, *Z* and *N* denote the mass number, the atomic number and the neutron number for a given nucleus respectively. Then,

a. isobars have same *A* but different *Z* and *N*
b. isotopes have same *Z* but different *N* and *A*
c. isotones have same *N* but different *A* and *Z*
d. mirror nucleus have different *N* and *Z*

15 Which of the following is/are correct?

a. Size of atom increases with increase in *Z*
b. Isotopes have different nuclear properties
c. Isobars have similar nuclear properties
d. For nuclei with $A < 20$ have maximum charge density at centre whereas nuclei with $A > 40$, the charge density is nearly constant with in the nucleus

16 Consider the nuclei ${}^1H_1$, ${}^2H_1$, ${}^3He_2$, ${}^{12}C_6$, ${}^{13}C_6$, ${}^{30}P_{15}$, ${}^{30}Si_{14}$, ${}^{10}B_5$ and ${}^{11}B_5$. Then, mark the incorrect options.

a. ${}^1H_1$ and ${}^2H_1$ are isotopes
b. ${}^{12}C_6$ and ${}^{13}C_6$ are isobars
c. ${}^{30}P_{15}$ and ${}^{30}Si_{14}$ are isotones
d. ${}^{10}B_5$ and ${}^{11}B_5$ are mirror nuclei

17 Stability of a nucleus is also depends on even and odd number of protons and neutrons. Then,

a. nucleus with even *N* and even *Z* are most stable
b. an even *N*, odd *Z* or odd *N* even *Z* nucleus is somewhat lesser stable than even-even nucleus
c. an odd *N*, odd *Z* nucleus is least stable
d. there are few odd *N*, odd *Z* nuclides which are stable

Statement Type

Working Space

Direction (Q. Nos. 18 and 19) *This section is based on Statement I and Statement II. Select the correct answer from the codes given below.*

a. Both Statement I and Statement II are correct and Statement II is the correct explanation of Statement I
b. Both Statement I and Statement II are correct but Statement II is not the correct explanation of Statement I
c. Statement I is correct but Statement II is incorrect
d. Statement II is correct but Statement I is incorrect

18 **Statement I** Positively charged nucleus of an atom has a radius of almost 10^{-15}m.

Statement II In α-particle scattering experiment, the distance of closest approach is around 10^{-15}m.

19 **Statement I** Atomic mass of an element may not be an integer.

Statement II Observed atomic mass of the element is the weighted average of atomic masses of the individual isotopes.

Comprehension Type

Direction (Q. Nos. 20 and 21) *This section contains a paragraph, describing theory, experiments, data, etc. Two questions related to the paragraph have been given. Each question has only one correct answer among the four given options (a), (b), (c) and (d).*

Passage

One of the process of separating the isotopes ^{235}U and ^{238}U is based on the difference of radii of their paths in a magnetic field.

Assume that, singly ionised atoms of Uranium starts from a common source and move perpendicular to a uniform magnetic field.

20 Maximum spatial separation of beams, when the two isotopes have same energies is

a. $\frac{2\sqrt{2}K}{BZ}(\sqrt{m_2}-\sqrt{m_1})$

b. $\frac{2\sqrt{2K}}{BZ}(\sqrt{m_2}-\sqrt{m_1})$

c. $\frac{2\sqrt{2K(m_2-m_1)}}{BZ}$

d. zero

21 Beam separation when both isotopes have same speed is

a. $\frac{v(m_2-m_1)}{BZ}$

b. $\frac{v(m_2-m_1)}{2BZ}$

c. $\frac{2v(m_2-m_1)}{BZ}$

d. $\frac{\sqrt{2}v(m_2-m_1)}{BZ}$

DPP-2

Mass Defect, Binding Energy, Nuclear Stability, Criteria for Fission and Fusion Processes

Subjective Question

Direction (Q. No. 1) This question is subjective in nature, need to be solved completely on notebook.

Working Space

1 i. Following mass differences have been found experimentally :

a. $^{1}H — ^{2}H = 1.5434 \times 10^{-3}$ amu

b. $3^{2}H — \frac{1}{2}\,^{12}C = 4.2300 \times 10^{-2}$ u

c. $^{12}C\,^{1}H_4 — ^{16}O = 3.6364 \times 10^{-2}u$

Calculate the masses of ^{1}H ^{2}H and ^{16}O. (Take, $^{12}C = 12.000u$)

ii. Consider the fission of $^{235}_{92}U$ by the slow neutrons

$$^{1}_{0}n + ^{235}_{92}U \longrightarrow ^{236}_{92}U^{*} \longrightarrow ^{95}_{42}Mo + ^{139}_{57}La + 2n.$$

Calculate the energy released in the process.

For the following calculations use following data.

$m(^{235}U) = 235.0439$, $m(^{95}Mo) = 94.9058$, $m(^{139}La) = 138.9061$

$m(^{1}n) = 1.0090$,

Only One Option Correct Type

Direction (Q. Nos. 2-11) This section contains 10 multiple choice questions. Each question has four choices (a), (b), (c) and (d), out of which ONLY ONE is correct.

2 Binding energy of a system of masses (nucleons) is

a. energy released when the system is formed
b. energy that must be supplied to the system to separate it into its components
c. energy equivalence of mass defect
d. All of the above

3 Energy equivalence of masses of $^{40}_{20}Ca$ and $^{41}_{20}Ca$ are 37225.154 MeV and 38156.362 MeV respectively. Energy equivalence of mass of a neutron is 939.57145 MeV. From the given data, we can conclude that

a. energy required to separate a neutron from $^{41}_{20}Ca$ is 939.57145 MeV
b. energy required to separate a neutron from $^{41}_{20}Ca$ is 37216.791 MeV
c. energy required to separate a neutron from $^{41}_{20}Ca$ 931.208 MeV
d. energy required to separate a neutron from $^{41}_{40}Ca$ is 8.36 MeV

Working Space

4 Binding energy for nuclei P, Q and R are E_P, E_Q and E_R respectively.
In the fusion processes

$$3P \longrightarrow Q + \text{Energy}(E_1)$$
$$2Q \longrightarrow R + \text{Energy}(E_2)$$

then, total energy released in the fusion process

$$6P \longrightarrow R + \text{Energy}(E_3) \text{ is}$$

a. $E_1 + E_2$ b. $E_1 - E_2$ c. $E_1 - 2E_2$ d. $2E_1 + E_2$

5 In fission of a $^{235}_{92}U$ nucleus 200 MeV energy is released. To maintain a power output of 5W, fission rate of ^{235}U must be maintained at

a. $1.56 \times 10^{11}\ s^{-1}$ b. $1.56 \times 10^{10}\ s^{-1}$
c. $1.56 \times 10^{16}\ s^{-1}$ d. $1.56 \times 10^{7}\ s^{-1}$

6 In the reaction, $^2_1H + ^3_1H \longrightarrow ^4_2He + ^1_0n$
if the binding energies of ^2_1H, $^3H_1$ and ^4_2He are x, y and z respectively, then

a. $x + y + z$ = energy released in the reaction
b. $x + y - z$ = energy released in the reaction
c. $z - x - y$ = energy released in the reaction
d. $z + x - y$ = energy released in the reaction

7 If the mass of a proton, $m_p = 1.008u$, mass of a neutron $m_n = 1.009u$, then the binding energy of an α-particle of mass 4.003u will be

a. 28.8 MeV   b. 21.4 MeV
c. 8.2 MeV   d. 32 MeV

8 When fission of U^{235} takes place the nucleus splits roughly into two equal parts. Use the given graph to find the energy released.

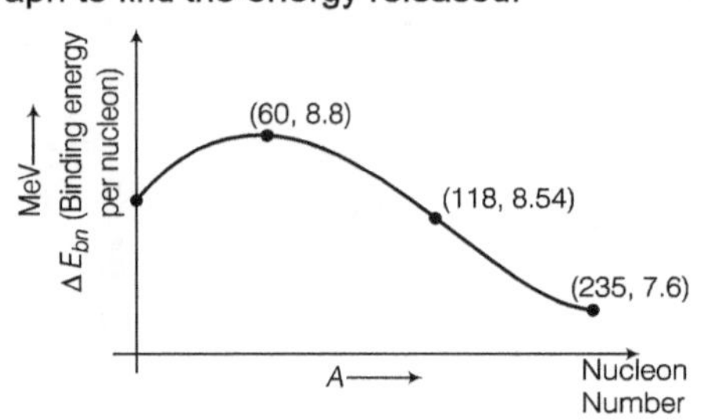

Energy released for each nucleus split of U^{235} is

a. more than 200 MeV   b. less than 200 MeV
c. equals to 200 MeV   d. Data insufficient

9 In order to fuse two nuclei, they must be brought at a separation of about 2 fm or less. Let two deuterium nuclei may be brought to fuse together by colliding them with equal and opposite velocities.
Minimum speed required for above process must be around.

a. $5.8 \times 10^{12}\ ms^{-1}$ b. $5.8 \times 10^{10}\ ms^{-1}$
c. $5.8 \times 10^{6}\ ms^{-1}$ d. $5.8 \times 10^{4}\ ms^{-1}$

10 Estimated number of nucleons in a 12th class student of mass 50 kg will be around

a. 4.5×10^{58} b. 7.5×10^{38} c. 2.5×10^{18} d. 1.5×10^{28}

11 What will be the possible order of magnitude of number of protons in a 12th class student of mass 70 kg?

a. 10^{18} b. 10^{58} c. 10^{38} d. 10^{28}

Working Space

One or More than One Options Correct Type

Direction *(Q. Nos. 12-16) This section contains 5 multiple choice questions. Each question has four choices (a), (b), (c) and (d), out of which ONE or MORE THAN ONE are correct.*

12 Mark the correct options from the following.

a. Like other light nuclei the $^4_2$He also have a low value of the binding energy per nucleon

b. The BE per nucleon decreases for nuclei with small as well as large atomic numbers

c. Energy required to remove one neutron from $^7_3$Li to transform it into $^6_3$Li is 5.6 MeV So, BE per nucleon of $^6_3$Li is also 5.6 MeV

d. When two deuterium nuclei fuse together, they give rise to tritium nucleus

13 Which of the following reactions are not possible?

a. $^1_1p \longrightarrow {}^1_0n + {}^0_{+1}e + \text{energy}$

b. $^1_0n \longrightarrow {}^1_1p + {}^0_{-1}e + \text{energy}$

c. $^{27}_{13}\text{Al} + {}^2_1\text{H} \longrightarrow {}^{25}_{12}\text{Mg} + {}^4_2\text{He}$

d. $^{235}_{92}\text{U} + {}^1_0n \longrightarrow {}^{140}_{54}\text{Xe} + {}^{95}_{38}\text{Sr} + 2{}^1_0n + \gamma$

14 Consider the following processes.

I. A proton is removed from a nucleus of $^{15}_7$N

II. A neutron is removed from a nucleus of $^{15}_7$N

Then, mark the correct options.

a. It takes less energy to perform I

b. Equal energy is required in both I and II

c. Binding energy per nucleon is not a constant quantity

d. Binding energy for each nucleon may have different value

15 Fusion process is not possible at room temperature because

a. nuclear forces are short ranged

b. nuclei are positively charged

c. original nuclei must be comletely ionised for fusion

d. original nuclei must break up into fragments for fusion

16 Which of the following are correct for binding energy per nucleon (ΔE_{bn}) curve plotted as a function of mass number (A)?

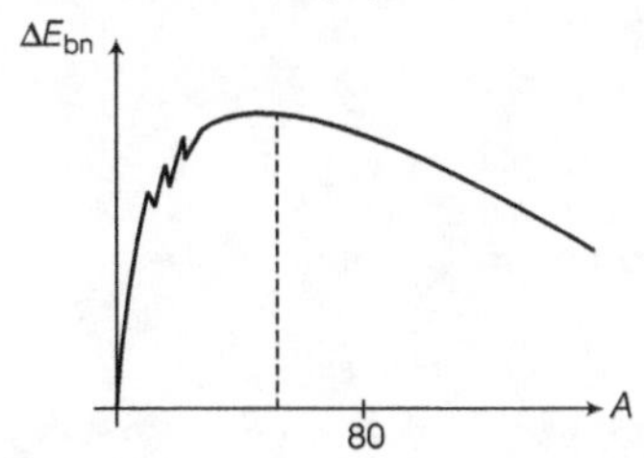

a. Maximum occurs at $A \sim 50$

b. When A increases from 0 to 30, there are considerable fluctuations as internucleon intraction is largely affected by pairing and symmetry effects

c. For $A > 30$, ΔE_{bn} varies only about 10%. This shows nucleons intracts with their neighbours only

d. For $A > 50$, the mean binding energy per nucleon decreases because of Coulomb repulsion

Statement Type

Direction (Q. Nos. 17-19) *This section is based on Statement I and Statement II. Select the correct answer from the codes given below.*

a. Both Statement I and Statement II are correct and Statement II is the correct explanation of Statement I
b. Both Statement I and Statement II are correct but Statement II is not the correct explanation of Statement I
c. Statement I is correct but Statement II is incorrect
d. Statement II is correct but Statement I is incorrect

17 **Statement I** The nuclear binding energy is nearly proportional to A^2.
Statement II Because of saturation property of nuclear forces, a nucleon can interact with its immediate neighbours and hence, with only a limited number of other nucleons.

18 **Statement I** Most stable nuclides are those for which $N = Z$.
Statement II Pauli's exclusion principle.

19 **Statement I** Sum of masses of nucleons is always greater than the mass of their nucleus.
Statement II A system is most stable in a configuration, when it has lowest possible energy.

Matching List Type

Direction (Q. No. 20) *Choices for the correct combination of elements from Column I and Column II are given as options (a), (b), (c) and (d), out of which one is correct.*

20 Match Column I in which a characteristic of an isotope atom is given with its correct match in Column II and choose the correct option from the codes given below.

	Column I		Column II
i.	Larger BE per nucleon	p.	More stable
ii.	Large mass defect	q.	Less stable
iii.	Large value of packing fraction		
iv.	Large value of number of electrons		
v.	Large value of number of neutrons		
vi.	Large nuclear radius		

Codes

	i	ii	iii	iv	v	vi
a.	p	q	p	q	p	q
b.	p	p	p	q	q	q
c.	q	p	p	q	q	q
d.	p	p	q	q	p	q

One Integer Value Correct Type

Direction (Q. No. 21) *This section contains a question. When worked out will result in an integer from 0 to 9 (both inclusive).*

21 When a slow moving neutron collides with a Uranium 235 atom ($^{235}_{92}U$), following reaction occurs.

$$^{1}_{0}n + ^{235}_{92}U \longrightarrow ^{236}_{92}U^{*} \longrightarrow ^{99}_{40}Zr + ^{134}_{52}Te + 3^{1}_{0}n$$

If kinetic energy of neutrons is neglected (as it is very small), then the Q value of reaction is $37 \times K$ MeV. Find the value of K. Use following data.

$$m(U^{235}) = 235.0439u, \quad m(n) = 1.0087u,$$
$$m(Zr^{99}) = 98.9165u, \quad m(Te^{134}) = 133.9115u.$$

DPP-3 Radioactivity and Law of Radioactive Disintegration

Subjective Question

Working Space

Direction (Q. No. 1) *This question is subjective in nature, need to be solved completely on notebook.*

1 Radioisotope of carbon ${}^{14}_{6}C$ has a half-life of 5730 yr. Normal activity of living carbon containing matter is found to be about 15 decays per minute per gram of carbon.

i. Let a bone specimen of an old civilisation gives an activity of 9 decays per minute per gram. Find the approximate age of civilisation.

ii. Can we find age of an old volcanic rock by same carbon-dating process?

Only One Option Correct Type

Direction (Q. Nos. 2-11) *This section contains 10 multiple choice questions. Each question has four choices (a), (b), (c) and (d), out of which ONLY ONE is correct.*

2 If ${}^{27}_{13}Al$ absorbs a neutron it becomes one of the following after stabilisation.

a. ${}^{28}_{13}Al$ b. ${}^{27}_{14}Al$

c. ${}^{28}_{14}Si$ d. ${}^{28}_{13}Si$

3 Consider the reaction,

$${}^{2}_{1}H + {}^{2}_{1}H \longrightarrow {}^{3}_{2}He + {}^{1}_{0}n$$

$m({}^{2}_{1}H) = 2.014082$ u, $m({}^{3}_{2}He) = 3.016029$, $m({}^{1}_{0}n) = 1.008665$

Then, mark the correct option.

a. Threshold energy of 3.23 MeV is required to initiate the reaction

b. Reaction occurs when total KE of reactants exceeds 3.23 MeV

c. Reaction occurs such that final total KE is 3.23 MeV lesser than total initial KE

d. No, threshold energy is required for the reaction

4 In the reaction, ${}^{1}_{0}n + {}^{24}_{12}Mg \longrightarrow {}^{23}_{11}Na + {}^{2}_{1}H$

$m({}^{24}_{12}Mg) = 23.985042$, $m({}^{23}_{11}Na) = 23.985042$

the bombarding neutrons have 16 MeV of kinetic energy. Then,

a. 6.55 MeV of extra energy is required for the reaction to occur

b. 6.55 MeV of energy is released in the reaction

c. reaction cannot occurs

d. 16 MeV of energy is released in the reaction

5 For collision of nuclei of ${}^{27}_{13}Al$ and ${}^{125}_{52}Te$, effective cross-sectional area will be

a. $92.16\,\pi\,(fm)^2$ b. $\frac{\pi}{8}(fm)^2$

c. $8\,\pi\,(fm)^2$ d. $64\,\pi\,(fm)^2$

Working Space

6 Let your town's daily requirement is of 950 MW power, which is fulfilled by a 38% efficient nuclear power plant. Mass of fuel $^{238}_{92}U$ required to meet demand for 1yr will be (Assume that a total of 200 MeV per fission is obtained)

a. more than 10 Ton
b. less than 1 Ton
c. less than 0.5 Ton
d. less than 0.1 Ton

7 A 1.0 MeV neutron emitted in a fission reaction loses one half of its kinetic energy in each collision with a moderator molecule. Number of collisions it must be make to reach thermal energy will be $\left(\text{Take, } \frac{3}{2}k_B T = 0.040 \text{ eV}\right)$

a. 5
b. 25
c. 125
d. 625

8 Three radioactive sources *A*, *B* and *C* of same activity 25 m Ci are dumped in a pond as waste. Source *A* emits 1.0 MeV γ-rays, source *B* emits 2.0 MeV γ-rays, and source *C* emits 2.0 MeV α-rays. Order of relative danger of *A*, *B* and *C* is

a. $A > B > C$
b. $B > A > C$
c. $C > B > A$
d. $A = B = C$

9 The threshold energy of the bombarding particle for a nuclear reaction is equal to (Take, m_b = mass of bombarding particle, m_{pr} = Total mass of products and Q = mass energy of reaction)

a. $\frac{Q \cdot m_b}{m_{pr}}$
b. $-\frac{Q \cdot m_b}{m_{pr}}$
c. $\frac{-Q \cdot m_{pr}}{m_{pr} - m_b}$
d. $\frac{Q \cdot m_{pr}}{m_{pr} - m_b}$

10 Technitium ($^{98}_{43}Tc$) has a half-life of 4.2×10^6 yr. Strontium has a half-life of 28.79 yr. Which of the following is correct?

a. Decay constant of Sr is greater than decay constant of Tc
b. Activity of 100 g of Sr is less than activity of 100 g of Tc
c. The long half-life of Tc shows that it decays by α-decay
d. A Tc atom has a higher probility of decaying in 1 yr than a Sr atom

11 In CO_2, obtained from atmospheric air, a small fraction of 1.3×10^{-12} is the radioactive isotope $^{14}_{6}C(\lambda = 3.83 \times 10^{-12})$. If 200g of carbon fragment is found in an animal's bone in an archaeological site shows an activity of 16 decays in 1second, then the age of bone will be

(Take, $\ln(25/8) \approx 1.1$)

a. about 10 yr
b. about 100 yr
c. about 1000 yr
d. about 10000 yr

One or More than One Options Correct Type

Direction (Q. Nos. 12-16) *This section contains 5 multiple choice questions. Each question has four choices (a), (b), (c) and (d), out of which ONE or MORE THAN ONE are correct.*

12 Which of the following are correct regarding α-decay?

a. α-decay occurs because the strong nuclear force is a short range force but the electric force is a long range force
b. If Q value for α-decay is negative, then the decay could not occur spontaneously
c. A nucleus can emit an α-particle but can not emit four separate nucleons as such mass of α-particle is lesser than the sum of masses of such four nucleons
d. Classical physics cannot explain α-emission

13 Which of the following are correct regarding a β-decay?

a. The electron energies observed in β-decay of a particular nuclide are found to vary continuously from zero to a maximum value KE_{max}
b. In a β-decay direction of recoil of product nucleus is almost exactly oppsite to the direction of emission of β-particle
c. In ordinary β-decay (β^- or electron), an antinutrino is emitted as
$${}_0^1n \longrightarrow {}_1^1P + {}_{-1}^{0}e + \nu$$
d. Neutrino and antinutrino are massless and they do not interact with matter

14 Which of the following are correct regarding γ-decay?

a. A γ-decay occurs only after an α-decay or a β-decay
b. A γ-decay may occur spontaneously (without any α or β-decay at first)
c. An excited nucleus can emits only one γ-photon in its transition to ground state
d. An excited nucleus can emits more than one γ-photon in its transition to ground state

15 Consider the process,

$${}_{92}^{232}U \longrightarrow {}_{90}^{228}Th + {}_2^4He$$

Masses of reactants and products are

$$m\,({}_{90}^{228}Th) = 228.0998 \text{ u},$$
$$m\,({}_2^4He) = 4.0039 \text{ u},$$
$$m({}_{92}^{232}U) = 232.1095 \text{ u}.$$

Then,
a. process may occur spontaneously
b. kinetic energy of an α-particle emitted is 5.30 MeV
c. kinetic energy of an α-particle emitted is always less than or equal to 5.30 MeV
d. process requires a total of 5.40 MeV of energy to proceed

16 The polonium isotope ${}_{84}^{210}Po$ is unstable, it emits an α-particle with energy of 5.30 MeV. The atomic mass of ${}_{84}^{210}Po$ is 209.9829 u and atomic mass of ${}_2^4He$ is 4.0026 u. Then,
a. Daughter nucleus is ${}_{82}^{206}Pb$
b. Disintegration energy of α-decay is 5.30 MeV
c. Disintegration energy of α-decay is 5.40 MeV
d. Mass of daughter nucleus is 205.9745 u

Comprehension Type

Direction *(Q. Nos. 17-20) This section contains 2 paragraphs, each describing theory, experiments, data, etc. Four questions related to the paragraphs have been given. Each question has only one correct answer among the four given options (a), (b), (c) and (d).*

Passage I

Let N_0 = initial number of parent nuclei of a radioactive substance (at t = 0s). Also, decay constant for this radioactive substances is λ.

17 Probability of a nucleus for survial upto time t is

a. $1-e^{-\lambda t}$
b. $N_0 \cdot e^{-\lambda t}$
c. $\dfrac{1}{e^{-\lambda t}}$
d. $e^{-\lambda t}$

18 Probability that a randomly choosen nucleus is disintegrated in time t is

a. $1-e^{-\lambda t}$
b. $N_0 \cdot e^{-\lambda t}$
c. $\dfrac{1}{e^{-\lambda t}}$
d. $e^{-\lambda t}$

Working Space

Passage II

Consider the following decay processes.

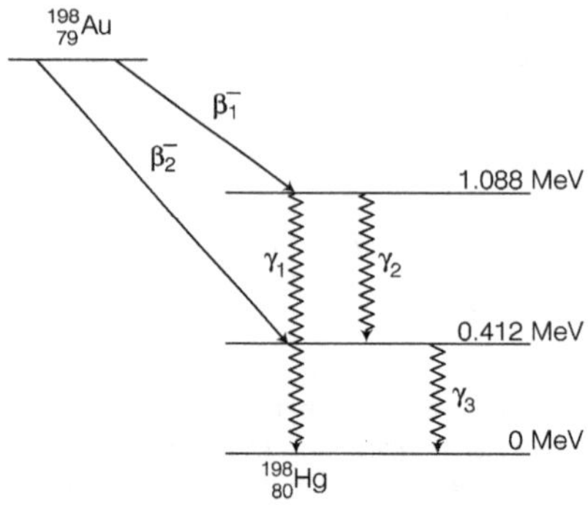

Masses of involved nuclei are

$m(^{198}_{79}Au) = 197.968233$ u

$m(^{198}_{80}Hg) = 197.966760$ u

19 Frequencies of emitted gamma radiations are in order

a. $\gamma_1 > \gamma_2 > \gamma_3$ b. $\gamma_3 > \gamma_2 > \gamma_1$
c. $\gamma_1 > \gamma_3 > \gamma_2$ d. $\gamma_3 > \gamma_1 > \gamma_2$

20 Kinetic energies of emitted β-particles are

a. $K_{\beta_1}(\max) = 0.281$ MeV, $K_{\beta_2}(\max) = 0.281$ MeV
b. $K_{\beta_1}(\max) = 0.957$ MeV; $K_{\beta_2}(\max) = 0.957$ MeV
c. $K_{\beta_1}(\max) = 0.281$ MeV, $K_{\beta_2}(\max) = 0.957$ MeV
d. $K_{\beta_1}(\max) - K_{\beta_2}(\max) = 0.281$ MeV

Statement Type

Direction (Q. Nos. 21-24) *This section is based on Statement I and Statement II. Select the correct answer from the codes given below.*

a. Both Statement I and Statement II are correct and Statement II is the correct explanation of Statement I
b. Both Statement I and Statement II are correct but Statement II is not the correct explanation of Statement I
c. Statement I is correct but Statement II is incorrect
d. Statement II is correct but Statement I is incorrect

21 **Statement I** In case of a β^- decay, when a β^--particle of maximum energy is emitted, then the energy carried by the daughter nucleus and the anti-neutrino is zero.

Statement II $^1_0n \longrightarrow {}^1_1p + {}^0_{-1}e + \bar{\nu}$.

22 **Statement I** Sea water is a suitable moderator for a nuclear reactor.

Statement II Sea water has much higher specific heat than ordinary water.

23 **Statement I** Moderator in a nuclear reactor is intimately mixed with uranium fuel to slow down fast neutrons (2 MeV) to the thermal energies (0.04 eV).

Statement II Neutrons transfer their energy to moderator molecules by partially elastic collisions.

Working Space

24 **Statement I** For a steady power operation, multiplication factor (K) must be kept lower than unity ($K < 1$).

Statement II Multiplication factor (K) is the ratio of number of neutrons present at the beginning of a particular generation to the number present at the begining of next generation.

Matching List Type

Direction (Q. No. 25) Choices for the correct combination of elements from Column I and Column II are given as options (a), (b), (c) and (d), out of which one is correct.

25 Match the Column I with Column II and mark the correct option from the codes given below.

	Column I		Column II
i.	α-decay	p.	Nucleus too large
ii.	β^+-decay	q.	Neutron number too large
iii.	β^--decay	r.	Proton number too large
iv.	γ-decay	s.	Nucleus has excess energy
v.	K-capture	t.	Proton number is more than 83

Codes

	i	ii	iii	iv	v
a.	p,t	r,t	q,t	s	r,t
b.	p	r	q	s	r,t
c.	p	q	r	t	s
d.	p	r	t	s	q

SET - 1

Revisal Problems
for JEE Main

Only One Option Correct Type

Working Space

Direction (*Q. Nos. 1-16*) *This section contains 16 multiple choice questions. Each question has four choices (a), (b), (c) and (d), out of which ONLY ONE option is correct.*

1 Atomic masses of many elements are not in whole numbers because of occurance of

a. isotopes
b. isobars
c. isotones
d. mirror nuclei

2 Most abundent element in our universe is

a. hydrogen
b. helium
c. iron
d. silicon

3 A very stable nucleus has a surface area $A_1 = 466.56\,\pi \times 10^{-30}\ m^2$.
Another nucleus X has an area four times the area of A_1. The mass number of X has a value

a. 108
b. 16
c. 216
d. 238

4 A radioactive substance for which rate of decay $\dfrac{dN}{dt}\ \alpha - N$, the mean life time is

a. about 69% more than half-life time
b. about 69% less than half-life time
c. about 44% more than half-life time
d. about 44% less then half-life time

5 1.0 kg of water is heated to raise its temperature by 1K. Then,

a. its mass decreases due to increase of volume
b. its mass decreases due to decrease of density
c. its mass may increases by 4.7×10^{-14} kg
d. its mass increases due to conversion of protons into neutrons

6 Consider the reaction,

$$^{14}N\,(\alpha, p)^{17}O$$

The masses involved are

$$^{14}N = 14.00307\ u,\quad ^4He = 4.00260\ u$$
$$^1H = 1.00783\ u,\quad ^{17}O = 16.99913\ u$$

Then,

a. reaction is spontaneous
b. reaction is possible when α-particle has a KE of 1.54 MeV
c. reaction is possible when α-particle has a KE of 0.54 MeV
d. Q-value of reaction is 1.54 MeV

7 A proton is bombarded over a lithium nucleus. Due to collision two α-particles are produced.

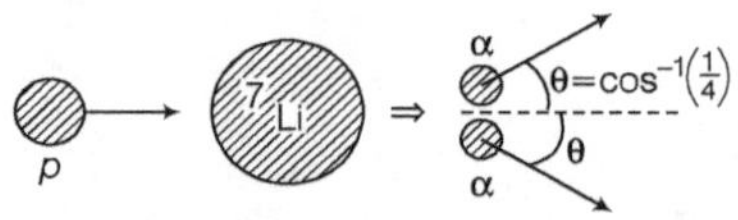

If direction of motion of the α-particles with the direction of bombarding proton is at an angle of $\cos^{-1}(1/4)$, then the correct relation between KE of proton (K_p) and KE of α-particle (K_α) is

a. $K_p = 2K_\alpha$
b. $K_p = \dfrac{1}{2}K_\alpha$
c. $2K_p = 3K_\alpha$
d. $K_p = K_\alpha$

Working Space

8 Mean lives of a radioactive substance are 1620 yr and 405 yr for α-emission and β emission respectively.

If a sample of this substance decays by emitting both α and β emissions simultaneously, then time in which $\frac{3}{4}$th of sample will decay is (Take, $\log_e 2 = 0.693$)

a. 1653 yr
b. 324 yr
c. 449 yr
d. 1468 yr

9 For two different radioactive substances, A and B let, half-life of A = mean life of B. If initially samples of both have same number of atoms, then

a. A and B both are decaying at same rate initially
b. A and B both have same decay rate at all time
c. A decays faster than B
d. B decays faster than A

10 Let two radioactive materials A and B have decay constants 10λ and λ respectively. Initially their samples have same number of atoms. The ratio of number of nuclei of A to that of B, after a time $(\lambda/9)$ seconds will be

a. 1 : 10
b. 10 : 1
c. e
d. e^{-1}

11 A more accurate value of nuclear binding energy is
(Use, m_p = mass of a proton, m_n = mass of a neutron, M = mass of neutral atom, and m_e = mass of an electron)

a. $(Zm_p + Nm_n - M)\cdot c^2$
b. $(Z(m_e + m_p) + Nm_n - M)\cdot c^2$
c. $(Zm_p + Nm_n - M + Z\cdot m_e)\cdot c^2$
d. $(Zm_p + Nm_n - M - Ze)\cdot c^2$

12 α-decay arises only for nuclides with mass numbers.

a. $A > 150$
b. $A > 83$
c. $A > 200$
d. $A > 216$

13 Which of the following is not a requirement for a fusion reaction?

a. high temperature
b. high density
c. high time
d. high mass number

14 Half-life of a neutron is about 693s and mass of neutron is 1.6×10^{-27} kg. A beam of 10000 high speed neutrons with kinetic energy 0.08 eV is fired in space. Number of neutrons that possibly decay in travelling through a distance of 4 km will be

a. 10
b. 100
c. 1000
d. 10000

15 Only $\frac{1}{8}$th of total initial number of active radioactive nuclei of a sample is left after 6 days.

Then, in 10 days fraction that decays is

a. $\frac{15}{16}$
b. $\frac{31}{32}$
c. $\frac{77}{80}$
d. $\frac{71}{80}$

16 You have a sample of deutron gas. Minimum temperature upto which you have to heat the gas to produce fusion will be

a. 2.8×10^5 K
b. 2.8×10^{12} K
c. 2.8×10^9 K
d. 2.8×10^2 K

Statement Type

Working Space

Direction *(Q. Nos. 17-20) This section is based on Statement I and Statement II. Select the correct answer from the codes given below.*

a. Both Statement I and Statement II are correct and Statement II is the correct explanation of Statement I
b. Both Statement I and Statement II are correct but Statement II is not the correct explanation of Statement I
c. Statement I is correct but Statement II is incorrect
d. Statement II is correct but Statement I is incorrect

17 **Statement I** α-particles are emitted with nearly same energy but β-particles are emitted with widely varrying energies.

Statement II α-decay is a two particle decay whereas β-decay is a three particle decay.

18 **Statement I** When stream of a α-particles (obtained from a decaying radioactive sample) is made to pass through a region of perpendicular magnetic field, they follow circular paths of some fixed values of radius.

Statement II Some of the daughter nuclei may be produced in their excited states.

19 **Statement I** Helium isotope ${}_2^6\text{He}$ is unstable and can achieve stability by emitting a β^- particle.

Statement II A β^--decay, ${}_0^1n \longrightarrow {}_1^1p + {}_{-1}^{0}e + \bar{\nu}$, results in increase of proton number of nucleus.

20 **Statement I** Most of the commerical nuclear reactors uses water both as moderator and as a coolant.

Statement II Water is easily available liquid with high specific heat.

SET - 2

Revisal Problems
for JEE Advanced

Only One Option Correct Type

Working Space

Direction (Q. Nos. 1-3) *This section contains 3 multiple choice questions. Each question has four choices (a), (b), (c) and (d), out of which ONLY ONE is correct.*

1 If 10% of a radioactive material decays in 5 days, then the probability of a randomly chosen nucleus to remain und]ecayed is around

a. 90% b. 81% c. 73% d. 65%

2 Binding energy per nucleon for deutron ${}_1^2H$ and helium ${}_2^4He$ are 1.1 MeV and 7.0 MeV. Energy released when two deutrons fuses to form a helium nucleus is

a. 2.2 MeV b. 23.6 MeV
c. 28.0 MeV d. 30.2 MeV

3 Four hypothetical nucleus *A, B, C* and *D* having mass numbers 2, 4, 6 and 8 respectively, are having same total binding energy, then

a. *A* is most stable b. *B* is most stable
c. *C* is most stable d. *D* is most stable

One or More than One Options Correct Type

Direction (Q. Nos. 4-7) *This section contains 4 multiple choice questions. Each question has four choices (a), (b), (c) and (d), out of which ONE or MORE THAN ONE are correct.*

4 Energy released Q values for various modes of decays of ${}^{232}U$ are listed below.

Emitted particle	Energy released (MeV)
${}_0^1n$	−7.26
${}_1^1H$	− 6.12
${}_2^4He$	+5.14
${}_1^2H$	−10.70

Mark the correct options.

a. Spontaneous decay is possible only for α-particles
b. Neutron decay is spontaneously possible
c. Spontaneous proton decay is not possible
d. All reactions except α-decay are endothermic

5 A free neutron decays into a proton and an electron. Then, mark the correct options.

a. Emitted electrons have a continuous distribution of energy due to existence of other particle
b. Antineutrinos have negligible rest mass and spin $+\frac{1}{2}$
c. β^+-decay of a free proton is not allowed
d. In a stable nucleus, the two reciprocal processes neutron decay and proton decay are in dynamical equilibrium

6 Consider the process,

$$e^- + {}_Z^AX \rightarrow {}_{Z-1}^{A}Y + \nu + Q_1$$

and

$$ {}_Z^AX \rightarrow {}_{Z-1}^{A}Y + e^+ + \nu + Q_2$$

Then, mark the correct options.

a. $Q_1 > 0$ implies $Q_2 > 0$ b. $Q_2 > 0$ implies $Q_1 > 0$
c. $Q_1 > 0$ does not implies $Q_2 > 0$ d. $Q_2 > 0$ does not implies $Q_1 > 0$

Working Space

7 Consider the decay as shown below.

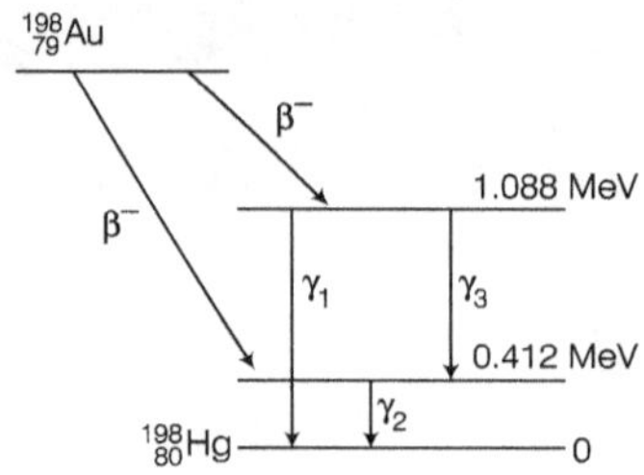

Mark the correct options.

a. Radiation frequencies of γ-decays are not constant

b. β^--particles are emitted with constant kinetic energy

c. $^{198}_{80}Hg$ is formed in its ground state only

d. γ-decays have frequencies of 2.608×10^{20} Hz, 9.950×10^{19} Hz and 1.633×10^{20} Hz

Comprehension Type

Direction (Q. Nos. 8-18) *This section contains 4 paragraphs, each describing theory, experiments, data, etc. Eleven questions related to the paragraphs have been given. Each question has only one correct answer among the four given options (a), (b), (c) and (d).*

Passage I

Let a radioactive substance A has N_0 active nuclei at time $t = 0$ s. Let A-decays into B with decay constant λ_1 and B-decays into a stable substance C with decay constant λ_2.

8 Number of nuclei of A at any instantaneous time t is

a. $N_0 e^{-\lambda_1 t}$

b. $N_0(e^{-\lambda_1 t} + e^{-\lambda_2 t})$

c. $N_0 e^{-\lambda_2 t}$

d. $N_0(1 - e^{-\lambda_1 t})$

9 Number of nuclei of B at any instantaneous time t is

a. $\dfrac{\lambda_1 N_0}{\lambda_2 - \lambda_1} \cdot (e^{-\lambda_1 t} - e^{-\lambda_2 t})$

b. $\dfrac{\lambda_2 N_0}{\lambda_2 - \lambda_1} \cdot (e^{-\lambda_1 t} - e^{-\lambda_2 t})$

c. $\dfrac{\lambda_1 \lambda_2 N_0}{\lambda_2 - \lambda_1} \cdot (e^{-\lambda_1 t} - e^{-\lambda_2 t})$

d. $\dfrac{\lambda_1 N_0}{\lambda_2 - \lambda_1} \cdot e^{-\lambda_1 t}$

10 Number of nuclei of C at any instant of time t is

a. $N_0 - \left\{\dfrac{\lambda_1 N_0}{\lambda_2 - \lambda_1} (e^{-\lambda_1 t} - e^{-\lambda_2 t}\right\}$

b. $N_0 - N_0 e^{-\lambda_1 t}$

c. $N_0 - N_0 e^{-\lambda_1 t} - \dfrac{\lambda_1 N_0}{\lambda_2 - \lambda_1} \cdot (e^{-\lambda_1 t} - e^{-\lambda_2 t})$

d. $N_0 e^{-\lambda_1 t} - \dfrac{N_0 \lambda_1}{\lambda_2 - \lambda_1} \cdot (e^{-\lambda_1 t} - e^{-\lambda_2 t})$

Passage II

An electrical power plant operates on the basis of thermal energy generated in a pressurised heavy water reactor. Electrical output of plant is 1 GW, efficiency of plant is 30% and energy per fission is 200 MeV.

11 Rate of fission events in the reactor core is

a. 0.72×10^{20} events/s

b. 0.31×10^{20} events/s

c. 1.04×10^{20} events/s

d. 5.02×10^{20} events/s

12 Mass of ^{235}U used in one year will be

a. 1.28×10^3 kg/yr

b. 0.5×10^3 kg/yr

c. 1000 kg/yr

d. 50 kg/yr

Working Space

Passage III

Unstable $^{236}_{92}U$ often decays as,

$$^{236}_{92}U \rightarrow ^{140}_{54}Xe + ^{94}_{38}Sr + \text{(other particles)}.$$

Reaction is energetically allowed as smaller nuclei have higher binding energy per nucleon. Inside the nucleus strong nuclear force holds the nucleons together. The nuclear force is charge independent.

13 As BE per nucleon of products is more than BE per nucleon of reactants, that means

a. energy is required to carry the reaction in forward direction
b. reaction is endothermic
c. reaction cannot occur spontaneously
d. reaction is exothermic

14 In above reaction, other particles are

a. an α-particle
b. two protons
c. one proton and one neutron
d. two neutrons

15 A proton and a neutron both are shot towards a $^{12}_{6}C$ nucleus with a speed of 100 ms^{-1}. Then, mark the correct option.

a. Both particles are equally likely to be absorbed
b. Proton has a higher probability of being absorbed
c. The neutron has a higher probability of being absorbed
d. Neither of particles will be absorbed

Passage IV

An α-decay spontaneously occurs as,

$$^{A}_{Z}X \rightarrow ^{A-4}_{Z-2}Y + \alpha$$

Let initial decaying nucleus X is at rest and energy released is Q.

16 Kinetic energy T_α of α-particle is given by

a. $T_\alpha = Q\left(\frac{m_\alpha}{m_Y}\right)$
b. $T_\alpha = Q\left(\frac{m_Y}{m_\alpha}\right)$
c. $T_\alpha = Q\left(1 - \frac{4}{A}\right)$
d. $T_\alpha = Q\left(1 + \frac{4}{A}\right)$

17 If original nucleus X is at rest, then which of the following are correct?

a. X and α move in opposite directions after reaction
b. Y and α moves in opposite directions
c. Y and α moves in random directions but not along the same line
d. X, Y and α moves along medians of an equilateral triangle

18 Relative potential energy V of α-particle and daughter nucleus system as a function of their separation is given : (radius of nucleus of X is a)

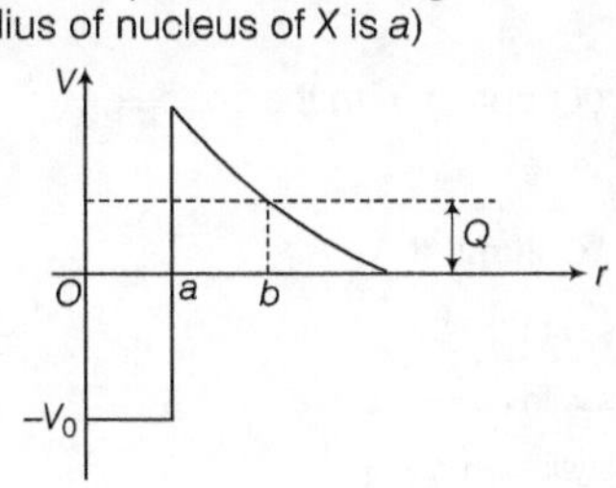

If $Q = (m_X - m_Y - m_\alpha)c^2$, then region in which α-particle cannot be found is

a. $r = 0$ to $r = \infty$
b. $r = a$ to $r = b$
c. $r = 0$ to $r = a$
d. $r = b$ to $r = \infty$

Matching List Type

Working Space

Direction (Q. Nos. 19-21) *Choices for the correct combination of elements from Column I and Column II are given as options (a), (b), (c) and (d), out of which one is correct.*

19 There are 3-series of naturally occurring radioactive nuclei and 1 artificial radioactive series. Match the Column I with Column II and mark the correct option from the codes given below.

	Column I (Starting Isotope)		Column II (Stable End Product)
i.	$^{238}_{92}U$	p.	$^{206}_{82}Pb$
ii.	$^{235}_{92}U$	q.	$^{207}_{82}Pb$
iii.	$^{232}_{90}Th$	r.	$^{208}_{82}Pb$
iv.	$^{237}_{93}Np$	s.	$^{209}_{83}Bi$

Codes

	i	ii	iii	iv		i	ii	iii	iv
a.	p	q	r	s	b.	s	r	q	p
c.	p	r	q	s	d.	s	r	p	q

20 Match the Column I with Column II and mark the correct option from the codes given below.

	Column I		Column II
I.	Mass of products formed is less than that of reactants	p.	α-decay
ii.	Binding energy per nucleon increases	q.	β^--decay
iii.	Emitted particles have variable KE	r.	β^+-decay
iv.	Emitted particles have same KE	s.	γ-decay

Codes

	i	ii	iii	iv
a.	p,q,r	p,q,r	q,r,s	p
b.	p	q	r	s
c.	p,q,r,s	p,q,r	q,r,s	p,s
d.	p,q,r,s	p,q,r,s	q,r	p

21 Match the Column I with Column II and mark the correct option from the codes given below.

	Column I		Column II
i.	γ-decay	p.	Nucleus with $P > N$
ii.	α-decay	q.	Nucleus with $N > P$
iii.	β-decay	r.	Nucleus with large A
iv.	Electron capture	s.	Nucleus has excess energy
v.	Positron emission		

Codes

	i	ii	iii	iv	v
a.	s	q	r	p	p
b.	p	q	q	r	s
c.	s	s	r	q	p
d.	s	r	q	p	p

JEE Main & AIEEE Archive

(Compilation of Last 13 Years' Questions)

1 Assume that a neutron breaks into a proton and an electron. The energy released during this process is (mass of neutron = 1.6725×10^{-27} kg, mass of proton = 1.6725×10^{-27} kg, mass of electron = 9×10^{-31} kg) **(AIEEE 2012)**

a. 0.73 MeV b. 7.10 MeV
c. 6.30 MeV d. 5.4 MeV

2 **Statement I** A nucleus having energy E_1 decays be β^- emission to daughter nucleus having energy E_2, but β^- rays are emitted with a continuous energy spectrum having end point energy $E_1 - E_2$.

Statement II To conserve energy and momentum in β-decay, atleast three particles must take part in the transformation. **(AIEEE 2011)**

a. Both Statement I and Statement II are correct and Statement II is the correct explanation of Statement I
b. Both Statement I and Statement II are correct but Statement II is not the correct explanation of Statement I
c. Statement I is correct but Statement II is incorrect
d. Statement II is correct but Statement I is incorrect

3 The half-life of a radioactive substance is 20 min. The approximate time interval $(t_2 - t_1)$ between the time t_2 when 2/3 of it has decayed and time t_1 when 1/3 of it had decayed is **(AIEEE 2011)**

a. 14 min b. 20 min
c. 28 min d. 7 min

Passage Q. Nos. (4 and 5)

A number of mass $M + \Delta m$ is at rest and decays into two daughter nuclei of equal mass $\frac{M}{2}$ each. Speed of light is C. **(AIEEE 2010)**

4 The binding energy per nucleon for the parent nucleus is E_1 and that for the daughter nuclei is E_2. Then,

a. $E_2 = 2E_1$ b. $E_1 > E_2$
c. $E_2 > E_1$ d. $E_1 = 2E_2$

5 The speed of daughter nuclei is

a. $c\dfrac{\Delta m}{M + \Delta m}$ b. $c\sqrt{\dfrac{2\Delta m}{M}}$
c. $c\sqrt{\dfrac{\Delta m}{M}}$ d. $c\sqrt{\dfrac{\Delta m}{M + \Delta m}}$

6 A radioactive nucleus (initial mass number A and atomic number Z) emits 3α-particles and 2 positrons. The ratio of number of neutrons to that of protons in the final nucleus will be **(AIEEE 2010)**

a. $\dfrac{A - Z - 8}{Z - 4}$ b. $\dfrac{A - Z - 4}{Z - 8}$
c. $\dfrac{A - Z - 12}{Z - 4}$ d. $\dfrac{A - Z - 4}{Z - 2}$

7 The graph given below is a plot of binding energy per nucleon E_b, against the nuclear mass M. A, B, C, D, E, F correspond to different nuclei. Consider the four reactions.

i. $A + B \rightarrow C + \varepsilon$
ii. $C \rightarrow A + B + \varepsilon$
iii. $D + E \rightarrow F + \varepsilon$ and
iv. $F \rightarrow D + E + \varepsilon$

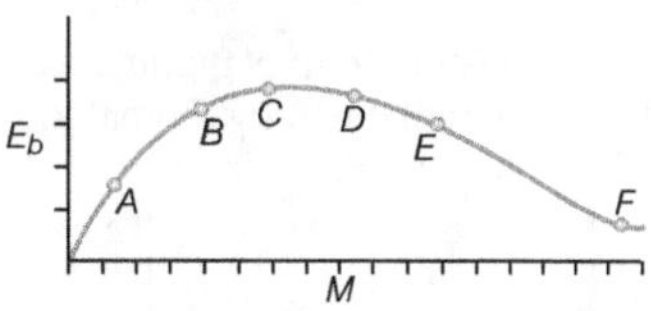

where, ε is the energy released. In which reactions ε is positive? **(AIEEE 2009)**

a. (i) and (iv) b. (i) and (iii)
c. (ii) and (iv) d. (ii) and (iii)

8 **Statement I** Energy is released when heavy nuclei undergo fission or light nuclei undergo fusion.

Statement II For heavy nuclei, binding energy per nucleon increases with increasing Z while for light nuclei, it decreases with increasing Z. **(AIEEE 2008)**

a. Both Statement I and Statement II are correct and Statement II is the correct explanation of Statement I
b. Both Statement I and Statement II are correct but Statement II is not the correct explanation of Statement I
c. Statement I is correct but Statement II is incorrect
d. Statement II is correct but Statement I is incorrect

9 The half-life period of a radioactive element X is same as the mean life time of another radioactive element Y. Initially they have the same number of atoms. Then, **(AIEEE 2007)**

a. X will decay faster than Y
b. Y will decay faster than X
c. Y and X have same decay rate initially
d. X and Y decay at same rate always

10 If M_o is the mass of an oxygen isotope $_8O^{17}$, M_p and M_n are the masses of a proton and a neutron respectively, the nuclear binding energy of the isotope is **(AIEEE 2007)**

a. $(M_o - 8M_p)c^2$
b. $(M_o - 8M_p - 9M_n)c^2$
c. $M_o c^2$
d. $(M_o - 17M_n)c^2$

11 An alpha nucleus of energy $\frac{1}{2}mv^2$ bombards a heavy nuclear target of charge Ze. Then, the distance of closest approach for the alpha nucleus will be proportional to **(AIEEE 2006)**

a. v^2 b. $\frac{1}{m}$ c. $\frac{1}{v^4}$ d. $\frac{1}{Ze}$

12 If the binding energy per nucleon in ${}_3^7\text{Li}$ and ${}_2^4\text{He}$ nuclei are 5.60 MeV and 7.06 MeV respectively, then in the reaction

$$p + {}_3^7\text{Li} \longrightarrow 2\,{}_2^4\text{He}$$

energy of proton must be **(AIEEE 2006)**

a. 28.24 MeV b. 17.28 MeV
c. 1.46 MeV d. 39.2 MeV

13 The energy spectrum of β-particles (number $N(E)$ as a function of β-energy E emitted from a radioactive source is **(AIEEE 2006)**

a. 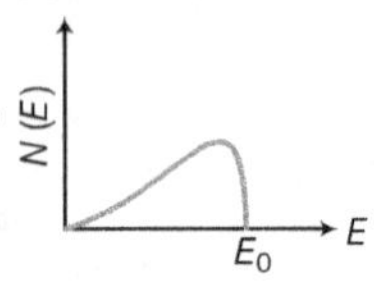

b. 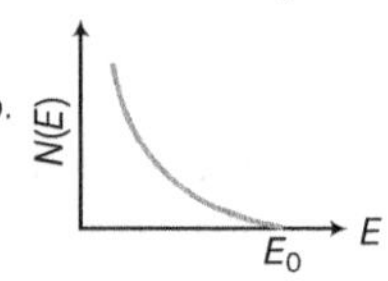

c. 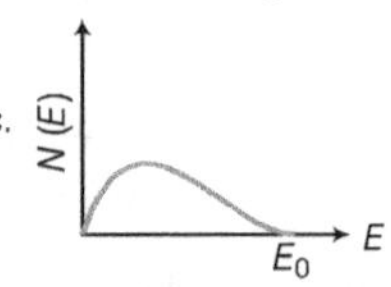

d. 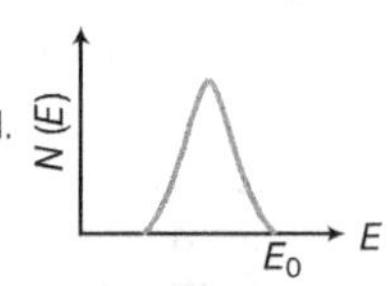

14 When ${}_3^7\text{Li}$ nuclei are bombarded by protons, and the resultant nuclei are ${}_4^8\text{Be}$, the emitted particles will be **(AIEEE 2006)**

a. alpha particles b. beta particles
c. gamma photons d. neutrons

15 Starting with a sample of pure ${}^{66}\text{Cu}$, 7/8 of it decays into Zn in 15 min. The corresponding half-life is **(AIEEE 2005)**

a. 10 min b. 15 min
c. 5 min d. $7\frac{1}{2}$ min

16 If radius of the ${}_{13}^{27}\text{Al}$ nucleus is estimated to be 3.6 fm, then the radius of ${}_{52}^{125}\text{Te}$ nucleus be nearly **(AIEEE 2005)**

a. 6 fm b. 8 fm c. 4 fm d. 5 fm

17 A nuclear transformation is denoted by $X(n, \alpha) \rightarrow {}_3^7\text{Li}$. Which of the following is the nucleus of element X? **(AIEEE 2005)**

a. ${}_6^{12}\text{C}$ b. ${}_5^{10}\text{B}$
c. ${}_5^9\text{B}$ d. ${}_4^{11}\text{Be}$

18 A nucleus disintegrates into two nuclear parts, which have their velocities in the ratio 2 : 1. The ratio of their nuclear sizes will be **(AIEEE 2004)**

a. $2^{1/3}:1$ b. $1:3^{1/2}$ c. $3^{1/2}:1$ d. $1:2^{1/3}$

19 The binding energy per nucleon of deuteron (${}_1^2\text{H}$) and helium nucleus (${}_2^4\text{He}$) is 1.1 MeV and 7 MeV respectively. If two deuteron nuclei react to form a single helium nucleus, then the energy released is **(AIEEE 2004)**

a. 13.9 MeV b. 26.9 MeV
c. 23.6 MeV d. 19.2 MeV

20 An α-particle of energy 5 MeV is scattered through 180° by a fixed uranium nucleus. The distance of the closest approach is of the order of **(AIEEE 2004)**

a. 1 Å b. 10^{-10} cm
c. 10^{-12} cm d. 10^{-15} cm

21 When U^{238} nucleus originally at rest, decays by emitting an α - particle having a speed u, the recoil speed of the residual nucleus is **(AIEEE 2003)**

a. $\frac{4u}{238}$ b. $-\frac{4u}{234}$
c. $\frac{4u}{234}$ d. $-\frac{4u}{238}$

22 A radioactive sample at any instant has its disintegration rate 5000 disintegrations/min. After 5 min, the rate is 1250 disintegrations/min. Then, the decay constant (per minute) is **(AIEEE 2003)**

a. 0.4 ln 2 b. 0.2 ln 2
c. 0.1 ln 2 d. 0.8 ln 2

23 A nucleus with $Z = 92$ emits the following in a sequence to

$\alpha, \alpha, \beta^-, \beta^-, \alpha, \alpha, \alpha, \alpha; \beta^-, \beta^-, \alpha, \beta^+, \beta^+, \alpha.$

The Z of the resulting nucleus is **(AIEEE 2003)**

a. 76 b. 78
c. 82 d. 74

24 Which of the following cannot be emitted by radioactive substances during their decay? **(AIEEE 2003)**

a. Protons b. Neutrinos
c. Helium nuclei d. Electrons

25 In the nuclear fusion reaction,

$${}_1^2\text{H} + {}_1^3\text{H} \longrightarrow {}_2^4\text{He} + n$$

given that the repulsive potential energy between the two nuclei is 7.7×10^{-14} J, the temperature at which the gases must be heated to initiate the reaction is nearly (Boltzmann's constant, $k = 1.38 \times 10^{-23}$ J/K) **(AIEEE 2003)**

a. 10^7 K b. 10^5 K
c. 10^3 K d. 10^9 K

26 If N_0 is the original mass of the substance of half-life period $t_{1/2} = 5$ yr, then the amount of substance left after 15 yr is **(AIEEE 2002)**

a. $\frac{N_0}{8}$ b. $\frac{N_0}{16}$
c. $\frac{N_0}{2}$ d. $\frac{N_0}{4}$

JEE Advanced & IIT JEE Archive

(Compilation of Last 10 Years' Questions)

Passage for Q. Nos. (1 and 2)

The mass of a nucleus ${}_Z^AX$ is less that the sum of the masses of $(A-Z)$ number of neutrons and Z number of protons in the nucleus. The energy equivalent to the corresponding mass difference is known as the binding energy of the nucleus. A heavy nucleus of mass M can break into two light nuclei of masses m_1 and m_2 only, if $(m_1 + m_2) < M$. Also, two light nuclei of masses m_3 and m_4 can undergo complete fusion and form a heavy nucleus of mass M' only, if $(m_3 + m_4) > M'$. The masses of some neutral atoms are given in the table below.

(2013 Adv., Comprehension type)

${}_1^1H$	1.007825 u	${}_1^2H$	2.014102 u
${}_3^6Li$	6.01513 u	${}_3^7Li$	7.016004 u
${}_{64}^{152}Gd$	151.919803 u	${}_{82}^{206}Pb$	205.974455 u
${}_1^3H$	3.016050 u	${}_2^4He$	4.002603 u
${}_{30}^{70}Zn$	69.925325 u	${}_{34}^{82}Se$	81.916709 u

1 The correct statement is

a. The nucleus ${}_3^6Li$ can emit an α- particle

b. The nucleus ${}_{84}^{210}Po$ can emit a proton

c. Deuteron and α- particle can undergo complete fusion

d. The nuclei ${}_{30}^{70}Zn$ and ${}_{34}^{82}Se$ can undergo complete fusion

2 The kinetic energy (in keV) of the α- particle, when the nucleus ${}_{84}^{210}Po$ (Polonium) at rest undergoes α- decay, is

a. 5319
b. 5422
c. 5707
d. 5818

3 A freshly prepared sample of a radioisotope of half-life 1386 s has activity 10^3 disintegrations per second. Given that $\ln 2 = 0.693$, the fraction of the initial number of nuclei (expressed in nearest integer percentage) that will decay in the first 80 s after preparation of the sample is **(2013 Adv., Integer Type)**

Passage for Q. Nos. (4 and 5)

The β-decay process, discovered around 1900, is basically the decay of a neutron (n). In the laboratory, a proton (p) and an electron (e^-) are observed as the decay products of the neutron.

Therefore, considering the decay of a neutron as a two-body decay process, it was predicted theoretically that the kinetic energy of the electron should be a constant. But experimentally, it was observed that the electron kinetic energy has a continuous spectrum. Considering a three-body decay process, i.e. $n \to p + e^- + \bar{\nu}_e$, around 1930, Pauli explained the observed electron energy spectrum.

Assuming the anti-neutrino $(\bar{\nu}_e)$ to be massless and possessing negligible energy, and the neutron to be at rest, momentum and energy conservation principles are applied. From this calculation, the maximum kinetic energy of the electron is 0.8×10^6 eV. The kinetic energy carried by the proton is only the recoil energy.

(2012, Comprehension Type)

4 If the anti-neutrino had a mass of 3 eV/c^2 (where, c is the speed of light) instead of zero mass, what should be the range of the kinetic energy K, of the electron?

a. $0 \le K \le 0.8 \times 10^6$ eV
b. $3.0 \text{ eV} \le K \le 0.8 \times 10^6$ eV
c. $3.0 \text{ eV} \le K < 0.8 \times 10^6$ eV
d. $0 \le K < 0.8 \times 10^6$ eV

5 What is the maximum energy of the anti-neutrino?

a. Zero

b. Much less than 0.8×10^6 eV

c. Nearly 0.8×10^6 eV

d. Much larger than 0.8×10^6 eV

6 The activity of a freshly prepared radioactive sample is 10^{10} disintegrations per second, whose mean life is 10^{-9} s. The mass of an atom of this radioisotope is 10^{-25} kg. The mass (in mg) of the radioactive sample is **(2011, Subjective Type)**

7 To determine the half-life of a radioactive element, a student plots a graph of $\ln\left|\frac{dN(t)}{dt}\right|$ versus t. Here $\frac{d\,N(t)}{dt}$ is the rate of radioactive decay at time t. If the number of radioactive nuclei of this element decreases by a factor of p after 4.16 yr, the value of p is **(2010, Subjective Type)**

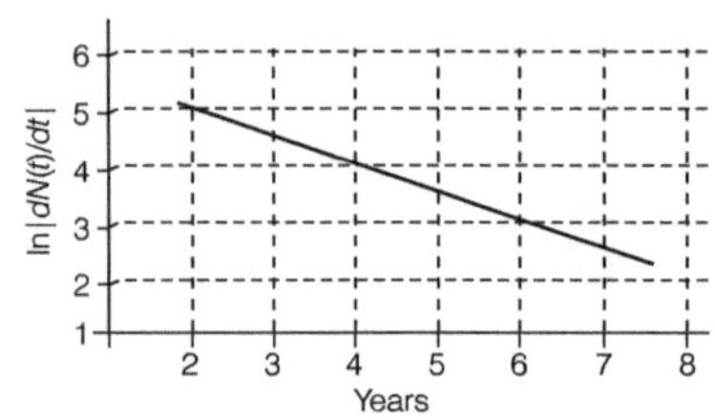

Passage for Q. Nos. (8-10)

Scientists are working hard to develop nuclear fusion reactor. Nuclei of heavy hydrogen, $^2_1$H known as deuteron and denoted by D can be thought of as a candidate for fusion reactor. The D-D reaction is $^2_1\text{H} + {^2_1}\text{H} \rightarrow {^3_2}\text{He} + n +$ energy. In the core of fusion reactor, a gas of heavy hydrogen is fully ionised into deuteron nuclei and electrons. This collection of $^4_1$H nuclei and electrons is known as plasma. The nuclei move randomly in the reactor core and occasionally come close enough for nuclear fusion to take place.

Usually, the temperatures in the reactor core are too high and no material wall can be used to confine the plasma. Special techniques are used, which confine the plasma for a time t_0 before the particles fly away from the core. If n is the density (number/volume) of deuterons, the product nt_0 is called Lawson number. In one of the criteria, a reactor is termed successful, if Lawson number is greater than 5×10^{14} s cm^{-3}.

Use, Boltzmann constant $k = 8.6 \times 10^{-5}$ eV/K, $\frac{e^2}{4\pi\varepsilon_0} = 1.44 \times 10^{-9}$ eV-m. **(2009, Comprehension Type)**

8 In the core of nuclear fusion reactor, the gas becomes plasma because of

a. strong nuclear force acting between the deuterons

b. Coulomb force acting between the deuterons

c. Coulomb force acting between deuteron-electron pairs

d. the high temperature maintained inside the reactor core

9 Assume that two deuteron nuclei in the core of fusion reactor at temperature T are moving towards each other, each with kinetic energy $1.5\,kT$, when the separation between them is large enough to neglect Coulomb potential energy. Also, neglect any interaction from other particles in the core. The minimum temperature T required for them to reach a separation of 4×10^{-15} m is in the range

a. 1.0×10^9 K $< T < 2.0 \times 10^9$ K

b. 2.0×10^9 K $< T < 3.0 \times 10^9$ K

c. 3.0×10^9 K $< T < 4.0 \times 10^9$ K

d. 4.0×10^9 K $< T < 5.0 \times 10^9$ K

10 Results of calculations for four different designs of a fusion reactor using D-D reaction are given below. Which of these is most promising based on Lawson criterion?

a. Deuteron density $= 2.0 \times 10^{12}$ cm^{-3}, confinement time $= 5.0 \times 10^{-3}$ s

b. Deuteron density $= 8.0 \times 10^{14}$ cm^{-3}, confinement time $= 9.0 \times 10^{-1}$ s

c. Deuteron density $= 4.0 \times 10^{23}$ cm^{-3}, confinement time $= 1.0 \times 10^{-11}$ s

d. Deuteron $= 1.0 \times 10^{24}$ cm^{-3}, confinement time $= 4.0 \times 10^{-12}$ s

11 A radioactive sample S_1 having an activity of 5μCi has twice the number of nuclei as another sample S_2 which has an activity of 10 μCi. The half-lives of S_1 and S_2 can be **(2008, Only One Option Correct Type)**

a. 20 yr and 5 yr, respectively
b. 20 yr and 10 yr, respectively
c. 10 yr each
d. 5 yr each

12 Match Column I of the nuclear process with Column II containing parent nucleus and one of the end products of each process and mark the correct option from the codes given below. **(2008, Matching Type)**

	Column I		Column II
i.	α-decay	p.	${}^{15}_{8}O \longrightarrow {}^{17}_{7}N + \ldots$
ii.	β^+-decay	q.	${}^{238}_{92}U \rightarrow {}^{234}_{90}Th + \ldots$
iii.	Fission	r.	${}^{185}_{83}Bi \rightarrow {}^{184}_{82}Pb + \ldots$
iv.	Proton emission	s.	${}^{239}_{94}Pu \rightarrow {}^{140}_{57}La +$

Codes

	i	ii	iii	iv
a.	s	q	p	r
b.	p	r	q	s
c.	q	p	s	r
d.	s	r	q	p

13 Assume that the nuclear binding energy per nucleon (B/A) *versus* mass number (A). is as shown in the figure. Use this plot to mark the correct choice(s) given below. **(2008, One or More than One Options Correct Type)**

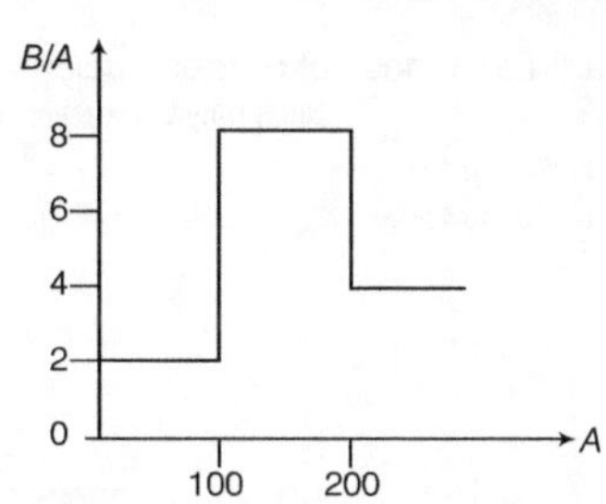

a. Fusion of two nuclei with mass numbers lying in the range of $1 < A < 50$ will release energy

b. Fusion of two nuclei with mass numbers lying in the range of $51 < A < 100$ will release energy

c. Fission of a nucleus lying in the mass range of $100 < A < 200$ will release energy, when broken into two equal fragments

d. Fission of a nucleus lying in the mass range of $200 < A < 260$ will release energy, when broken into two equal fragments

14 In the options given below, let E denote the rest mass energy of a nucleus and n a neutron. The correct option is **(2007, Only One Option Correct Type)**

a. $E\,({}^{236}_{92}\text{U}) > E\,({}^{137}_{53}\text{I}) + E\,({}^{97}_{39}\text{Y}) + 2E\,(n)$

b. $E\,({}^{236}_{92}\text{U}) < E\,({}^{137}_{53}\text{I}) + E\,({}^{97}_{39}\text{Y}) + 2E\,(n)$

c. $E\,({}^{236}_{92}\text{U}) < E\,({}^{140}_{56}\text{Ba}) + E\,({}^{94}_{36}\text{Kr}) + 2E\,(n)$

d. $E\,({}^{235}_{92}\text{U}) < E\,({}^{140}_{56}\text{Ba}) + E\,({}^{94}_{36}\text{Kr}) + E\,(n)$

15 Half-life of a radioactive substance A is 4 days. The probability that a nucleus will decay in two half-lives is **(2006, Only One Option Correct Type)**

a. $\frac{1}{4}$

b. $\frac{3}{4}$

c. $\frac{1}{2}$

d. 1

16 Some laws/processes are given in Column I. Match with the physical phenomena given in Column II and mark the correct option from the codes given below. **(2006,Matching Type)**

	Column I		Column II
i.	Nuclear fusion	p.	Converts some matter into energy
ii.	Nuclear fission	q.	Generally possible for nuclei with low atomic number
iii.	β-decay	r.	Generally possible for nuclei with higher atomic number
iv.	Exothermic nuclear reaction	s.	Essentially proceeds by weak nuclear forces

Codes

	i	ii	iii	iv
a.	p,q	p,r	p,s	p,q,r
b.	q,r	p	s	q
c.	s	r	r,s	p,r
d.	q	q,p	q,r	s

17 If a star can convert all the He nuclei completely into oxygen nuclei. The energy released per oxygen nuclei is
(mass of the helium nucleus is 4.0026 amu and mass of oxygen nucleus is 15.9994 amu) **(2005, Only One Option Correct Type)**

a. 7.6 MeV

b. 56.12 MeV

c. 10.24 MeV

d. 23.4 MeV

JEE Archive

4. Electronic Devices

DPP-1 Semiconductors

Only One Option Correct Type

Working Space

Direction (Q. Nos. 1-10) *This section contains 10 multiple choice questions. Each question has four choices (a), (b), (c) and (d), out of which ONLY ONE option is correct.*

1 Which of the following is potential energy curve for two point charges of opposite signs at a distance r from each other?

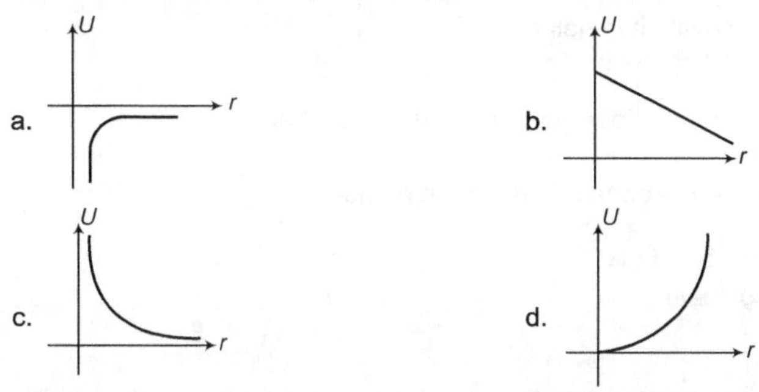

2 Potential energy curve for H_2 molecule or a function of separation r between two H-atoms is as shown in the figure.

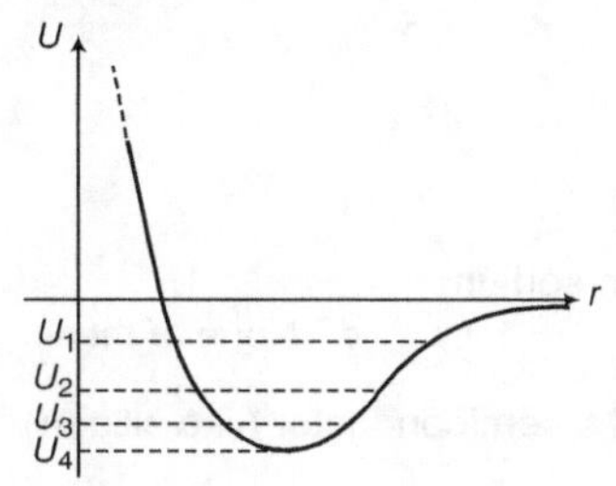

Where,

U_3 = Energy corresponding to the lowest quantum energy state

U_4 = Energy corresponding to the lowest value of potential enegy function

Then, binding energy for H_2 molecule will be

a. $|U_4|$ b. $|U_3|$

c. $|U_3 - U_4|$ d. $|U_4 - U_1|$

3 Let a diatomic molecule has a potential energy given by

$$U = -K\left(\frac{e^2}{r}\right) + \frac{B}{r^6}$$

where, $K = 9 \times 10^9$ Nm^2C^{-2}, $B = 1.0 \times 10^{-78}$ Jm^6 and

r = distance between centres by two atoms.

Then, bondlength of molecule is nearly to

a. 2 fm b. 1 fm

c. 2.5 fm d. 1.2 fm

Working Space

4 A Fermi level is

a. a possible energy value, which an electron can have in free state
b. a possible energy value, which an electron can have in free state at zero kelvin
c. an unfilled energy level which can be occupied by two electrons of opposite spin
d. maximum possible level at zero kelvin upto which every possible energy levels are filled

5 Fermi energy of an electron is

a. energy of electron
b. energy of electron at zero kelvin
c. energy of transition of electron
d. maximum possible allowed value of energy for an electron at zero kelvin

6 Most solids are crystalline with atoms, ions or molecules falling into regular repeated three dimensional patterns. Crystalline solids are

a. ionic in nature
b. covalent in nature
c. metallic in nature
d. All of these

7 When a large number (N) of identical atoms come closer uniformly, we get a band consisting of closely spaced level which are

a. N in number
b. $2N$ in number
c. N^2 in number
d. 2^N in number

8 A forbidden energy gap between two energy bands is

a. interval of energies that an electron cannot have
b. interval of energies of electrons
c. energy value taken by an excited electron
d. difference of enegy values of conduction and valence bands

9 Energy band in solid sodium metal is as shown in the figure.

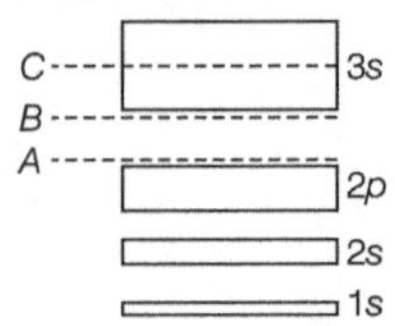

What will be the position of fermi energy level in sodium

a. A
b. B
c. C
d. None of these

10 Where does a fermi energy level lies in case of a semiconductor (like silicon) atom?

a. At the top of valence band
b. At the bottom of conduction band
c. In middle of energy gap between conduction and valence band
d. At the top of conduction band

One or More than One Options Correct Type

Direction (Q. Nos. 11-15) *This section contains 5 multiple choice questions. Each question has four choices (a), (b), (c) and (d), out of which ONE or MORE THAN ONE are correct.*

11 A small amount of impurity can drastically change the conductivity of a semiconductor. By adding an impurity, few of impurity atom electrons have their own energy level.

Then, mark the correct option(s).

a. Impurity atoms have their energy level lie above valence band
b. Impurity atoms have their energy level lie below conduction band
c. Impuirty atoms have their energy level below valence band
d. Impurity atoms have their energy level above conduction band

Working Space

12 Which of the following statements are correct statements?

a. Good conductors of electricity are the solids in which uppermost occupied band is not completely filled
b. Substances whose atoms have complete shells but which in the solid state, have the uppermost filled band overlapping an empty band are conductors
c. Insulators are substances whose uppermost valence band is completely filled and separated by an energy gap of a few eV from the next empty band
d. Semiconductors are the substances in which the energy gap between the completely filled valence band and the conduction band is about 3 eV or less

13 Aluminium metal has three valence electrons per atom, an atomic weight of 0.02698 kg/mol, a density of 2700 kgm^{-3} and a conductivity of $3.54 \times 10^7\ sm^{-1}$ at room temperature. Now, choose the correct options.

a. Number of free electrons increases with increase in temperature
b. Electron mobility for sample at room temperature is $3.67 \times 10^{-3}\ (m^2 V^{-1} s^{-1})$
c. Electron mobility for sample increases with temperature
d. Electron mobility for sample decreases with temperature

14 An intrinsic semiconductors is a crystalline solid, not contaminated with impurity atoms. Which of the following are correct regarding intrinsic conduction?

a. At any temperature a stable state is achieved when rate of creation of electron-hole pairs is equal to the recombination rate
b. Number of free electrons (at any temperature) = number of holes
c. Number of intrinsic pairs, $n_i = n_0 T^{3/2} \cdot e^{\left(\frac{-V_G}{2V_T}\right)}$

   where, n_0 = constant, T = absolute temperature,
   V_G = semiconductor band gap voltage, V_T = thermal voltage = $\left(\frac{k_B T}{q}\right)$.

d. Number of intrinsic pairs, $n_i \propto e^{\left(\frac{-E_g}{2k_B T}\right)}$,

   where, k_B = Boltzmann's constant.

15 A pure semiconductor can be doped to increase its conductivity. Which of the following statement are correct?

a. A pentavalent donor impurity produces n-type semiconductor
b. An acceptor trivalent impurity produces p-type semiconductor
c. Conductivity of a doped semiconductor sample is $\sigma = (n_e \mu_e + n_n \mu_n) q$
d. Conductivity of a doped semiconductor sample is $\sigma = n_i(\mu_e + \mu_n) q$

Comprehension Type

Direction (Q. Nos. 16 and 17) *This section contains a paragraph, which describing theory, experiments, data, etc. Two questions related to the paragraph have been given. Each question has only one correct answer among the four given options (a), (b), (c) and (d).*

Passage

Potential energy of an ionic crystal is given by $U = -\alpha \frac{1}{4\pi\varepsilon_0} \cdot \frac{e^2}{r} + \frac{B}{r^m}$

where, r = distance between a positive and negative ion and $-\alpha \frac{1}{4\pi\varepsilon_0} \cdot \frac{e^2}{r}$ in the coulomb energy when summation is taken for all the positive and negative ions in a crystal.

Quantity α is called Madelung constant and α = 1.7 for simple crystal structures.

Also, $\frac{B}{r^m} \propto$ repulsion force, m = index of dependence a difference constant for each crystal energy due to repulsion between positive-positive or negative-negative ions in the crystal.

Working Space

16 Minimum potential energy of each ion due to its intraction with other ions in the crystal is where, (r_0 = equilibrium separation of positive and negative ions)

a. $U_0 = \frac{-\alpha e^2}{4\pi\varepsilon_0 r_0}$
b. $U_0 = \frac{-\alpha e^2}{4\pi\varepsilon_0 r_0 m}$
c. $U_0 = \frac{-\alpha e^2}{4\pi\varepsilon_0 r_0 m^2}$
d. $U_0 = \frac{-\alpha e^2}{4\pi\varepsilon_0 r_0}\left(1-\frac{1}{m}\right)$

17 At minimum potential energy separation ($r = r_0$), force on each ion is

a. $F_0 = \frac{\alpha e^2}{4\pi\varepsilon_0 r_0^2}$
b. $F_0 = \frac{e^2}{4\pi\varepsilon_0 r_0^2}$
c. $F_0 = \frac{-\alpha e^2}{4\pi\varepsilon_0 r_0^2}$
d. zero

Matching List Type

Direction (Q. No. 18) *Choices for the correct combination of elements from Column I and Column II are given as options (a), (b), (c) and (d), out of which one is correct.*

18 In Column I, different type of crystalline materials are listed. In Column II, values of energy gap E_g in eV and values of resistivities in Ω m are given.

Match the Column I with Column II and mark the correct option from the codes given below.

	Column I		Column II
i.	Insulator	p.	10^8
ii.	Semicondutor	q.	10^5
iii.	Optically sensitive semiconductor	r.	1
iv.	Intrinsic semiconductor	s.	6
v.	Extrinsic semiconductor	t.	2

Codes

	i	ii	iii	iv	v
a.	p	q,r,s	t	s	t
b.	p,s	q,r,s	r,t	q,r,s	r,t
c.	p	q	r	s	t
d.	p	r,s	q	p,t	t

Statement Type

Direction (Q. Nos. 19 and 20) *This section is based on Statement I and Statement II. Select the correct answer from the codes given below.*

a. Both Statement I and Statement II are correct and Statement II is the correct explanation of Statement I
b. Both Statement I and Statement II are correct but Statement II is not the correct explanation of Statement I
c. Statement I is correct but Statement II is incorrect
d. Statement II is correct but Statement I is incorrect

19 **Statement I** Conductivity of a semiconductor increases with temperature.

Statement II When temperature of a semiconductor is increased, energy gap between valence band and conduction band decreases.

20 **Statement I** When a light beam of wavelength 600 nm is incident over a semiconductor with 2.8 eV energy gap, conductivity of semiconductor increases.

Statement II If an electron absorbs a photon of energy hf such that $hf > E_g$, then it is transferred to the conduction band.

DPP-2 Semiconductor Diodes

Only One Option Correct Type

Direction (Q. Nos. 1-10) *This section contains 10 multiple choice questions. Each question has four choices (a), (b), (c) and (d), out of which ONLY ONE option is correct.*

Working Space

1 Choose the incorrect option for a p - n junction diode.

a. A p - n junction diode is formed by joining together n-type and p-type silicon crystals, using a suitable glue

b. To form a p - n junction, as n-type silicon crystal is being grown the process is abruptly altered to grow a p-type silicon crystal

c. Depletion zone appears in a p-n junction, when p-type silicon is given a negative potential and n-type silicon is given a positive potential

d. Diffusion of electrons in a newly formed p - n junction creates a junction field that stops further diffusion of electrons from n to p-side

2 To draw the characteristics curves for a p - n junction, a student gives you two circuit diagrams.

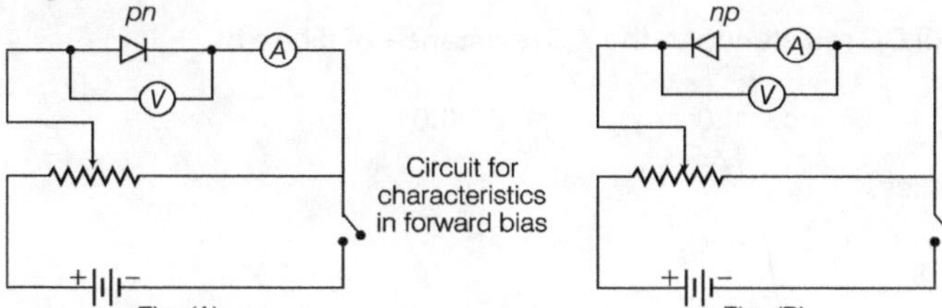

Circuit for characteristics in forward bias

Fig. (A)

Circuit for characteristics in reverse bias

Fig. (B)

What you would conclude?

a. Both Fig. (A) and Fig. (B) are correct

b. In Fig. (A) voltmeter is connected wrongly

c. In Fig. (B) voltemeter is connected wrongly

d. In both Fig. (A) and Fig. (B) potential divider is connected wrongly

3 Characteristics of a p - n junction diode are non-linear as shown in the figure.

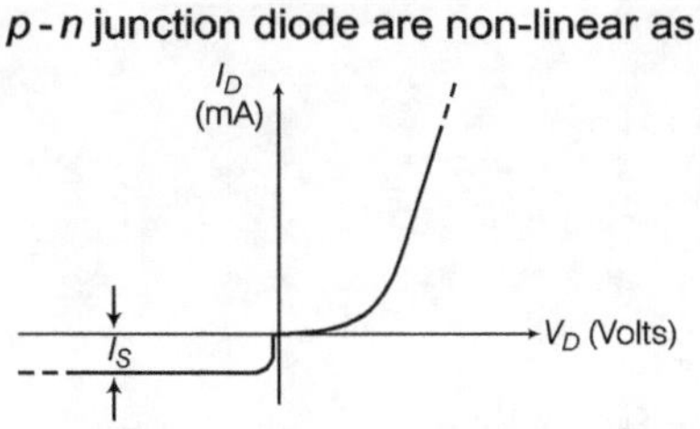

A general purpose p - n junction diode follows one of the given rules. Then, choose the correct options.

a. $I_D = I_S e^{kV_D/T}$ for $V_D > 0$ and $I_D = -I_S$ for $V_D \le 0$

b. $I_D = I_S e^{kV_D/T}$

c. $I_D = I_S \ln\left(\frac{k}{V_D T}\right)$

d. $I_D = I_S(e^{V_D \cdot k/T} - 1)$

Working Space

4 A typical diode characteristics curve A is drawn at temperature T_2(K).

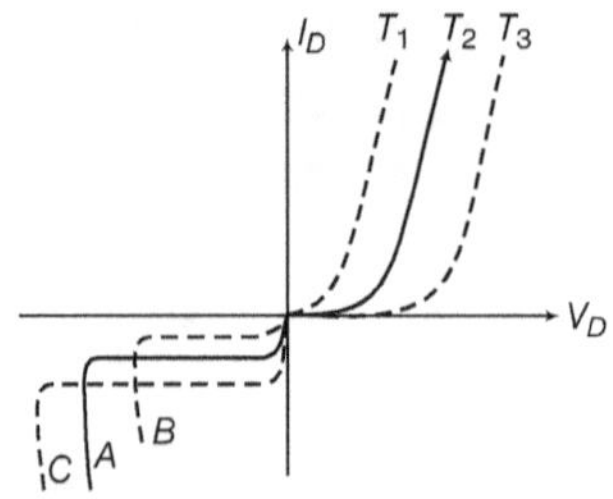

Let curves B and C are drawn at temperatures T_3 and T_1 respectively (under same set up and for same diode). What will be the possible relation between temperatures T_1, T_2 and T_3?

a. $T_1 > T_2 > T_3$ b. $T_1 < T_2 < T_3$

c. $T_2 = \frac{T_1 + T_3}{2}$ d. $T_1 = T_2 = T_3$

5 Your school lab's diode characteristics equipment gives following graph.

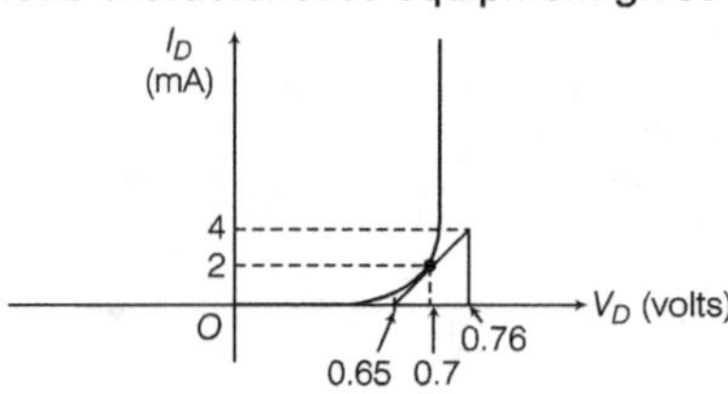

Then, choose the correct ratio of DC resistance to the AC resistance of diode at 2 mA diode current.

a. 12.73 b. 0.078 c. 120 d. 0.01

6 A diode has a given characteristics.

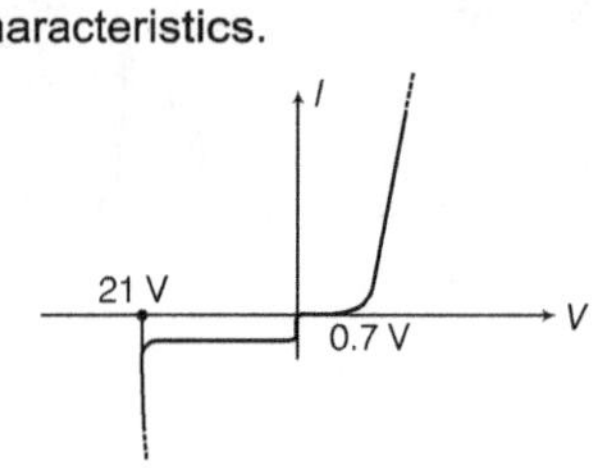

This diode is connected in a circuit as

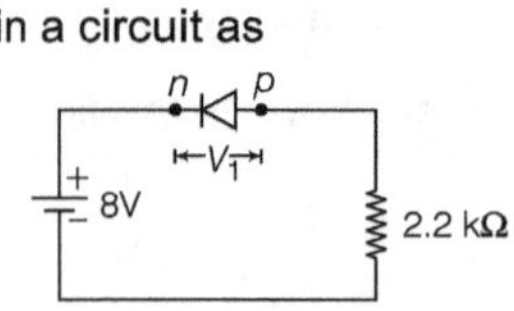

and then as

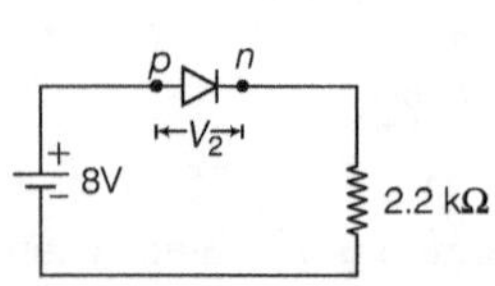

Ratio of potential drops $V_1 : V_2$ (in given circuits) will be

a. 0 b. $\frac{80}{70}$

c. $\frac{80}{7}$ d. $\frac{79}{1}$

Working Space

7 A silicon diode has a "knee voltage" or "threshold voltage" value of 0.7 V. This Si diode is connected in a circuit as shown in the figure.

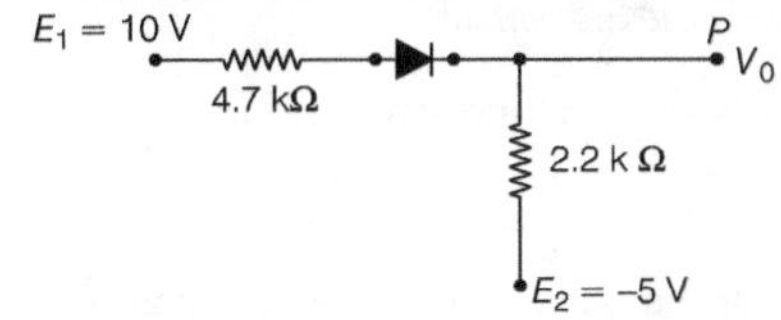

Potential V_0 of point *P* in the circuit will be

a. − 0.44 V   b. 0.44 V
c. 4.4 V   d. − 4.4 V

8 Knee voltage or threshold voltage values for Si and Ge diodes are 0.7 V and 0.3 V respectively. Let a Si and a Ge diode are connected in a circuit as shown in the figure.

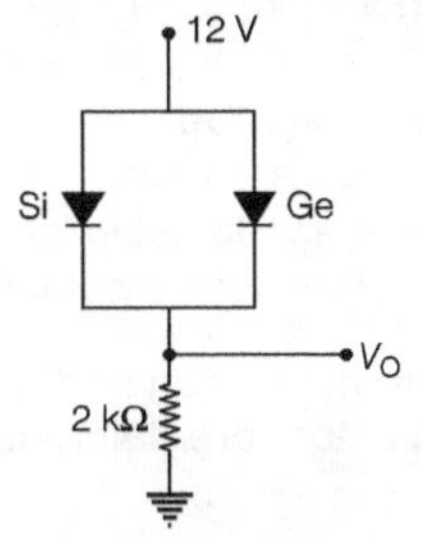

Then, output voltage is

a. 12 V   b. 11.3 V
c. 11.7 V   d. 11 V

9 An AC input is made to an ideal diode. Both input and characteristics of diode are shown below.

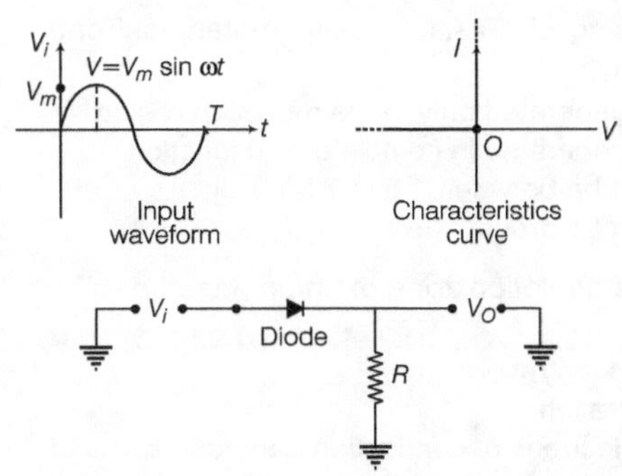

The average value of output voltage V_O over one complete time period of input cycle is

a. V_m b. $\frac{V_m}{\sqrt{2}}$ c. $\frac{2V_m}{\pi}$ d. $\frac{V_m}{\pi}$

10 PIV or peak inverse voltage rating is of primary importance in design of rectifier system. Let similar silicon diodes are used for half wave rectification and for full wave rectification (using a centre tap transformer). If supply voltage for both half-wave and full wave rectification is $V_i = V_m \sin \omega t$, then ratio

$$\frac{\text{PIV (Half-wave rectifier)}}{\text{PIV (Full wave rectifier, CT transformer)}}$$ will be

a. 2   b. $\frac{1}{\sqrt{2}}$ c. $\frac{\pi}{2}$ d. $\frac{1}{2}$

One or More than One Options Correct Type

Working Space

Direction (Q. Nos. 11-15) *This section contains 5 multiple choice questions. Each question has four choices (a), (b), (c) and (d), out of which ONE or MORE THAN ONE are correct.*

11 An LED (light emitting diode) is a forwardly biased p - n junction used for electroluminescence.

Mark the correct options.

a. In an LED, a recombination of holes and electrons occurs in the depletion zone
b. Silicon and Germenium diodes cannot be used as LED
c. Energy gap for material used for fabricating an LED must be in the interval of 1.8 and 2.8 eV
d. Energy gap for material used for red LED is in range $E_g = [1.6 \text{ to } 2.0]$ eV and for violet LED it is $E_g = [2.50 \text{ to } 2.80]$ eV

12 Despite two different mechanisms the avalanche breakdown and the zener breakdown, breakdown diodes are usually called zener diodes. Choose the correct behaviour of two types at high temperatures.

a. Temperature coefficient of zener breakdown voltage is negative
b. Temperature coefficient of avalanche breakdown voltage is positive
c. Avalanche breakdown takes place, when junction have a wide depletion layer
d. Zener breakdown takes place when the p - n junction is moderately or lightly doped

13 Photodetectors are an essential part of an optical communication link. With the advent of technology, semiconductor based photodetectors are being used.
In a p - n junction, photodetector

a. change in intensity of incident light is measured by change in conductivity of photodetectors
b. reverse bias is used
c. forward bias is used
d. change in intensity of incident light produces a change in reverse saturation current

14 A solar cell is a p - n junction fabricated using Si, GaAs, Cds etc. materials. For a solar cell, which of the following are correct?

a. Photogenerated electron-hole pairs are generated only in the depletion region
b. Photogenerated electron-hole pairs are generated in complete p - n junction
c. Energy gap for material of solar cell must be between 1.8 to 2.8 eV
d. Energy gap for material of solar cell must be around 1 eV

15 Which of these are correct concerning the depletion zone of an unbiased p - n junction?

a. Zone width is proportional to impurity concentration
b. Zone width depends on impurity concentration
c. Electric field in the zone is provided by electrons of conduction band and holes of valence band
d. Electric field in the zone is provided by the ionised dopant atoms

Comprehension Type

Direction (Q. Nos. 16-19) *This section contains 2 paragraphs, each describing theory, experiments, data, etc. Four questions related to the paragraphs have been given. Each question has only one correct answer among the four given options (a), (b), (c) and (d).*

Passage I

Resistivity of pure silicon is 3000 Ωm with electron and hole mobilities are $0.12 \text{ m}^2\text{V}^{-1}\text{s}^{-1}$ and $0.045 \text{ m}^2\text{V}^{-1}\text{s}^{-1}$ respectively.

16 When 10^{19} atoms of phosphorus are added (specimen is doped) to specimen of silicon, its resistivity is around

a. 5.2 Ωm
b. 25.2 Ωm
c. 1.2 Ωm
d. 2.5 Ωm

17 Resistivity of specimen, when 2×10^{19} boron atoms per m^3 are further added is

a. 5.2 Ωm
b. 25 Ωm
c. 1.2 Ωm
d. 2.5 Ωm

Passage II

Consider the given circuit.

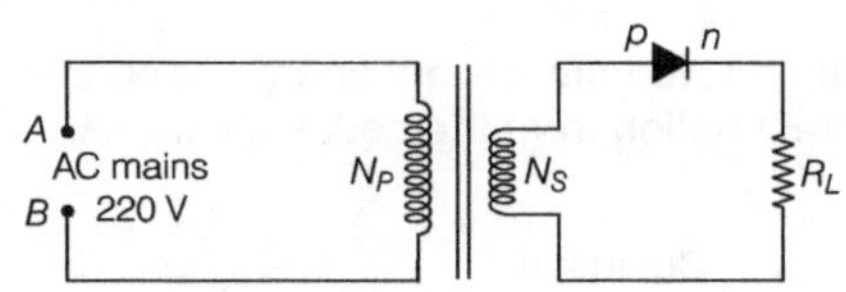

Given, supply voltage (AC mains) = 220 V
Number of turns of primary coil = 1000
Number of turns of secondary coil = 50
Current in primary coil = 0.25 A
Diode is ideal.

18 Average value of current flowing through load resistance R_L is

a. 5 A
b. $\frac{5}{\pi}$ A
c. $\frac{5}{\sqrt{2}}$ A
d. $\frac{5\sqrt{2}}{\pi}$ A

19 RMS value of current through load resistance R_L is

a. $5\sqrt{2}$ A
b. $\frac{5}{\pi}$ A
c. $\frac{5}{\sqrt{2}}$ A
d. 2.5 A

Statement Type

Direction (Q. Nos. 20-22) This section is based on Statement I and Statement II. Select the correct answer from the codes given below.

a. Both Statement I and Statement II are correct and Statement II is the correct explanation of Statement I
b. Both Statement I and Statement II are correct but Statement II is not the correct explanation of Statement I
c. Statement I is correct but Statement II is incorrect
d. Statement II is correct but Statement I is incorrect

20 **Statement I** Ionisation energy of isolated pentavalent atom of phosphorus is very large so, in a *n*-type semiconductor, there are no free charge carries at room temperature.

Statement II When current flows through a *p*-*n* junction diode, it becomes hot.

21 **Statement I** Rate of recombination for both intrinsic and extrinsic semiconductors are equal.

Statement II For an extrinsic semiconductor, $n_e n_h = n_i^2$.

Working Space

22 **Statement I** When a current flows through a p-n junction, an electron moves from a lower energy level to higher energy level by absorbing energy from the source (cell).

Statement II To move from a lower quantum state to a higher quantum state, an electron must absorb an energy which is equals to difference in energy of higher state and energy of lower state.

Matching List Type

Direction (Q. No. 23) *Choices for the correct combination of elements from Column I and Column II are given as options (a), (b), (c) and (d), out of which one is correct.*

23 Match type of semiconductors given in Column I with the correct energy band and gap diagram in Column II and mark the correct option from the codes given below.

	Column I		Column II
i.	Intrinsic semiconductor	p.	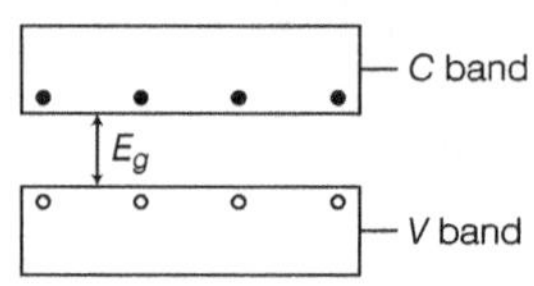
ii.	Extrinsic semiconductor	q.	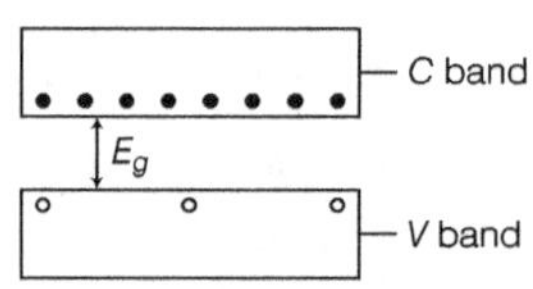
iii.	p-type semiconductor	r.	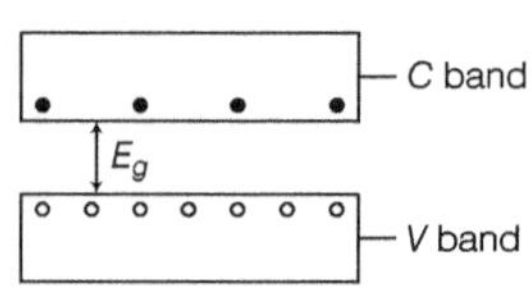
iv.	p-n junction	s.	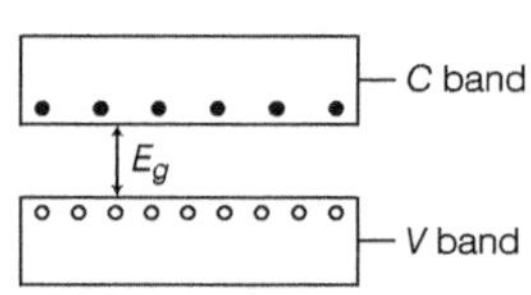
v.	Zener diode	t.	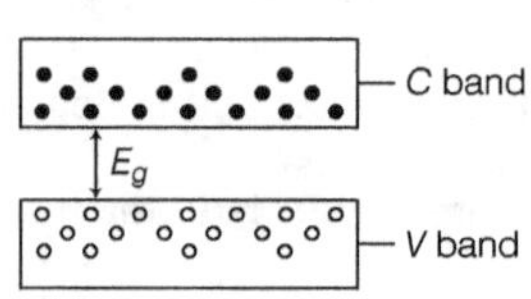

Codes

	i	ii	iii	iv	v
a.	p	q	r	s	t
b.	t	s	r	p	q
c.	p	q,r,s,t	r	s,t	t
d.	p	p,	r	s,t	t

DPP-3 Junction Transistors

Only One Option Correct Type

Working Space

Direction (Q. Nos. 1-10) *This section contains 10 multiple choice questions. Each question has four choices (a), (b), (c) and (d), out of which ONLY ONE option is correct.*

1 A transistor is equivalent to

a. one *p-n* junction
b. two *p-n* junction
c. three *p-n* junction
d. four *p-n* junction

2 A transistor has two depletion layers, first depletion layer is smaller and second is larger with no biasing. With a proper biasing, size of first depletion layer is

a. equal to the size of second depletion layer
b. larger than the size of second depletion layer
c. smaller than the size of second depletion layer
d. keep on increasing and decreasing

3 A transistor is analogous to a combination of two diodes.

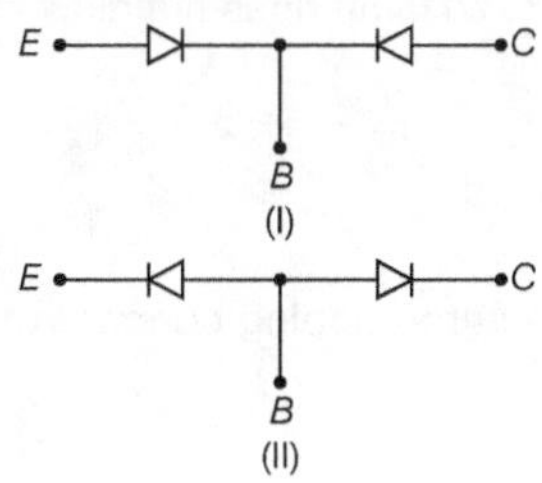

Now, mark the correct option.

a. I is for *p-n-p* and II is *n-p-n* transistor
b. I is for *p-n-p* and II for *p-n-p* transistor
c. Both I and II are for *n-p-n* transistor
d. Both I and II are for *p-n-p* transistor

4 Which of the following is correct option? (Consider a *n-p-n* type transistor).

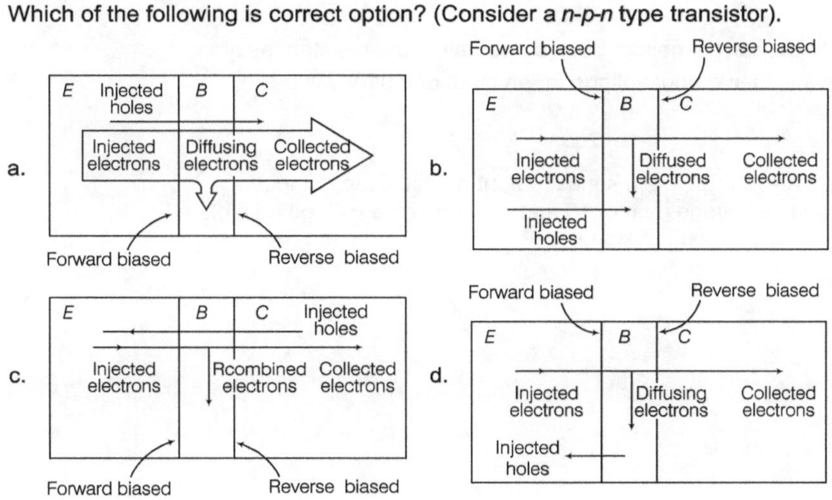

Working Space

5 In an *n-p-n* transistor, collector current is 10 mA. If 90% of charge carriers reach the collector, then emitter current I_E and base current I_B are

a. $I_E = 1\text{mA}, I_B = 10\text{ mA}$
b. $I_E = 10\text{ mA}, I_B = 1\text{mA}$
c. $I_E = 9\text{ mA}, I_B = 1\text{mA}$
d. $I_E = 11\text{mA}, I_B = 1\text{mA}$

6 Consider the given circuits,

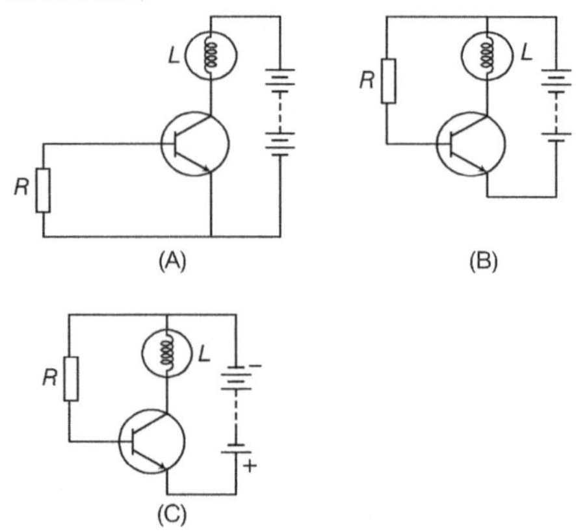

L is a lamp and R is a resistance. Choose the correct option.

a. As there is only one biasing battery in all 3 circuits, so lamp does not glow in any of the three
b. Lamp will glow in case A
c. Lamp will glow in case B
d. Lamp will glow in case C

7 Emitter and base regions of a *n-p-n* transistor have same doping concentrations. Then, which of the following is correct?

a. $I_E = I_B + I_C$
b. $I_E = I_C$
c. $I_E = I_B$
d. $I_B = I_E + I_C$

8 For a common emitter transistor current gain is 50. When same transistor is used in common base configuration, current gain will be

a. 50
b. 102
c. $\frac{1}{50}$
d. 0.98

9 An *n-p-n* CE amplifier has a current gain of 50. If output and input resistances are $1\text{k}\Omega$ and $5\text{ k}\Omega$ respectively, then output voltage for an input of 0.01 V will be

a. 0.25 V
b. 2.5 V
c. 25 V
d. 250 V

10 Two *n-p-n* CE amplifiers are connected in series, output of first is fed to input of second. First amplifier has a voltage gain of 10 and second has a voltage gain of 20. For an input signal of 0.01 V, output signal is of

a. 200 amplification
b. 2 amplification
c. 30 amplification
d. 0.5 amplification

Working Space

One or More than One Options Correct Type

Direction (Q. Nos. 11-15) *This section contains 5 multiple choice questions. Each question has four choices (a), (b), (c) and (d), out of which ONE or MORE THAN ONE are correct.*

11 Given are the transfer characteristics of a base biased CE transistor.

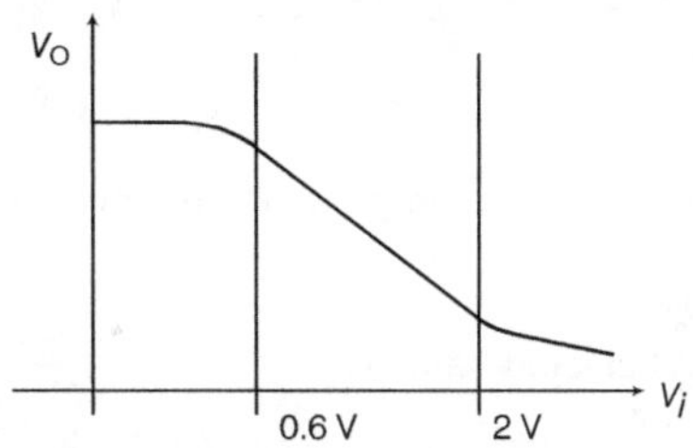

Which of these are correct ?

a. At $V_i = 0.4\,V$, transistor is in active state
b. At $v_i = 1V$, transistor can be used as an amplifier
c. At $V_i = 0.5\,V$, it can acts like switch in ON condition
d. At $V_i = 2.5\,V$, it can acts like a switch turned OFF

12 Consider the oscillatory circuit as shown in the figure.

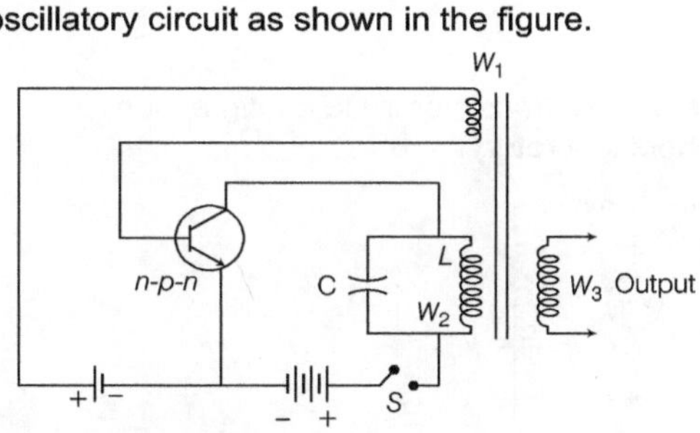

Then, mark the correct options.

a. When voltage in winding W_1 is maximum, voltage in winding W_3 is also maximum
b. When voltage in winding W_2 is maximum, voltage in windings W_1 and W_3 is minimum
c. When voltage in W_2 is growing, voltage in both W_1 and W_3 is also growing
d. When voltage in W_2 is maximum, potential drop across capacitor C is minimum

13 Consider an amplifier circuit using an *n-p-n* CE transistor,

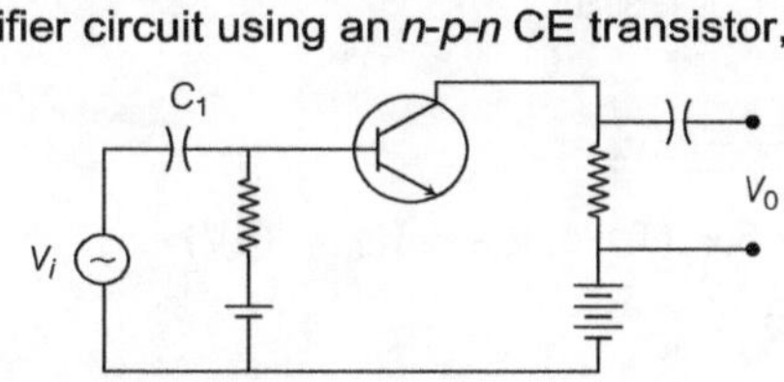

Now, mark the correct option.

a. Output voltage in DC
b. Output voltage in AC in phase with input
c. Output voltage is AC out of phase with input
d. Capacitors are used to block DC biasing voltages

Working Space

14 Output characteristics of an *n-p-n* CE transistor are plotted as shown in the figure.
For the given transistor,

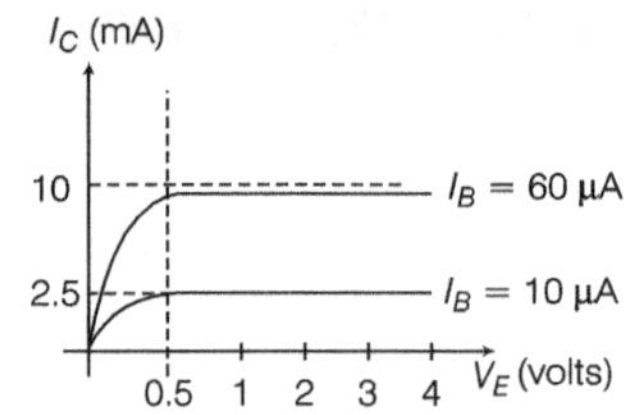

a. current amplification is of 50
b. current amplification is of 140
c. output resistance of transistor is initially 500 Ω
d. output resistance of transistor is initially 50 Ω

15 A transistor amplifier produces a current gain and a voltage gain. Then, choose the correct option.
a. Amplifier violate law of conservation of energy
b. Power gain is produced on the expense of DC biasing batteries
c. Voltage gain increases with increase in frequency
d. Voltage gain decreases both at low as well as high frequencies

Comprehension Type

Direction (Q. Nos. 16-20) This section contains 2 paragraphs, each describing theory, experiments, data, etc. Five questions related to the paragraphs have been given. Each question has only one correct answer among the four given options (a), (b), (c) and (d).

Passage I

To study input and output characteristics of *n-p-n* transistor in CE configuration, a student uses following circuit in the school laboratory.

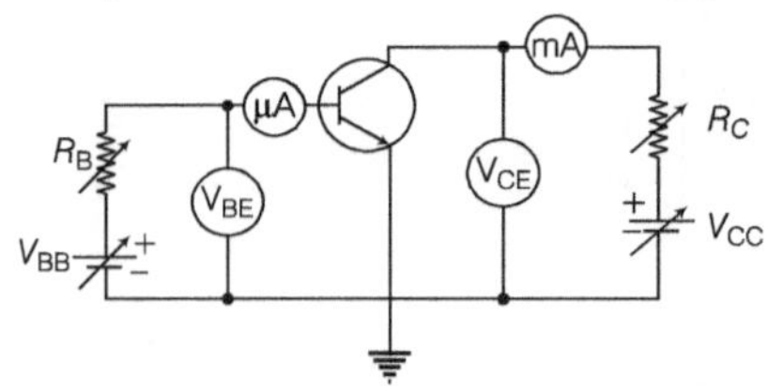

He also prepares following variation of I_C with V_{CE} at a constant base current of 30μA.

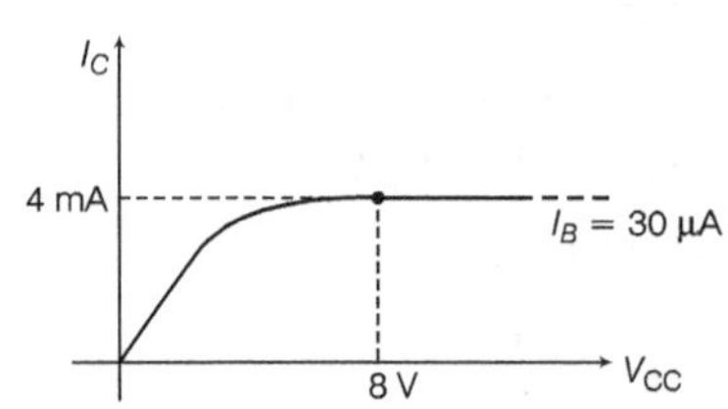

Let V_{BE} for the transistor be 0.7 V with V_{CE} = 8V. (Take, $V_{CC} = V_{BB}$ = 16V)

16 What is the ratio R_B/R_C for given condition ?
a. 50 : 1
b. 1 : 510
c. 1 : 255
d. 225 : 1

17 What is current gain β?
a. 510
b. 225
c. 166
d. 133

18 What will be the power gain?
a. 19
b. 49
c. 69
d. 79

Working Space

Passage II

Consider the transistor given,

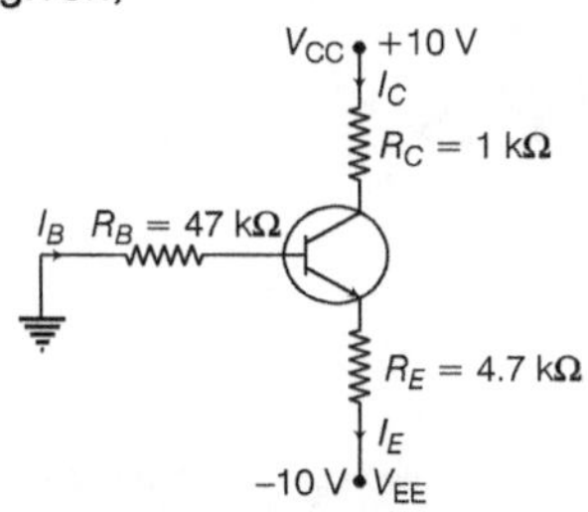

19 In steady state, current through the collector is

a. 1.8 mA
b. 2.0 mA
c. 4 mA
d. 0.9 mA

20 V_{CE} is

a. 5.67 V
b. 10.28 V
c. 9.74 V
d. 5.35 V

Statement Type

Direction (*Q. Nos.* 21 *and* 22) *This section is based on Statement I and Statement II. Select the correct answer from the codes given below.*

a. Both Statement I and Statement II are correct and Statement II is the correct explanation of Statement I
b. Both Statement I and Statement II are correct but Statement II is not the correct explanation of Statement I
c. Statement I is correct but Statement II is incorrect
d. Statement II is correct but Statement I is incorrect

21 **Statement I** For voltage amplification, a transformer can be used in place of transistor.

Statement II Transformer does not work with a DC supply.

22 **Statement I** For a positive feed back to an *L-C* oscillatory circuit, an *n-p-n* CE transistor is used.

Statement II A positive feed back requires a high voltage input.

DPP-4 Logic Gates

Subjective Question

Working Space

Direction (Q. No. 1) This question is subjective in nature, need to be solved completely on notebook.

1 i. What is Boolean algebra?
ii. State basic laws of Boolean algebra.
iii. How Boolean algebra helps us to express a logical function?
iv. How Boolean algebra is helpful to designing/modifying logic gates?

Only One Option Correct Type

Direction (Q. Nos. 2-12) This section contains 11 multiple choice questions. Each question has four choices (a), (b), (c) and (d), out of which ONLY ONE option is correct.

2 Identify the logic operation of given logic gate.

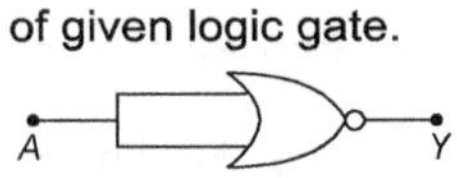

a. OR b. AND
c. NOT d. XOR

3 Consider the given circuit,

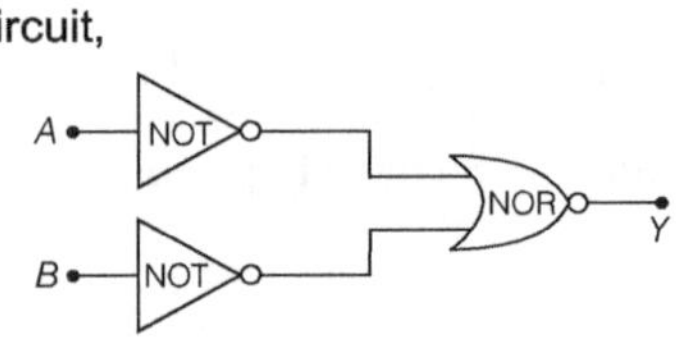

It is equivalent to

a. 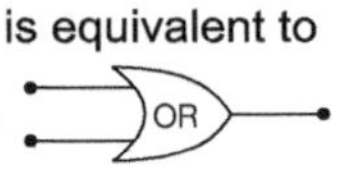

b. 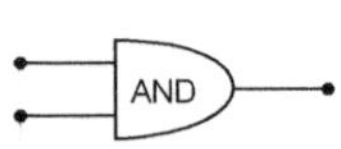

c. 

d. 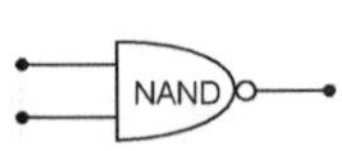

4 Two NOR gates are connected in series with each other. They result in the logic operation as

a. AND b. OR
c. NOT d. XOR

5 Two logical circuits are equivalent when they produce same outputs for same inputs. Now you have given following circuit.

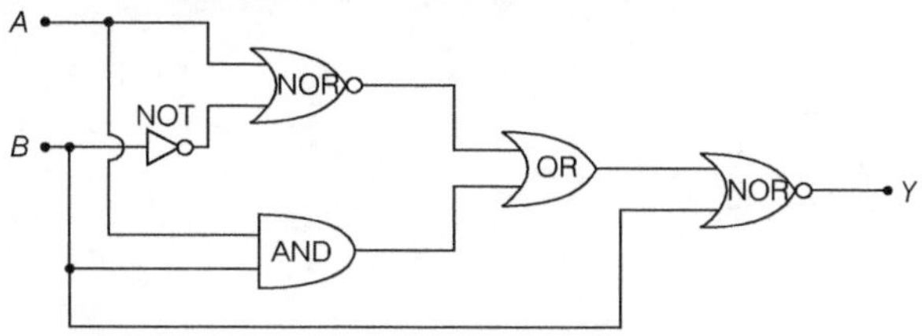

This is equivalent to

a. 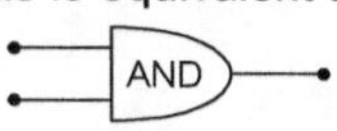

b. 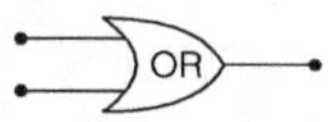

c. 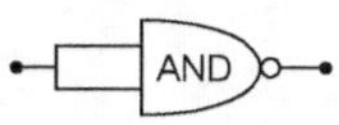

d. 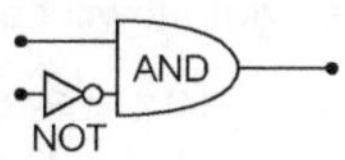

6 Identify the logical function performed by given circuit.

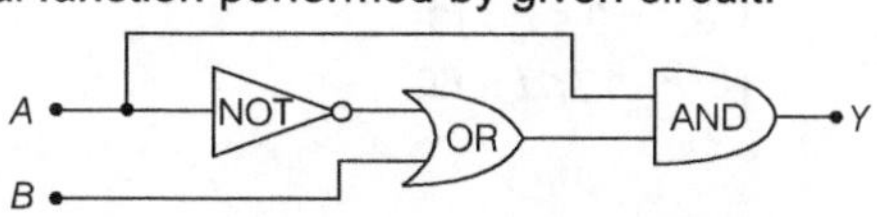

a. $Y \equiv A \cdot (A + B)$

b. $Y \equiv A(\bar{A} \cdot B)$

c. $Y \equiv A + (A + \bar{B})$

d. $Y \equiv A \cdot (\bar{A} + B)$

7 A logical circuit requires an input to perform. This input must be

a. an analog signal
b. a digital signal
c. a mixture of analog and digital signal
d. a constant signal

8 0 and 1 inputs for a logical circuit are

a. 0 V and 1 V
b. $0 \equiv$ Some low voltage level, $1 \equiv$ Some high voltage level
c. $0 \equiv$ A range of low voltage level, $1 \equiv$ A range of high voltage level
d. $0 \equiv$ High voltage level, $1 \equiv$ Low voltage level

9 A circuit is prepared by using two *p-n* junctions and a resistor as

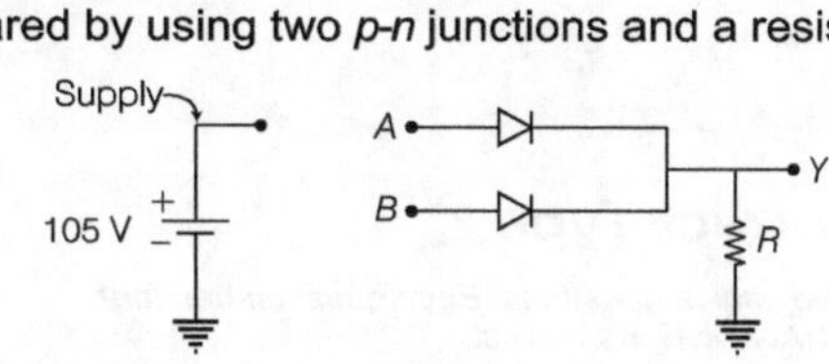

Logical operation performed by the circuit is

a. $Y \equiv A \cdot B$

b. $Y \equiv A + B$

c. $Y \equiv (A \cdot B)'$

d. $Y = (A + B)'$

10 Which of these can not be made using only *p-n* junctions?

a. AND gate

b. OR gate

c. NOT gate

d. All of these

Working Space

11 Two time varrying inputs A and B are

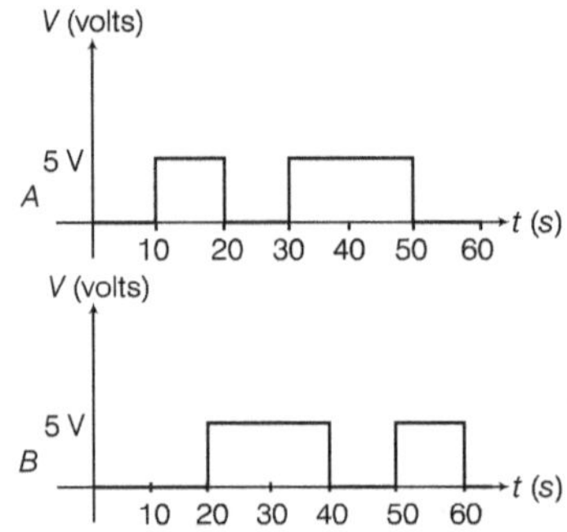

If there are fed to a two input OR gate, then output will be

a. $V = \begin{cases} 0, & 0 \le t \le 10 \\ 1, & 10 \le t \le 20 \\ 0, & 20 \le t \end{cases}$

b. $V = \begin{cases} 0, & 0 \le t \le 20 \\ 5, & 20 \le t \end{cases}$

c. $V = \begin{cases} 0, & 0 \le t \le 10 \\ 5\text{V} & 10 \le t \le 60 \end{cases}$

d. $V = \begin{cases} 0, & 0 \le t < 10 \\ 5\text{V} & 10 \le t \le 60 \end{cases}$

12 Two time varying functions are as follows :

$$A = \begin{cases} 0, & 0 \le t < 1, \\ 1, & 1 \le t \le 2, \\ 0, & 2 < t \le 3, \end{cases} \quad B = \begin{cases} 0, & 0 \le t < \frac{1}{2} \\ 1, & \frac{1}{2} \le t \le \frac{5}{2} \\ 0, & \frac{5}{2} < t \le 3 \end{cases}$$

Output of a NAND gate will be

a.

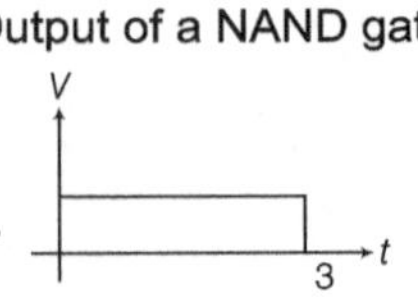

b.

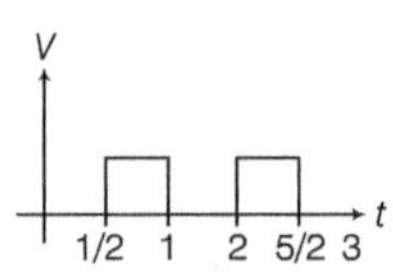

c.

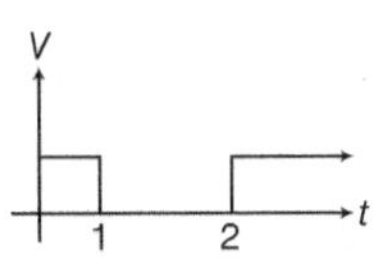

d.

One or More than One Options Correct Type

Direction (Q. Nos. 13-16) *This section contains 4 multiple choice questions. Each question has four choices (a), (b), (c) and (d), out of which ONE or MORE THAN ONE are correct.*

13 A universal gate is

a. XOR gate
b. NOR gate
c. NAND gate
d. integrated circuit

14 Concept of fabricating an entire circuit over a single block of a semiconductor has revolutionised the computer technology. Such circuits as we know as called Integrated circuits. Then, mark the correct options.

a. SSI means IC contains upto 10 logic gates
b. MIS means IC contains upto 100 logic gates
c. LSI means IC contains upto 1000 logic gates
d. VLSI means IC contains upto 2000 logic gates

15 In digital system such as computers 0 and 1 represent two different voltage levels. Then, choose the correct options.

a. For two voltage levels of 2 V and 5 V in a negative logic system, 2 V is 1 and 5V is 0 of the logic

b. For two voltage levels of 2 V and 5 V, in positive logic system 2 V is represented by 0 and 5V is represented by 1

c. A positive OR is a negative AND and *vice-versa*.

d. In a negative logic, the logic : 1010010111 is represented by waveform

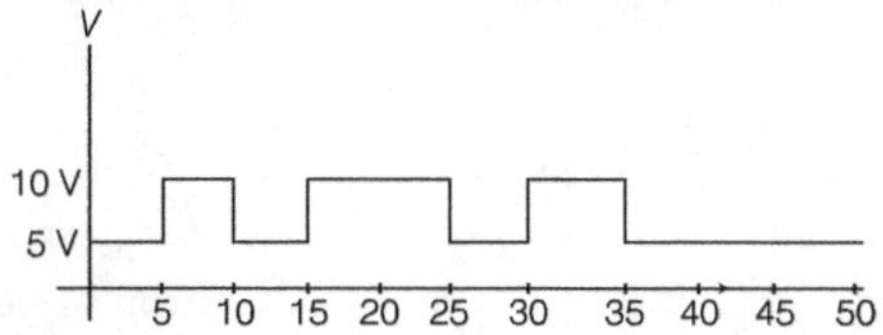

16 Mark the correct option(s).

a. A logic gate can have more than one output

b. A logic gate can have a maximum of two inputs

c. A logic gate can have two or more inputs

d. A logic gate cannot have more than one output

Comprehension Type

Direction *(Q. Nos. 17 and 18) This section contains a paragraph, describing theory, experiments, data, etc. Two questions related to the paragraph have been given. Each question has only one correct answer among the four given options (a), (b), (c) and (d).*

Passage

Consider given input waveforms *A* and *B*,

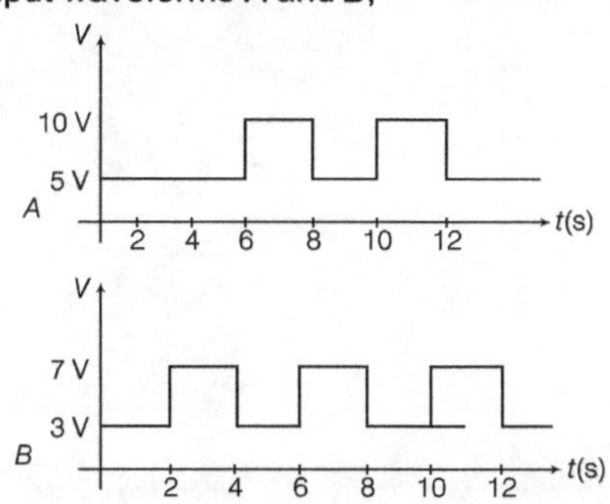

17 When the inputs are fed to a NAND gate with a negative logic, output in binary form is

a. 111010 b. 111111

c. 000101 d. 000100

18 Output of a NOR gate working with a positive logic on these inputs will be

a. 0, 1, 0, 1, 0, 1

b. 1, 0, 1, 0, 1, 0

c. 1, 0, 0, 1, 1, 1

d. 1, 1, 0, 0, 1, 0

Matching List Type

Direction *(Q. No. 19) Choices for the correct combination of elements from Column I and Column II are given as options (a), (b), (c) and (d), out of which one is correct.*

19 Match given gates in Column I with the logical Boolean operation they perform in Column II and mark the correct option from the codes given below.

	Column I		Column II
i.	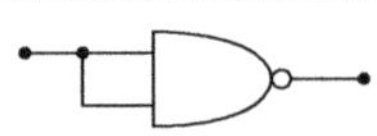	p.	$Y \equiv \overline{A}$
ii.	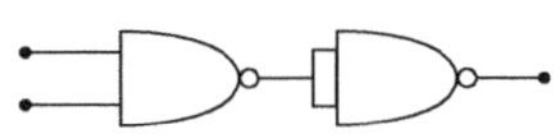		
iii.	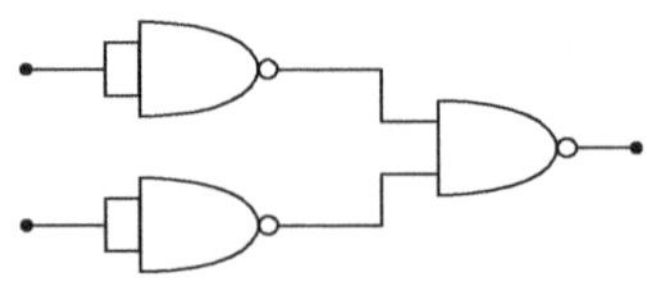		
iv.	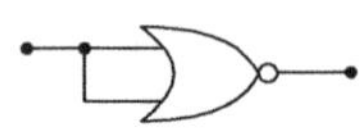	q.	$Y \equiv A + B$
v.	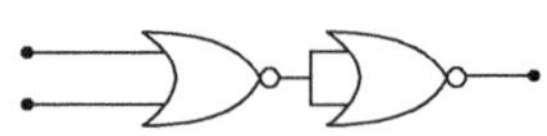		
vi	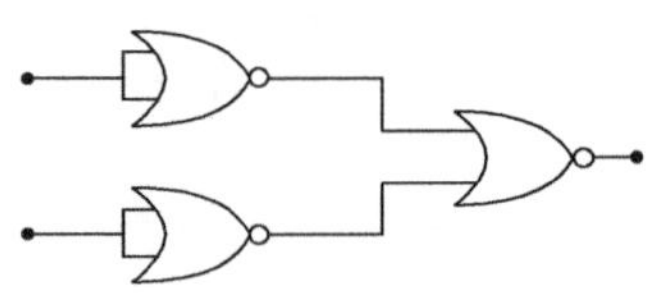	r.	$Y \equiv A \cdot B$

Codes

	i	ii	iii	iv	v	vi
a.	p	r	q	p	q	r
b.	p	q	r	p	r	q
c.	r	q	p	q	p	p
d.	q	p	r	q,p	r	p

Statement Type

Direction *(Q. Nos. 20 and 21) This section is based on Statement I and Statement II. Select the correct answer from the codes given below.*

a. Both Statement I and Statement II are correct and Statement II is the correct explanation of Statement I
b. Both Statement I and Statement II are correct but Statement II is not the correct explanation of Statement I
c. Statement I is correct but Statement II is incorrect
d. Statement II is correct but Statement I is incorrect

20 **Statement I** A logic gate can have one and only one output.

Statement II A Boolean function like any other function can have a unique output.

21 **Statement I** NAND or NOR gates are universal gates.

Statement II A combination of NAND gates or a combination of NOR gates can be used to perform or implement any Boolean function.

Revisal Problems for JEE Main

Only One Option Correct Type

Working Space

Direction *(Q. Nos. 1-17) This section contains 17 multiple choice questions. Each question has four choices (a), (b), (c) and (d), out of which ONLY ONE option is correct.*

1 In pure silicon sample of 5×10^{22} atoms at 300 K, electron hole concentration is $7 \times 10^{15}\ m^{-3}$. Antimony is doped into silicon in a proportion of 1 atom for every 10^7 atoms of silicon. Half of impurity atoms contribute band.

Number of electrons in the conduction band increases by a factor of

a. 3.6 b. 2.0
c. 1.8 d. 0.5

2 Let we have a half-wave rectifier using one *p-n* junction and a full wave rectifier using two *p-n* junctions and a centre tap transformer. All of *p-n* junctions used are identical and ideal. Ratio of rectifier efficiencies in half wave and full wave rectification is

a. $1 : \sqrt{2}$ b. $\sqrt{2} : 1$
c. $1 : 2$ d. $2 : 1$

3 Change in resistivity of semiconductor piece with wavelength of incident light is as shown in the figure. Intensity of light is kept constant while experiment. The band gap (in eV) for the semiconductor is

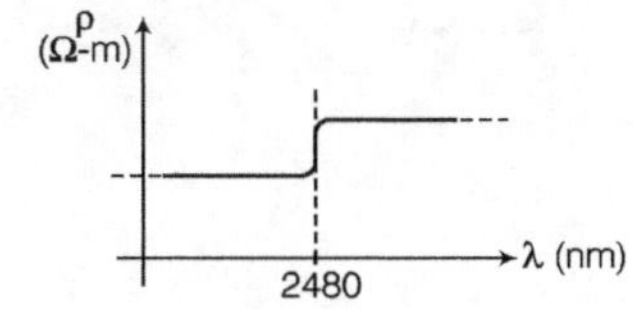

a. 0.9 b. 0.7
c. 0.5 d. 1.1

4 Consider the given circuit,

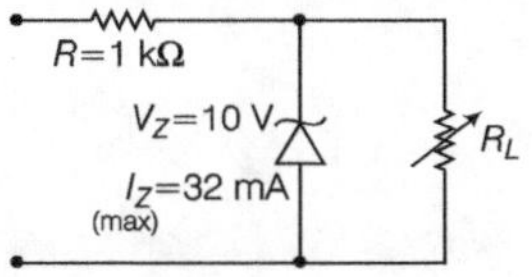

If voltage across load resistance R_L is to be maintained at 10 V, range of values of R_L will be

a. 50 Ω to 1 kΩ
b. 100 Ω to 250 Ω
c. 250 Ω to 1.25 kΩ
d. 500 Ω to 1.25 kΩ

5 Consider the given circuit,

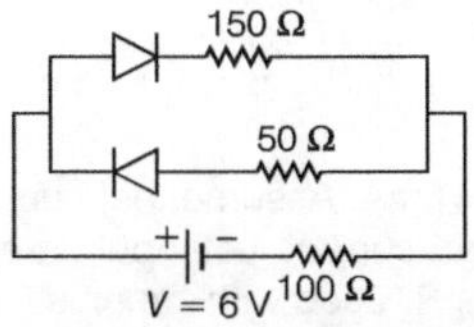

If forward resistance of each diode is 50Ω, then current through 100Ω resistor is

a. 0A b. 0.02 A
c. 0.03 A d. 0.036 A

Working Space

6 Which of the following statements is incorrect?

a. Resistance of intrinsic semiconductor decreases with rise of temperature
b. Resistance of extrinsic semiconductor decreases with rise of temperature
c. Doping pure Si with aluminium gives p-type semiconductor
d. Doping pure Si with antimony gives p-type semiconductor

7 Correct order of conductivity of Si, Ge and C is

a. Ge > Si > C
b. Si > Ge > C
c. C > Si > Ge
d. Si > C > Ge

8 In given circuit resistance of resistor is 1kΩ and threshold (knee) voltage for diode is 0.7 V.

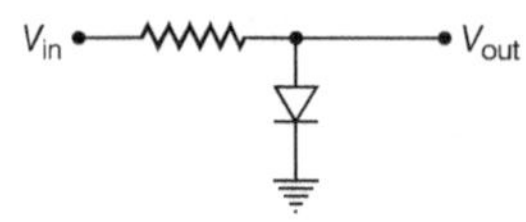

Correct variation of V_{out} with respect to V_{in} is

a.
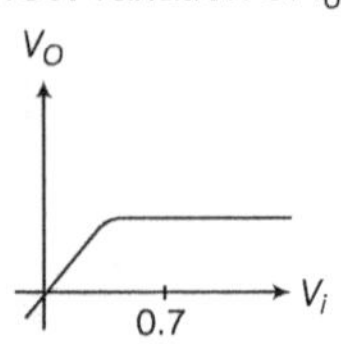

b.
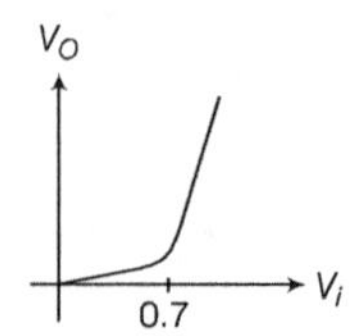

c.
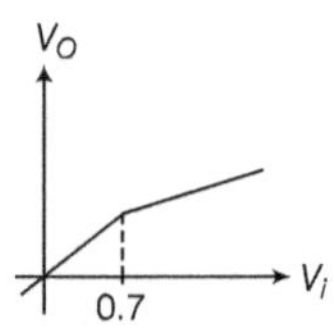

d.
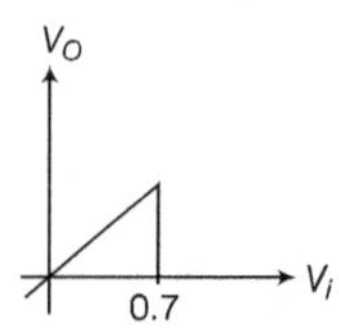

9 Consider the given circuit,

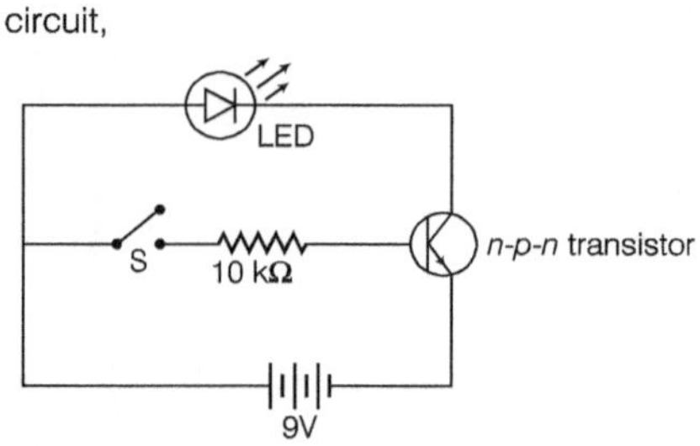

When switch is turned ON

a. LED changes its colour
b. LED becomes OFF
c. LED becomes ON
d. LED is short circuited

10 In a zener regulator,

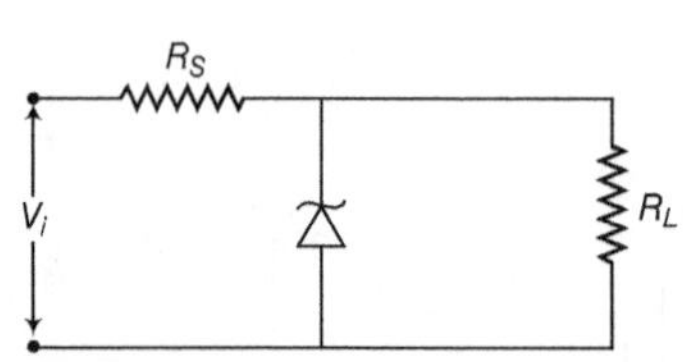

Let zener breakedowns at 6 V, permissible load current = 4mA. Assume that current through zener diode at breakdown is about 5 times of load current. Let input unregulated voltage can reaches upto 10 V. Value of series resistance R_s used with zener is

a. 166.67Ω
b. 1000 Ω
c. 3.33 Ω
d. 57.79 Ω

11 For a transistor,

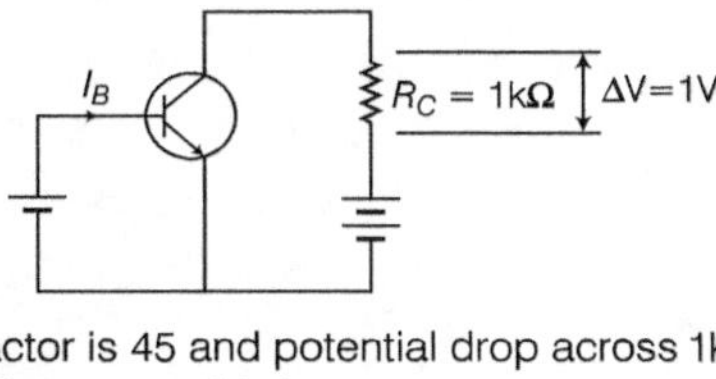

current amplification factor is 45 and potential drop across $1k\Omega$ resistor connected in the collector circuit is 1V. Base current I_B is

a. 0.044 mA
b. 0.033 mA
c. 0.022 mA
d. 0.011 mA

12 In the given transistor circuit, $V_{BE} = 0.7$ V.

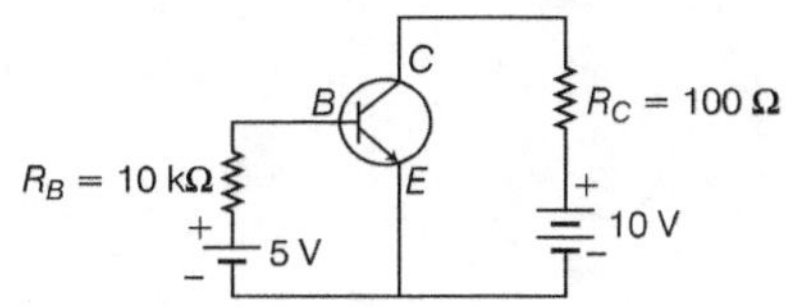

If $\beta = 150$, then V_{CB} is

a. −2.85V
b. +2.85 V
c. +3.55V
d. −3.55V

13 For the circuit shown,

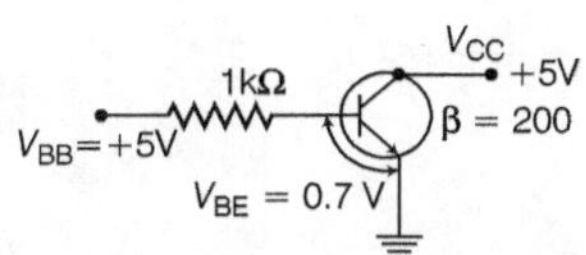

Power dissipation in the transistor will be

a. 4.3 W
b. 5.2 W
c. 7.2 W
d. 9.7 W

14 For an unbiased *n-p-n* transistor,

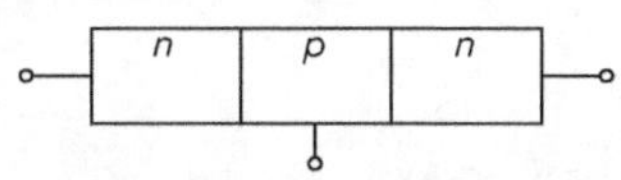

Which of the following is a correct figure?

a.
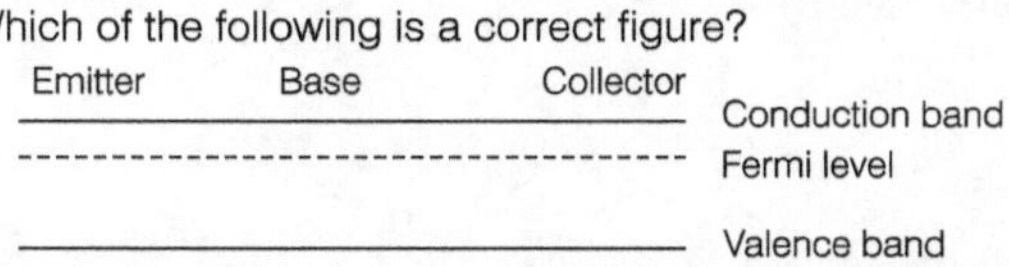

b.
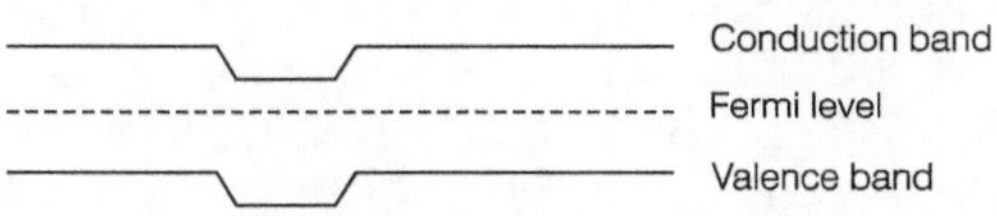

c.
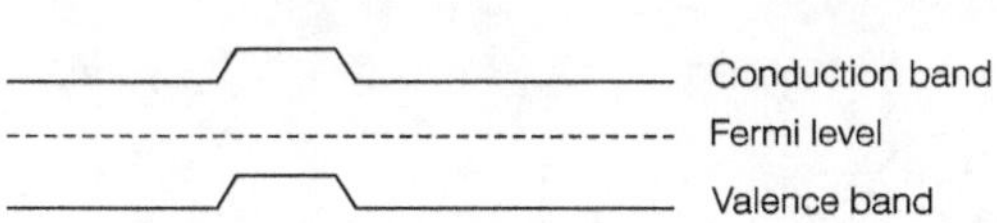

d.
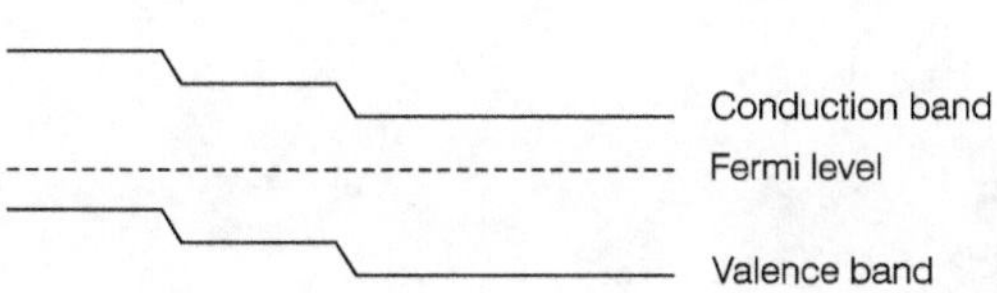

Working Space

15 Consider the given circuit,

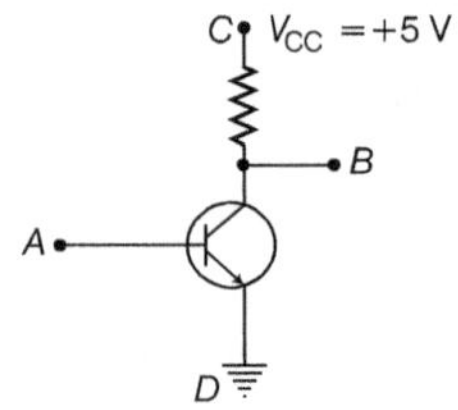

Point C is maintained at + 5V and point D is earthed.
If the following input is given at point A,

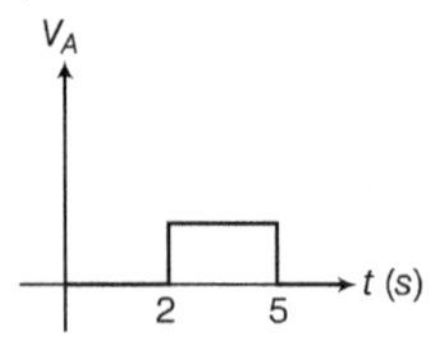

Then, output at point B is

a.
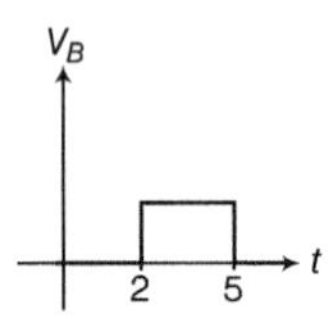

b.
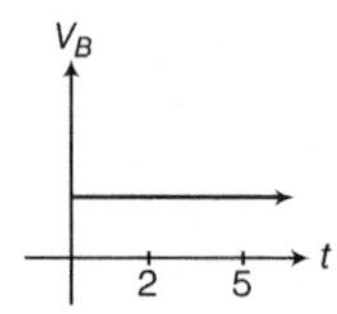

c.
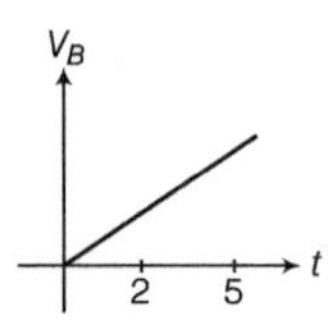

d.
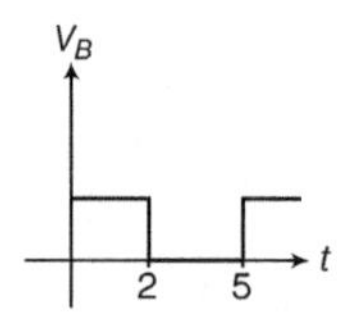

16 For the given logic circuit,

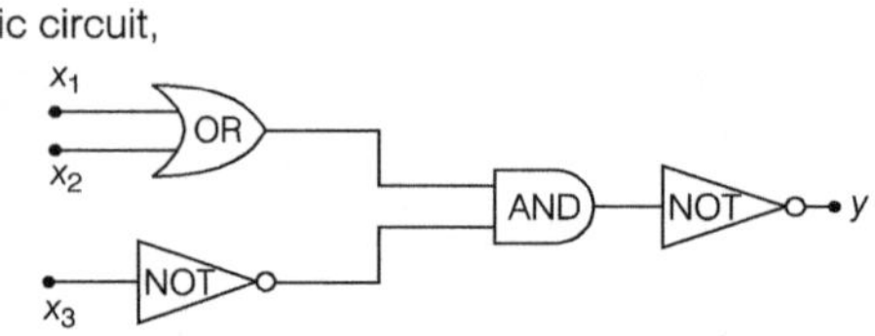

Which of these is correct?

a. $y = 0$ for $x_1 = x_3 = 0$ and $x_2 = 1$
b. $y = 0$ for $x_1 = x_2 = x_3 = 0$
c. $y = 1$ for $x_1 = x_2 = x_3 = 1$
d. $y = 1$ for $x_1 = x_2 = 1$ and $x_3 = 0$

17 Consider the circuit,

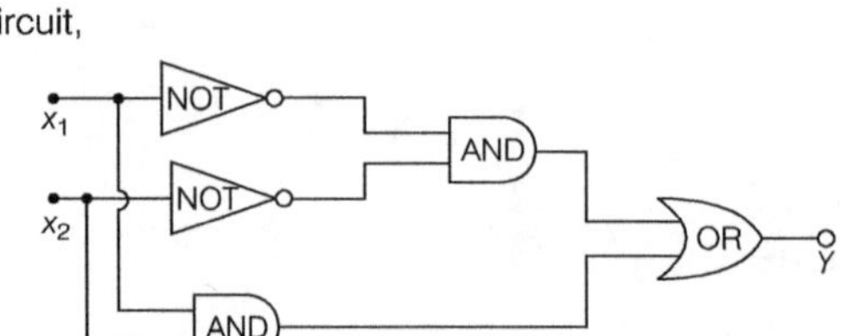

Logical operation performed by the circuit is

a. $(x_1 \cdot x_2) + (\bar{x}_1 \cdot \bar{x}_2)$
b. $(\overline{x_1 \cdot x_2})$
c. $(x_1 \cdot x_2) + (\overline{x_1} \cdot \overline{x_2})$
d. $(\overline{x_1} + \overline{x_2}) \cdot (x_1\ x_2)$

Matching List Type

Working Space

Direction *(Q. No. 18) Choices for the correct combination of elements from Column I and Column II are given as options (a), (b), (c) and (d), out of which one is correct.*

18 Match the Column I with Column II and mark the correct option from the codes given below.

	Column I		Column II
i.	Energy gap for Si (in eV)	p.	1.1
ii.	Energy gap for Ge (in eV)	q.	0.67
iii.	Threshold voltage for Si diode (in volts)	r.	0.7
iv.	Threshold voltage for Ge (in volts)	s.	0.3

Codes

	i	ii	iii	iv
a.	p	q	r	s
b.	s	r	q	p
c.	p	r	q	s
d.	p	s	q	r

Statement Type

Direction *(Q. Nos. 19-21) This section is based on Statement I and Statement II. Select the correct answer from the codes given below.*

a. Both Statement I and Statement II are correct and Statement II is the correct explanation of Statement I
b. Both Statement I and Statement II are correct but Statement II is not the correct explanation of Statement I
c. Statement I is correct but Statement II is incorrect
d. Statement II is correct but Statement I is incorrect

19 **Statement I** Output frequency of time varying DC voltage in a full wave rectifier is twice that of input.

Statement II A centre tap transformer increases the frequency of input.

20 **Statement I** In a transistor, I_C (the collector current) is almost independent of V_{CE} (potential difference of collector and emitter).

Statement II Base emitter junction of a transistor is in forward bias.

21 **Statement I** *A* NAND gate is not obtained by using NOR gates only.

Statement II NOR and NAND gates are called universal gates.

JEE Main & AIEEE Archive

(Compilation of Last 13 Years' Questions)

1 The forward biased diode connection is **(JEE Main 2014)**

a. 

b. 

c. 

d. 

2 The *I*-*V* characteristic of an LED is **(JEE Main 2013)**

a. 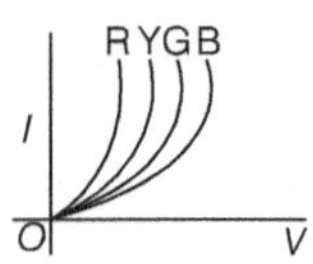

b. 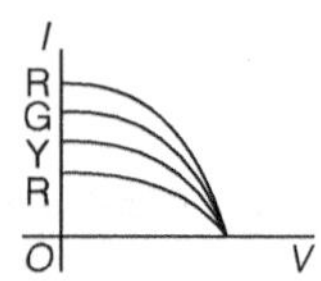

c. 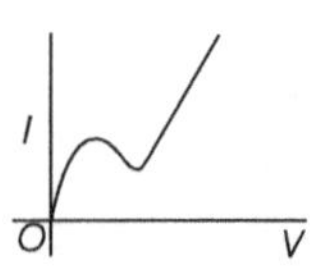

d. 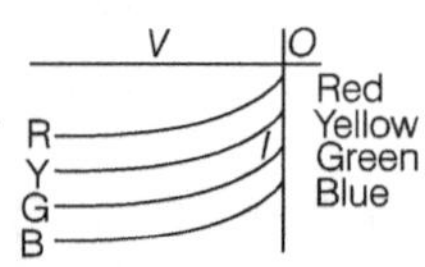

3 Truth table for system of four NAND gates as shown in figure is **(AIEEE 2012)**

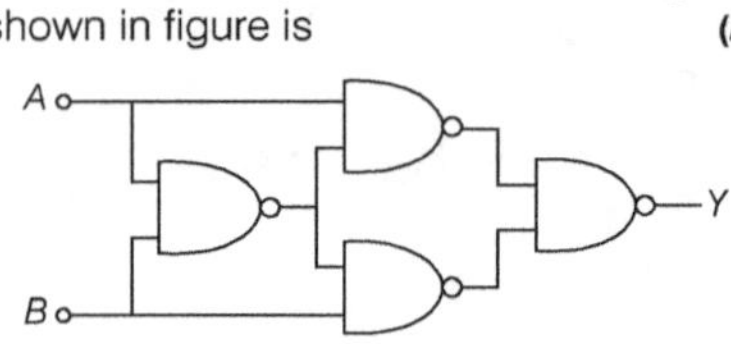

a.

A	B	Y
0	0	0
0	1	1
1	0	1
1	1	0

b.

A	B	Y
0	0	0
0	1	0
1	0	1
1	1	1

c.

A	B	Y
0	0	1
0	1	1
1	0	0
1	1	0

d.

A	B	Y
0	0	1
0	1	0
1	0	0
1	1	1

4 The output of an OR gate is connected to both the inputs of a NAND gate. The combination will serve as a **(AIEEE 2011)**

a. OR gate
b. NOT gate
c. NOR gate
d. AND gate

5 The combination of gates shown below yields **(AIEEE 2010)**

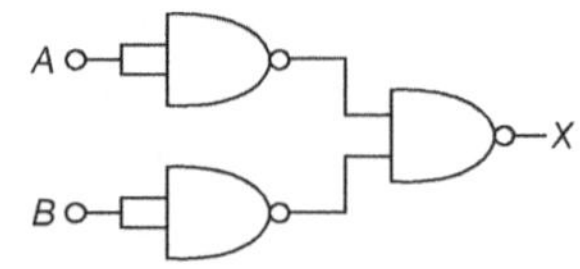

a. OR gate b. NOT gate
c. XOR gate d. NAND gate

6 A *p*-*n* junction (*D*) shown in the figure can act as a rectifier. An alternating current source (V) is connected in the circuit. **(AIEEE 2009)**

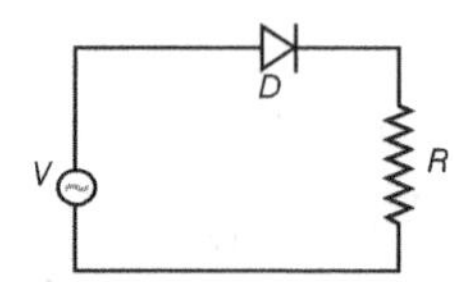

Then, which of the following graphs is correct?

a. 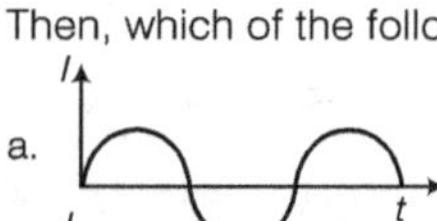

b. 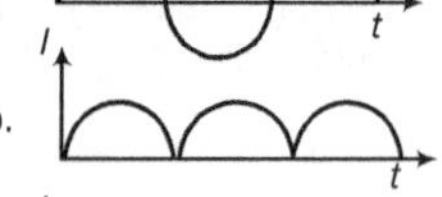

c. 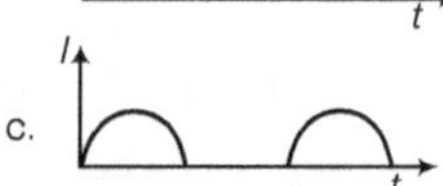

d. 

7 The logic circuit ashown below has the input waveforms *A* and *B* as shown. Pick out the correct output waveform. **(AIEEE 2009)**

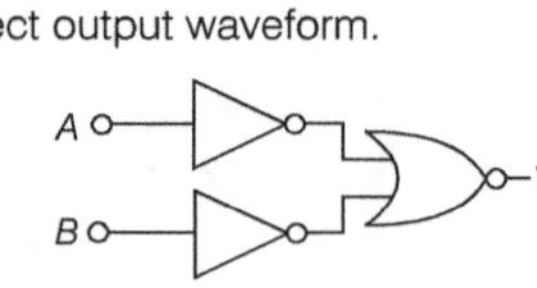

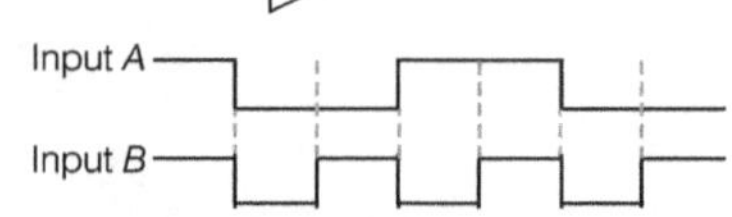

a. 

b. 

c. 

d. 

8 In the adjacent circuit, A and B represent two inputs and C represents the output.

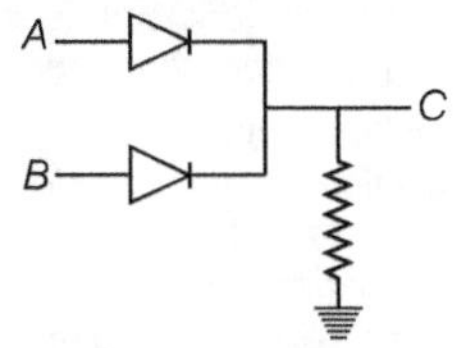

The circuit represents **(AIEEE 2008)**

a. NOR gate
b. AND gate
c. NAND gate
d. OR gate

9 If in a p-n junction diode, a square input signal of 10 V is applied as shown in the figure. **(AIEEE 2007)**

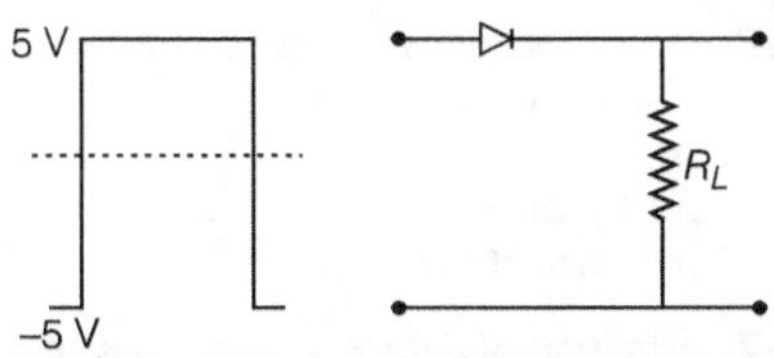

Then, the output signal across R_L will be

a. 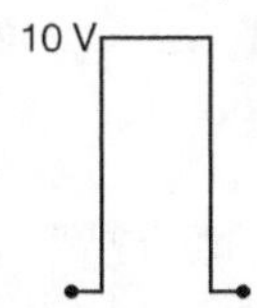

b. 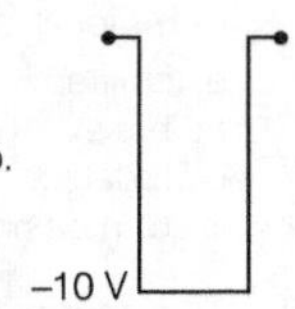

c. 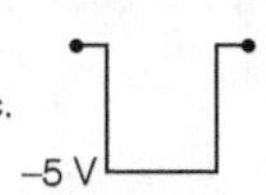

d. 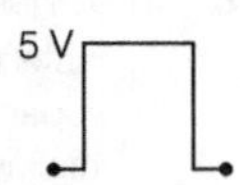

10 Carbon, silicon and germanium have four valence electrons each. At room temperature, which one of the following statements is most appropriate? **(AIEEE 2007)**

a. The number of free conduction electrons is significant in C but small in Si and Ge
b. The number of free conduction electrons is negligibly small in all the three
c. The number of free electrons for conduction is significant in all the three
d. The number of free electrons for conduction is significant only in Si and Ge but small is C

11 In a common-base mode of a transistor, the collector current is 5.488 mA for an emitter current of 5.60 mA. The value of the base current amplification factor (β) will be **(AIEEE 2006)**

a. 49 b. 50
c. 51 d. 48

12 If the lattice constant of this semiconductor is decreased, then which of the following is correct? **(AIEEE 2006)**

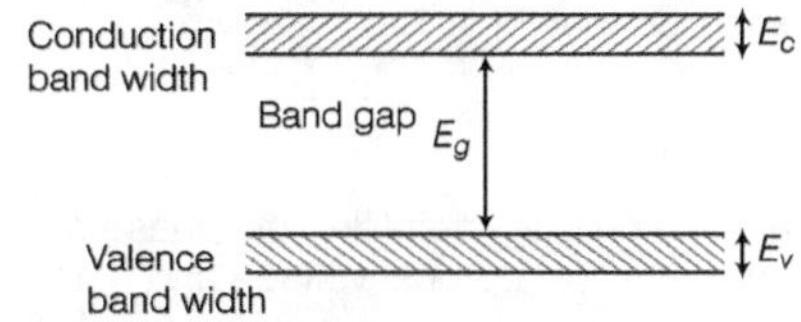

a. All E_c, E_g, E_v increase
b. E_c and E_v increase but E_g decreases
c. E_c and E_v decrease but E_g increases
d. All E_c, E_g, E_v decrease

13 A solid which is not transparent to visible light and whose conductivity increases with temperature is formed by **(AIEEE 2006)**

a. ionic bonding
b. covalent bonding
c. van der Waals' bonding
d. metallic bonding

14 If the ratio of the concentration of electrons to that of holes in a semiconductor is $\frac{7}{5}$ and the ratio of currents is $\frac{7}{4}$, then what is the ratio of their drift velocities? **(AIEEE 2006)**

a. $\frac{5}{8}$ b. $\frac{4}{5}$ c. $\frac{5}{4}$ d. $\frac{4}{7}$

15 The circuit has two oppositely connected ideal diodes in parallel. What is the current flowing in the circuit? **(AIEEE 2006)**

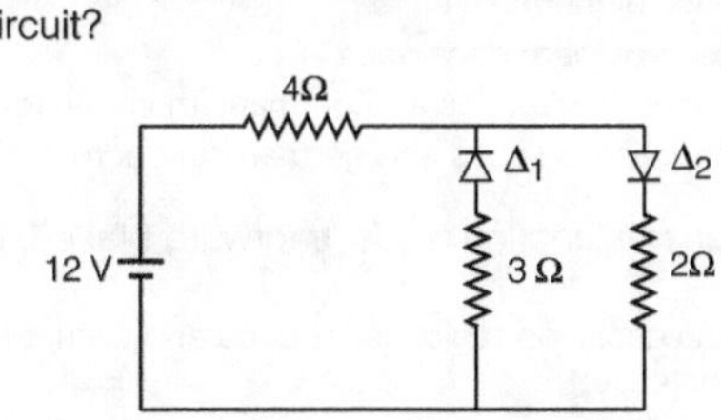

a. 1.71 A b. 2.00 A
c. 2.31 A d. 1.33 A

16 In the following, which one of the diodes is reverse biased? **(AIEEE 2006)**

a. 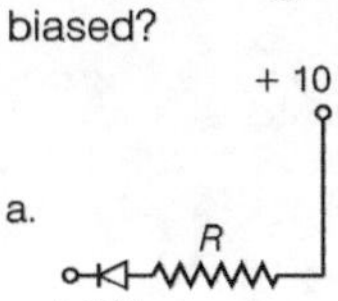

b. 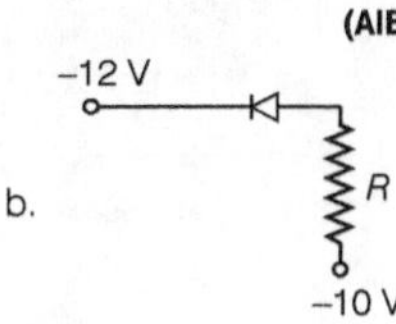

c. 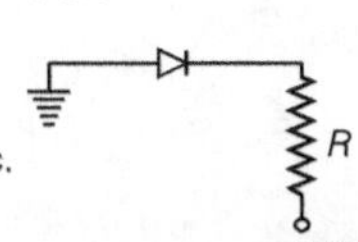

d. 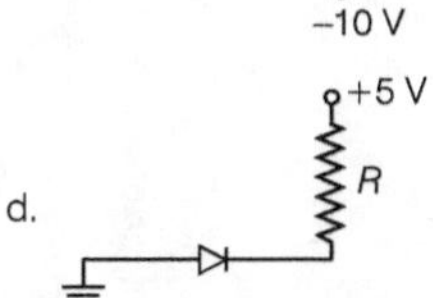

17 The electrical conductivity of a semiconductor increases when electromagnetic radiation of wavelength shorter than 2480 nm, is incident on it. The band gap (in eV) for the semiconductor is **(AIEEE 2005)**

a. 1.1 eV b. 2.5 eV
c. 0.5 eV d. 0.7 eV

18 In a common base amplifier, the phase difference between the input signal voltage and output voltage is **(AIEEE 2005)**

a. $\frac{\pi}{4}$ b. π c. zero d. $\frac{\pi}{2}$

19 In a full wave rectifier circuit operating from 50 Hz mains frequency, the fundamental frequency in the ripple would be **(AIEEE 2005)**

a. 50 Hz b. 25 Hz c. 100 Hz d. 70.7 Hz

20 When *n-p-n* transistor is used as an amplifier, **(AIEEE 2004)**

a. electrons move from base to collector
b. holes move from emitter to base
c. electrons move from collector to base
d. holes move from base to emitter

21 For a transistor amplifier in common emitter configuration for load impedance of 1 kΩ (h_{fe} = 50 and h_{oe} = 25 μ A/V), the current gain is **(AIEEE 2004)**

a. – 5.2 b. – 15.7
c. – 4.8 d. – 48.78

22 A piece of copper and another of germanium are cooled from room temperature to 77 K, the resistance of **(AIEEE 2004, 03)**

a. each of them increases
b. each of them decreases
c. copper decreases and germanium increases
d. copper increases and germanium decreases

23 When *p-n* junction diode is forward biased, then **(AIEEE 2004)**

a. the depletion region is reduced and barrier height is increased
b. the depletion region is widened and barrier height is reduced
c. both the depletion region and barrier height are reduced
d. both the depletion region and barrier height are increased

24 The difference in the variation of resistance with temperature in a metal and a semiconductor arises essentially due to the difference in the **(AIEEE 2003)**

a. crystal structure
b. variation of the number of charge carriers with temperature
c. type of bonding
d. variation of scattering mechanism with temperature

25 In the middle of the depletion layer of reverse biased *p-n* junction, the **(AIEEE 2003)**

a. electric field is zero
b. potential is maximum
c. electric field is maximum
d. potential is zero

26 At absolute zero, Si acts as

a. non-metal
b. metal
c. insulator
d. None of the above

27 The part of a transistor which is most heavily doped to produce large number of majority carriers is **(AIEEE 2002)**

a. emitter
b. base
c. collector
d. Can be any of the above three

28 The energy band gap is maximum in **(AIEEE 2002)**

a. metals
b. superconductors
c. insulators
d. semiconductors

29 By increasing the temperature, the specific resistance of a conductor and a semiconductor **(AIEEE 2002)**

a. increases for both
b. decreases for both
c. increases, decreases respectively
d. decreases, increases respectively

5. Communication System

DPP

DAILY PRACTICE PROBLEMS

DPP-1 Propagation of Electromagnetic Waves

Only One Option Correct Type

Working Space

Direction (Q. Nos. 1-10) *This section contains 10 multiple choice questions. Each question has four choices (a), (b), (c) and (d), out of which ONLY ONE is correct.*

1 A material has a dielectric constant of 9 for optical frequencies. Speed of light through this medium is most close to

a. $3 \times 10^8 \text{ ms}^{-1}$ b. $1 \times 10^8 \text{ ms}^{-1}$

c. $0.34 \times 10^8 \text{ ms}^{-1}$ d. Data insufficient

2 A scientist is using following setup to produce an electromagnetic wave.

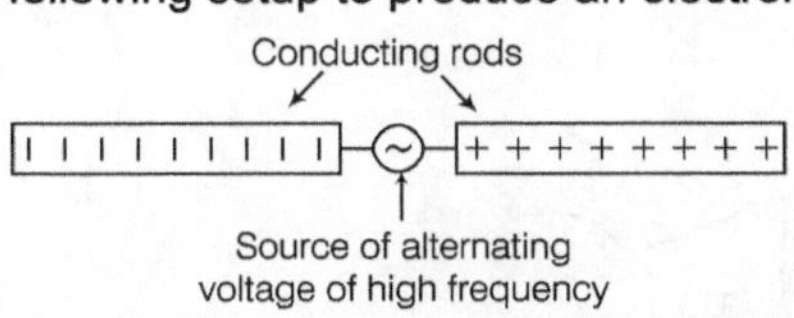

He places this setup in an evacuated chamber. If the length of each rod is nearly 0.5 m, then frequency of electromagnetic waves produced is closest to

a. $1.5 \times 10^5 \text{ Hz}$ b. $1.5 \times 10^8 \text{ Hz}$

c. $1.5 \times 10^{11} \text{ Hz}$ d. $1.5 \times 10^{14} \text{ Hz}$

3 A small fishing boat relays a SOS signal which cannot be received beyond a distance of 200 km over sea surface. A coast gaurd vessel received that signal, is using a vertical antenna because

a. ground waves are vertically polarised
b. ground waves are horizontally polarised
c. sky waves are vertically polarised
d. sky waves are horizontally polarised

4 Ground wave propagation is not suitable for high frequency waves and not suitable for very long range (even for low frequencies) due to

a. distortion of signal
b. interference of other signals
c. lack of signal security
d. high attenuation

5 Ground wave propagation is used in

a. FM radio transmission
b. AM radio transmission
c. mobile phone transmission
d. TV transmission

Working Space

6 If you have an equipment like a band radio which can be tuned to receive a particular frequency, then which of the following intervals you will tune it to receive a ground wave signal?

a. 2 kHz to 20 kHz
b. 500 kHz to 1500 kHz
c. 30 MHz to 300 MHz
d. 8 GHz to 8000 GHz

7 The ionospheric layer acts as a reflector for 3 MHz to 30 MHz frequency range. For less than 3 MHz frequencies ionospheric reflection is not used because

a. high attenuation
b. high power requirement
c. distance between earth surface and ionosphere is less than one complete wavelength
d. high interference due to man made noises (electromagnetic signals)

8 You are trying to send your signal (electromagnetic wave) up to a distance of 3500 kms by ionospheric reflection mode. Then, you prefer your signal to be reflected by the following layer of ionosphere.

a. D b. E
c. F_1 d. F_2

9 What is the approximate value of $|x_1 - x_2|$ in the given figure?

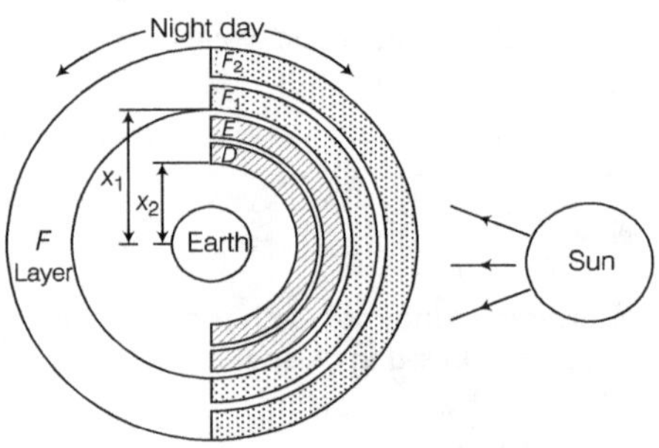

a. 30 km b. 100 km
c. 200 km d. 150 km

10 AM radio broadcast is relayed mostly by

a. ground wave mode
b. sky wave mode
c. space wave mode
d. satellite communication

One or More than One Options Correct Type

Direction (Q. Nos. 11-15) This section contains 5 multiple choice questions. Each question has four choices (a), (b), (c) and (d), out of which ONE or MORE THAN ONE are correct.

11 Ionosphere extends from ~ 70 km to ~ 300 km above earth's surface. For the ionosphere, following are correct.

a. We are most likely to lost (or get a fade signal), signal during early morning hours or at the evening
b. We are most likely to lost (or get a fade signal), signal during afternoon
c. We are most likely to get a fade signal during occurrence of Sun spots
d. There is not much affect due to a solar eclipse

Working Space

12 Geostationary satellites are used for weather forcasting, satellite TV, satellite radio and other types of global communications. A geostationary satellite

a. can be parked at any distance above the earth in an orbit with $T = 24\text{h}$
b. can be parked only in equatorial plane with an orbital radius of 35900 km
c. around 21st march noise in signal from the geostatic satellite is excessive
d. around September 22, noise in signal from the geostatic satellite is excessive

13 A particular mode of communication is preferred when it is reliable, economical and user friendly. Then, choose the correct options.

a. For AM radio broadcasting, ground waves are commonly used
b. For cellular network, LOS communication system are used
c. For sky TV transmission, satellite communication is used
d. For wireless used by armed forces, police, etc. ground waves are commonly used

14 Free space is an ideal condition for transmission of a signal without any absorption or adverse propagation effects. Let an electromagnetic wave is propagating in space such that flux density of signal is $100p\ \text{Wm}^{-2}$. Then,

a. effective value of electric field of signal $= 196\mu\text{Vm}^{-1}$
b. effective value of magnetic field is $\dfrac{10^{-5}}{\sqrt{60\pi}}\ \mu\text{Am}^{-1}$
c. impedance of free space is 120π or 377 ohms
d. Wave number for free space ε_0/c

15 Line of sight communication is one of the most widely used mode of signal transmission. Then, choose the correct options.

a. Area of coverage of transmission tower of height h is $A = \pi(d^2)$, where $d = 3.27\sqrt{h}$
b. Maximum distance between two repeater towers is $d = 3.57\,(\sqrt{h_1} + \sqrt{h_2})$
c. A repeater is a combination of a receiver and a transmitter
d. Frequency of signal received by a repeater is same as that retransmitted by the repeater

Matching List Type

Direction (Q. No. 16) Choices for the correct combination of elements from Column I and Column II are given as options (a), (b), (c) and (d), out of which one is correct.

16 Match the radio frequency bands in Column I with the mode in which they are employed in Column II and mark the correct option from the codes given below.

	Column I		Column II
i.	300 Hz, to 300 kHz	p.	Space wave
ii.	300 kHz to 3 MHz	q.	Sky wave
iii.	3 MHz to 30 MHz	r.	Ground wave
iv.	30 MHz to 3000 GHz		

Codes

	i	ii	iii	iv
a.	p	p,q	p,r	r
b.	p	q	r	r
c.	r	p	q	q,p
d.	r	r,q	q	p

Working Space

One Integer Value Correct Type

Direction (Q. No. 17) This section contains a question. When worked out will result in an integer from 0 to 9 (both inclusive).

17 The maximum frequency of sky wave that can be reflected from an ionospheric layer is given by $f_{max} = 9(N_{max})^{1/2}$

where, N_{max} = maximum electron density of that layer.

On a certain day, we have following observations.

i. Signals of frequencies higher than 5 MHz are not reflected from F_1 layer.

ii. Signals of frequencies higher than 8 MHz are not reflected from F_2 layer.

If the ratio of maximum electron densities of F_1 and F_2 layer at that day is equal to $\frac{N^2}{64}$, then find the value of N?

Statement Type

Direction (Q. Nos. 18-20) This section is based on Statement I and Statement II. Select the correct answer from the codes given below.

a. Both Statement I and Statement II are correct and Statement II is the correct explanation of Statement I
b. Both Statement I and Statement II are correct but Statement II is not the correct explanation of Statement I
c. Statement I is correct but Statement II is incorrect
d. Statement II is correct but Statement I is incorrect

18 **Statement I** For TV signal transmission sky wave propagation mode is used for long distance coverage.

Statement II For F_2 layer of ionosphere, hop or skip distance is large.

19 **Statement I** "Conductivity of atmosphere is higher at low altitudes."

Statement II Pressure at low altitudes is higher and so, the number of molecules per unit volume is more at low altitudes.

20 **Statement I** A short wave radio broadcast is actually a long distance broadcast.

Statement II Wavelength of a broadcast $\propto \frac{1}{\text{distance}}$.

DPP-2 Elements of Communication System

Only One Option Correct Type

Working Space

Direction (Q. Nos. 1-10) This section contains 10 multiple choice questions. Each question has four choices (a), (b), (c) and (d), out of which ONLY ONE is correct.

1 A block diagram of a basic communication system is as shown in the figure :

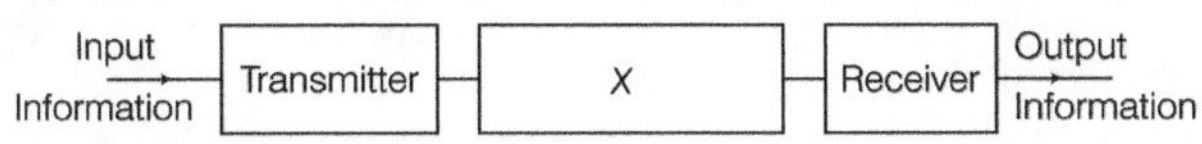

The bolck marked as X is

a. modem
b. transducer
c. communication channel
d. channel decoder

2 An analog communication system is one for which

a. input signal is similar to intelligent information
b. output signal from the system is analogous to intelligent information
c. input information contains few discreet levels
d. output signal from the system contains only two discreet levels

3 In a digital communication system,

a. input is in only 2 levels (0 and 1)
b. transmitted signal is in only 2 levels (0 and 1)
c. input is converted into discreet amplitude levels and then coded in binary form
d. transmitted signal is of very high frequency

4 For a communication system designed for multiple users (like cell phone), value of ratio $\frac{\text{Range of frequency}}{\text{Bandwidth per user}}$ must be

a. 0 b. 1
c. as high as possible d. as low as possible

5 You must be aware of old mobile phones (initially launched). They have a large collapsible antenna like a radio (or car radio). But in a new model of a mobile phones, there is no such feature. Which of the following is responsible for this change?

a. Frequency of signal
b. Phone battery
c. Number of transmission towers
d. Customer demand

6 To relay a signal of frequency 20 kHz, size of dipole antenna is around

a. 0.15 cm b. 1.5 cm
c. 15 cm d. 15 km

Working Space

7 A public addressing system, one of these may be in use in your school, microphone is an example of

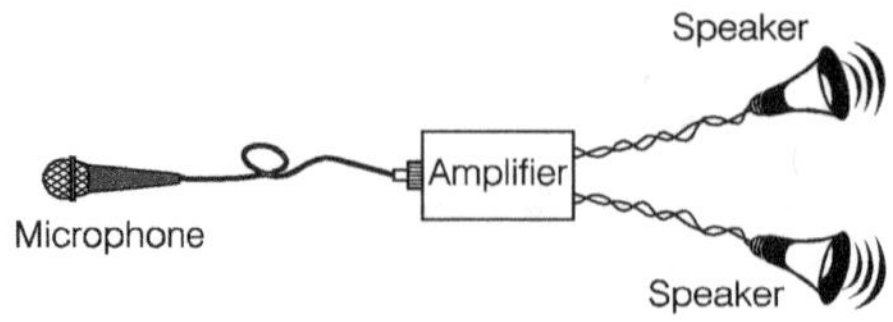

a. modem
b. channel decoder
c. transducer
d. repeater

8 If you want a low bandwidth for your communication system, you must incorporate a

a. channel decoder
b. channel encoder
c. coherent detector
d. modem

9 If a communication system uses a signal which is a single valued function of time, then the system must be

a. a digital system
b. an analog system
c. a combination of a digital and analog system
d. None of the above

10 A recent development of communication technology is a GPS, the global positioning system, which determines exact location and speed of a user's device. GPS system employs a total of

a. one satellite
b. three satellites
c. five satellites
d. twenty four satellites

One or More than One Options Correct Type

Direction *(Q. Nos. 11-15) This section contains 5 multiple choice questions. Each question has four choices (a), (b), (c) and (d), out of which ONE or MORE THAN ONE are correct.*

11 An optical communication system can give a data transmission capacity above 1 Tb/s (commercial systems of today uses 2.5 Gb/s). Then, mark the correct options.

a. Core of an optical fibre is more denser than cladding
b. For signal from user 1 to user 2 and from user 2 to user 1 we need only one optical fibre

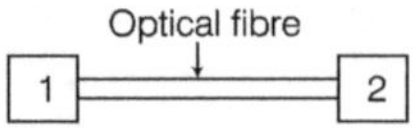

c. Photo detectors used to sense change intensity of signal are reverse biased *p-n* junctions
d. To convert an intelligent signal into optical signal, we uses an invar prism

12 Many communication systems are still using wires as transmission media. Then, mark the correct options.

a. A two wire twisted pair is not suitable above 30 kHz to 300 kHz band due to higher resistance (due to skin effect) and radiation loss
b. Coaxial cable (like that of a TV) can be used to carry medium (0.3 MHz to 3 MHz), high frequency (3 MHz to 30 MHz) and very high frequency(30 MHz to 300 MHz)
c. Optical fibres can carry signals of frequency upto 300 GHz
d. At very high frequency, a coaxial cable may undergo a dielectric breakdown

13 Which of the following are correct regarding radio systems

a. a space wave and a surface wave exists inside the troposphere
b. sky wave can exists below as well as above troposphere
c. bandwidth of a channel is the range of frequencies that it can transmit
d. several radio stations can broadcast simultaneously provided they all use a different frequency range

14 A basic communication system is as shown in the figure.

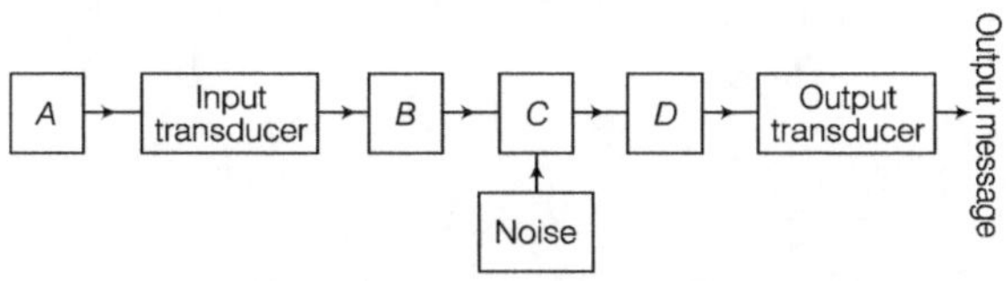

Then, mark the correct options.

a. *B* modifies the base band signal for efficient transmission
b. *C* carries output of *B* and delivers it to *D* with some attenuation
c. *D* may contain a modem
d. *A* creates a signal which can have more than one value at any instant of time

15 A ground receiver station is receiving a signal at 5 MHz and 100 MHz, transmitted from a ground transmitter at a height of 300 m located at a distance of 100 km. Radius of the earth is 6.4×10^6 m. N_{max} for ionosphere at that day is 10^{12} m^{-3}.
Then, mark the correct option(s).

a. 5 MHz frequency signal is not transmitted using space wave communication
b. 100 MHz frequency signal is not transmitted using space wave communication
c. 5 MHz frequency signal reaches *via* ionospheric mode
d. 100 MHz frequency signal reaches *via* ionospheric mode

Comprehension Type

Direction *(Q. Nos. 16-18) This section contains a paragraph, describing theory, experiments, data, etc. Three questions related to the paragraph have been given. Each question has only one correct answer among the four given options (a), (b), (c) and (d).*

Passage

Consider a communication system made for transmitting a message signal (200 Hz to 2000 Hz). Height of transmitting antenna and receiving antenna are given.

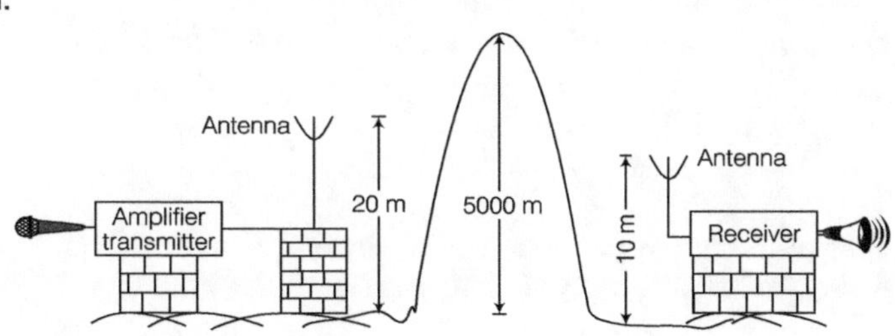

There is a high hill between the transmitter and receiver of height 5000 m. Distance between the 2 antennas is more than 1000 kms.

16 If minimum power condition must be fulfilled, which of the following corrections you must use for efficient and economical signal transmission.

a. A modem for satellite communication
b. LOS communication system with repeaters
c. A modem for ionospheric reflection for transmitting the signal
d. A coaxial cable with a suitable modem and amplifier

Working Space

17 When you put a modem to superimpose signal over a carrier wave of suitable frequency, modem must be set to produce a signal frequency of (Assume day time in summer)

a. less than 27 MHz
b. more than 27 MHz
c. less than 30 MHz
d. more than 30 MHz

18 Which of the following layers are used to reflect a transmitted signal? (Assume day time in summer)

a. D layer
b. E layer
c. F_1 layer
d. F_2 layer

Statement Type

Direction (Q. Nos. 19 and 20) *This section is based on Statement I and Statement II. Select the correct answer from the codes given below.*

a. Both Statement I and Statement II are correct and Statement II is the correct explanation of Statement I
b. Both Statement I and Statement II are correct but Statement II is not the correct explanation of Statement I
c. Statement I is correct but Statement II is incorrect
d. Statement II is correct but Statement I is incorrect

19 **Statement I** For a communication system using ionospheric reflection, maximum frequency can not be increased beyond a certain value.

Statement II Mean electron density of ionosphere is generally of order 10^9 to 10^{12} m^{-3}.

20 **Statement I** A Jammer (an instrument used to produce an interfering wave of same frequency) required to Jam an optical communication system, must be operate around frequency of 10^{15} Hz.

Statement II Frequency of light is 10^{15} Hz.

DPP-3 Modulation : AM and FM

Only One Option Correct Type

Working Space

Direction (Q. Nos. 1-10) *This section contains* 10 *multiple choice questions. Each question has four choices (a), (b), (c) and (d), out of which ONLY ONE is correct.*

1 In modulation process, following is done

a. carrier of high frequency is send instead of message signal

b. frequency of base band signal is changed to some higher frequency

c. amplitude of base band signal is changed to some higher value

d. some characteristics of a high frequency carrier is changed in direct proportional to the instantaneous amplitude of base band signal

2 Consider a carrier given by $V(t) = 4\sin(8\pi t)$ V. It is modulated by a intelligent signal

$$V(t) = \begin{cases} 1, & n \le t < n + \dfrac{1}{2} \\ -1, & n + \dfrac{1}{2} \le t \le n + 1 \end{cases}$$

where t is in seconds and $n = 0, 1, 2, \ldots$ etc. Then, the modulated signal will be

a.
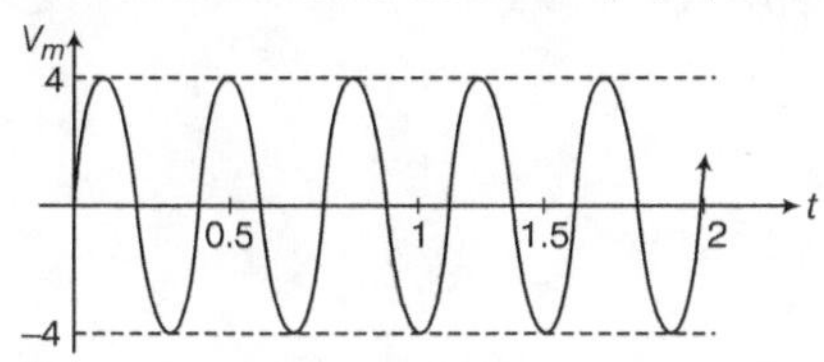

b.
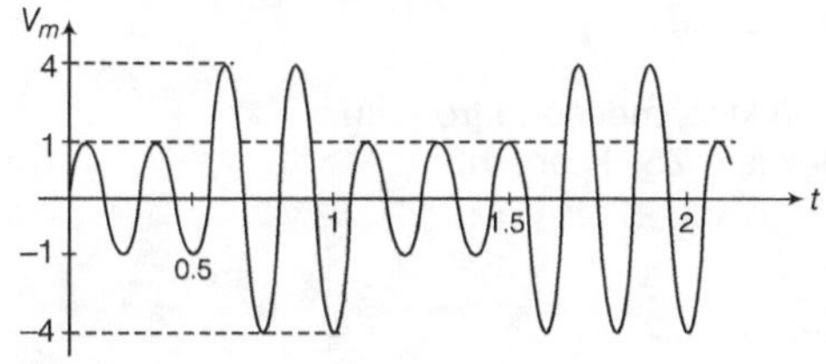

c.
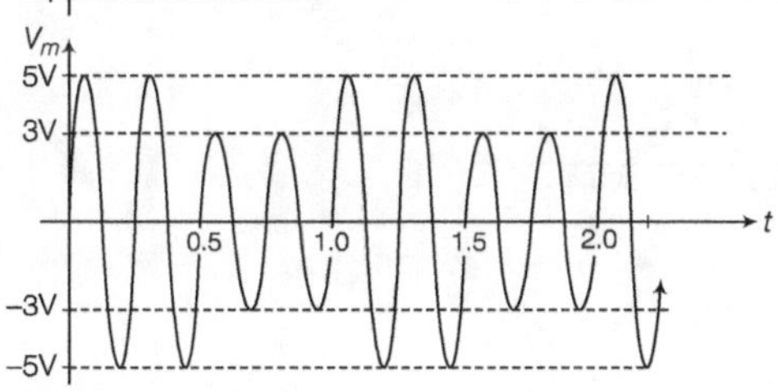

d.
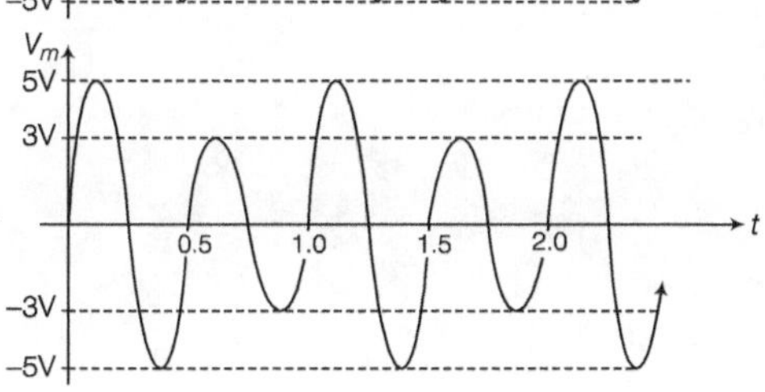

3 A sine carrier with amplitude V_c and frequency f_c is modulated with a modulating sinusoidal signal of amplitude V_m and frequency f_m. The result will be

a. $v = [(V_c + V_m) \sin 2\pi f_m t] \cdot \sin 2\pi f_c t$
b. $v = [V_c \sin 2\pi f_m t + V_m] \sin 2\pi f_c t$
c. $v = V_m \sin 2\pi f_m t + V_c \sin 2\pi f_c t$
d. $v = V_c \sin 2\pi f_c t + (V_m \sin 2\pi f_m t) \sin 2\pi f_c t$

4 Let a sinusoidal carrier with a periodic time of 0.25s is modulated by using an intelligent signal given by

$$m(t) = 2 \sin 2\pi t \text{ (V)}$$

Assuming ideal condition for modulation, the modulated signal will be

a. $v(t) = 2 \sin 8\pi t + (2 \sin 2\pi t) \sin 8\pi t$
b. $v(t) = 4 \sin 8\pi t + (2 \sin 2\pi t) \sin 8\pi t$
c. $v(t) = 4 \sin 8\pi t + 2 \sin 2\pi t$
d. Cannot be found or modulation index is not given

5 An amplitude modulated wave is as shown in the figure.

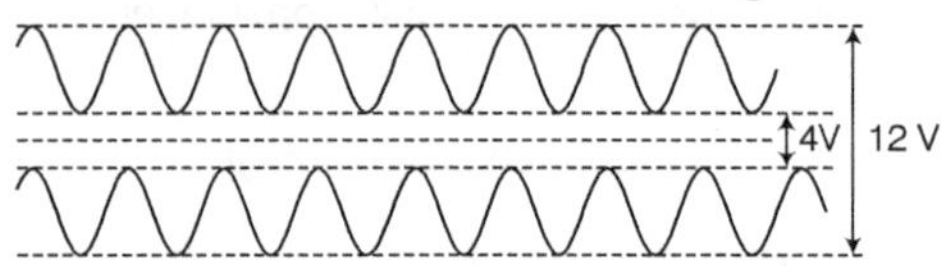

Modulation index for given wave is

a. 1
b. 0.5
c. 0.75
d. 0.25

6 A 400 Hz tone modulates a 300 kHz carrier. Amplitude of carrier is 2V and modulation index is 1.
Then, the modulated signal is given by

a. $2 \sin (6 \times 10^5 \pi t) + 2 \cos (600.8 \times 10^3 \pi t) + \cos (599.2 \times 10^3 \pi t)$
b. $2 \sin (6 \times 10^5 \pi t) + 2 \sin (600.8 \times 10^3 \pi t) + 2 \sin (599.2 \times 10^3 \pi t)$
c. $2 \sin (6 \times 10^5 \pi t) + 2 \sin (600.8 \times 10^3 \pi t) + 2 \cos (599.2 \times 10^3 \pi t)$
d. $2 \sin (6 \times 10^5 \pi t) + \cos ((600.8 \times 10^3) \pi t) - \cos ((599.2 \times 10^3) \pi t)$

7 Let in an amplitude modulation, carrier frequency is 300 kHz, message frequency is 400 Hz, carrier amplitude is 9V and modulation index is 0.75. Then, the frequency domain display for AM signal (voltage) is

a.
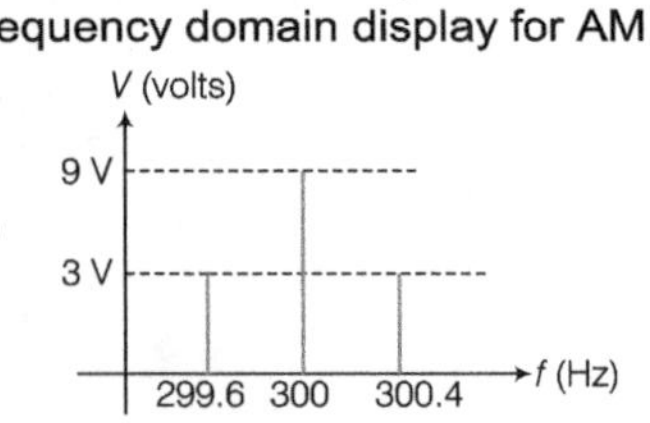

b.
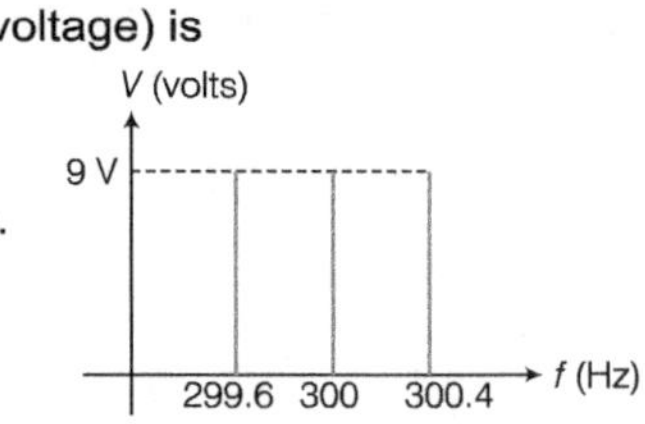

c.
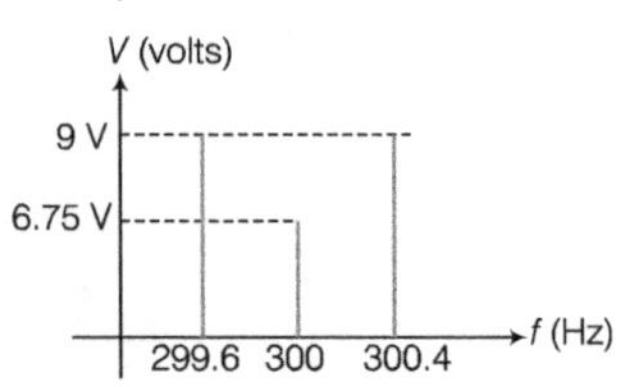

d.
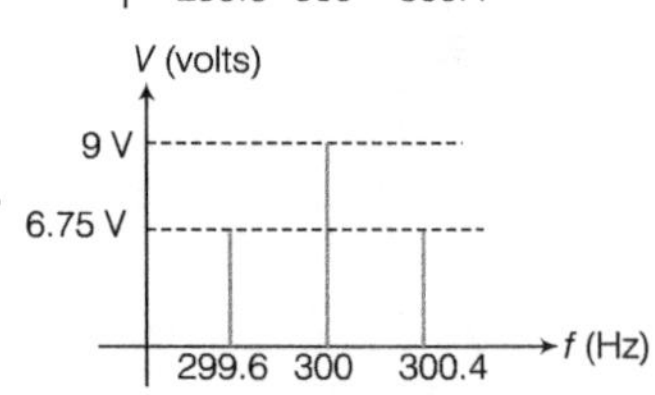

Working Space

Working Space

8 An AM signal is represented as

$$v_{AM} = V_c \cdot \sin 2\pi f_c t + \frac{V_m}{2}[\cos 2\pi t\,(f_c - f_m) - \cos 2\pi t\,(f_c + f_m)]$$

and R = resistance of load impedance (antenna)

Then, total power carried by the signal is

a. $\frac{V_c^2 + V_m^2}{R}$ b. $V_c^2 + 2V_m^2$

c. $\frac{V_c^2}{R} \times \frac{V_m^2}{R} \times 2$ d. $\frac{V_c^2}{2R} + \frac{V_m^2}{4R}$

9 For a 100% modulation, ratio of total side band power with the carrier power is

a. 1 : 4 b. 4 : 1

c. 1 : 2 d. 2 : 1

10 For an AM signal, if modulation index is reduced (from 1), then

a. power of side bands increases

b. power of side bands decreases

c. power of side bands remains same

d. power of carrier increases

Comprehension Type

Direction (Q. Nos. 11-18) *This section contains 3 paragraphs, each describing theory, experiments, data, etc. Eight questions related to the paragraphs have been given. Each question has only one correct answer among the four given options (a), (b), (c) and (d).*

Passage I

An antenna has an impedance of 40 Ω. An unmodulated AM signal produces a current of 4.8 A. The modulation is 90%.

11 Carrier power is

a. 921.6 W b. 1295 W

c. 373.4 W d. 1453 W

12 Total transmitted power is

a. 921.6 W b. 1295 W

c. 373.4 W d. 1453 W

13 Power carried by each side band is

a. 921.6 W b. 1295 W

c. 373.4 W d. 186.7 W

Passage II

In an AM transmission, carrier itself conveys no information. The carrier can be received and transmitted but without modulation, no information will be transmitted. When modulation occurs, side bands are produced. So, all of information is contained within the side bands.

One way of increasing efficiency is to suppress the carrier and eliminate one side band. This result in a single side band signal (SSB) transmission.

Frequency domain of a SSB signal is as shown in the figure.

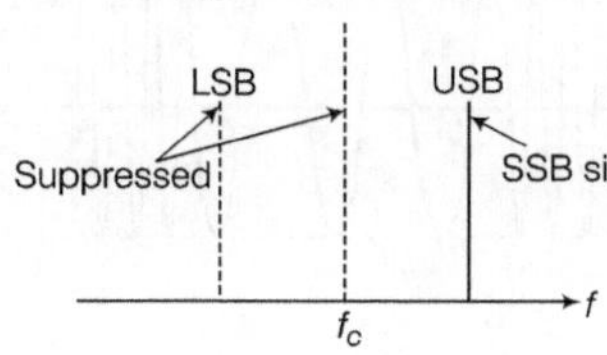

Working Space

14 In a standard AM broadcast, percentage of total power wastage is (Take, modulation index $m = 1$)

a. 57.57% b. 79.98%
c. 66.67% d. 25.52%

15 Power requirement ratio of SSB with standard AM is

a. 1 : 6 b. 1 : 4
c. 3 : 2 d. 2 : 3

Passage III

When frequency of carrier is changed in accordance with the amplitude of intelligent signal, it is called frequency modulation.

Amplitude of modulated wave remains constant (= carrier amplitude).

FM signals gives noiseless reception, they have a high range and transmission efficiency is higher.

16 One of the disadvantage of FM transmission is

a. decrease in noise or amplitude variations
b. increase in operating range
c. increase in bandwidth
d. a high modulation index

17 A carrier wave as shown in the figure.

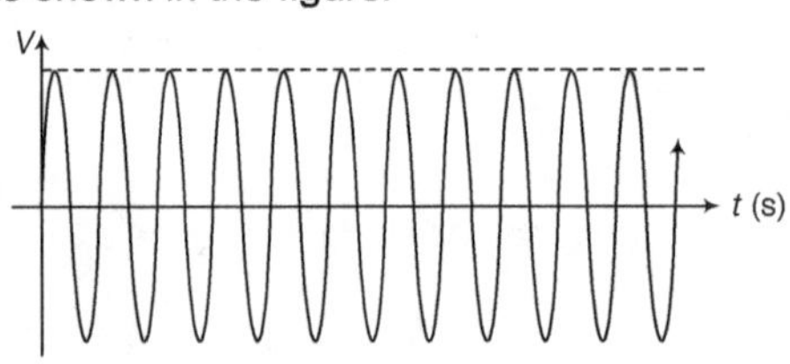

So, modulated by an intelligent signal as shown below.

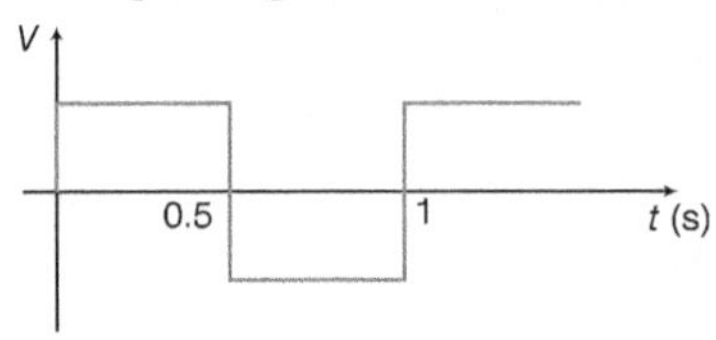

Which of the following represents a FM carrier wave?

a. 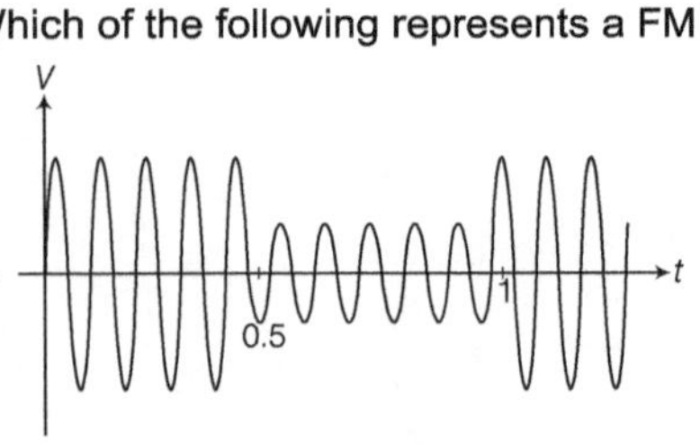

b. 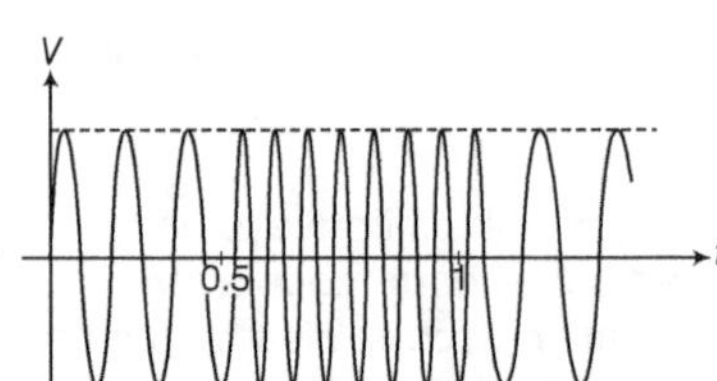

c. 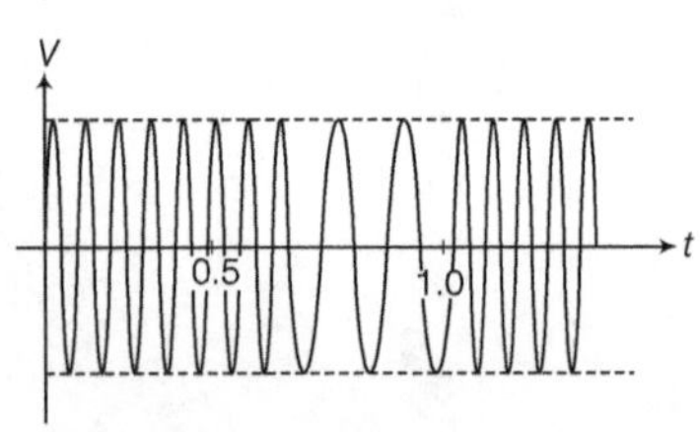

d. 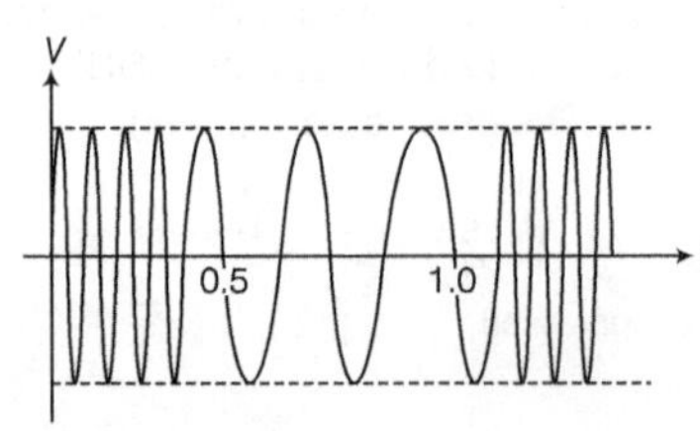

Working Space

18 For which type of transmissions FM is preferred?

a. Air control tower transmissions
b. Cable TV transmissions
c. Music channels transmissions
d. High speed data transmissions

One or More than One Options Correct Type

Direction (Q. Nos. 19-21) *This section contains 3 multiple choice questions. Each question has four choices (a), (b), (c) and (d), out of which ONE or MORE THAN ONE are correct.*

19 A simple technique to produce an AM wave is shown in the figure.

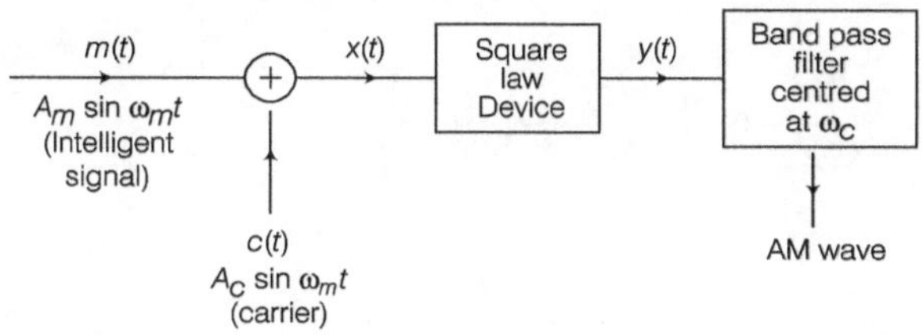

In above arrangement, a band pass filter,

a. rejects DC and sine frequencies ω_m, $2\omega_m$ and ω_c
b. retains frequencies ω_c and $\omega_c \pm \omega_m$
c. retains DC, ω_m, $2\omega_m$ and ω_c
d. rejects ω_c and $\omega_c \pm \omega_m$

20 An AM wave produced by an experimental set up by a student is

$$v_{AM} = 5(1 + 0.6 \cos 6280t) \sin (211 \times 10^4 t) \text{ (in V)}$$

Mark the correct options.

a. Maximum amplitude of AM wave is 8V and minimum amplitude is 2V
b. AM contains frequencies of 335 kHz, 336 kHz and 337 kHz
c. Amplitudes of three components of AM wave are 2.5V, 5V and 7.5V
d. Signal amplitude is 6V

21 An AM wave consists of

Carrier component : 5V peak value
LSB component : 2.5 peak value
USB component : 2.5 peak value

This AM wave is received by a receiver with a resistance of 2 kΩ. Then

a. power delivered by the carrier is 6.25 MW
b. total power delivered by both side band components is more than 3 MW
c. total power delivered by AM is less than 10 MW
d. only USB delivers power to receiver

Working Space

Matching List Type

Direction *(Q. No. 22) Choices for the correct combination of elements from Column I and Column II are given as options (a), (b), (c) and (d), out of which one is correct.*

22 In Column I, a combination of a carrier and a modulating signal is given. Match the combination with the output signal given in Column II and mark the correct option from the codes given below.

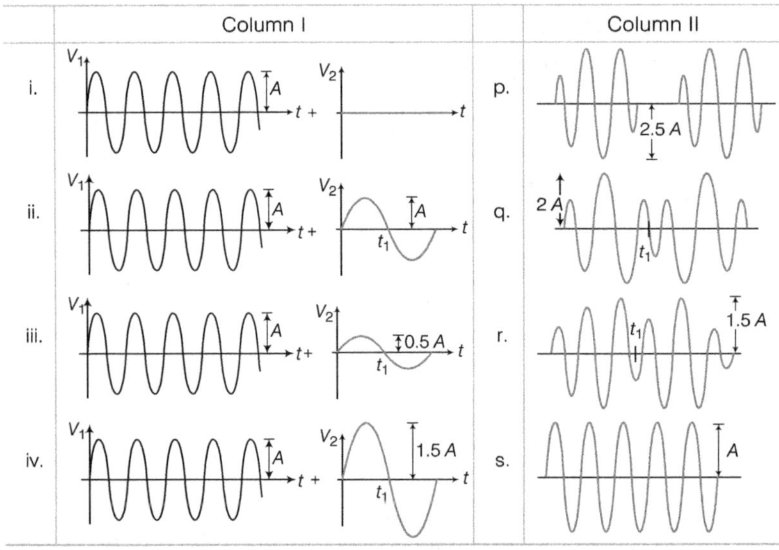

Codes

	i	ii	iii	iv
a.	p	q	r	s
b.	p	q	s	r
c.	s	q	p	r
d.	s	q	r	p

Revisal Problems
for JEE Main

Only One Option Correct Type

Working Space

Direction *(Q. Nos. 1-20) This section contains 20 multiple choice questions. Each question has four choices (a), (b), (c) and (d), out of which ONLY ONE option is correct.*

1 In case of communication between *A* and *B* as shown in the given figure.

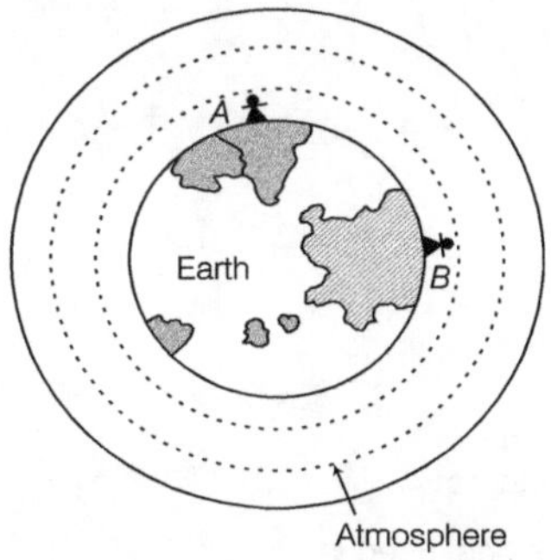

The frequency of signal must be

a. 10 kHz
b. 10 MHz
c. 1 GHz
d. 1000 GHz

2 Frequencies in the ultra high frequency range (UHF range) normally propagate by means of

a. ground waves
b. sky waves
c. surface waves
d. space waves

3 Which of the following is a digital signal?

a.
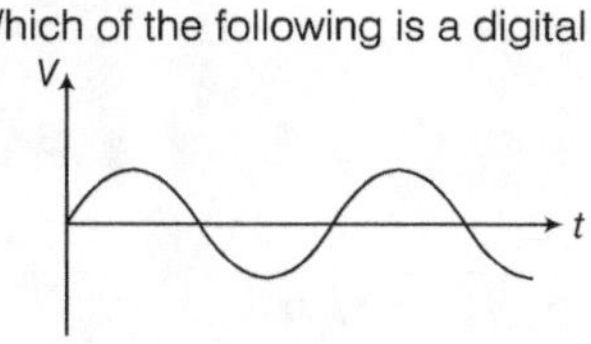

b.
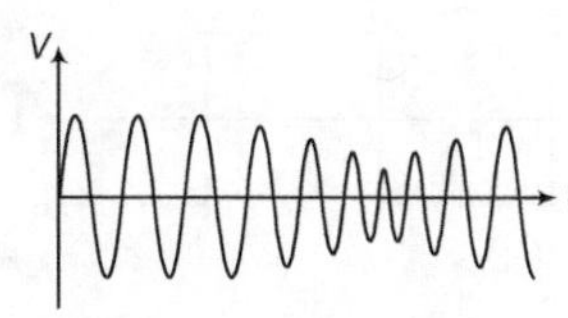

c.
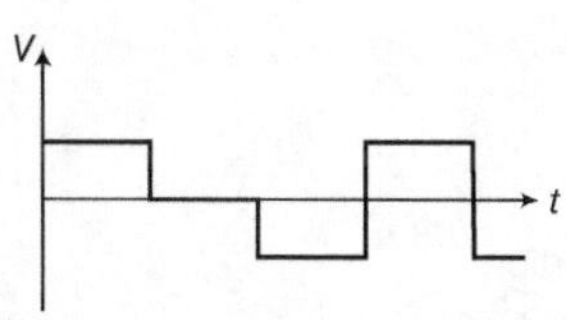

d.
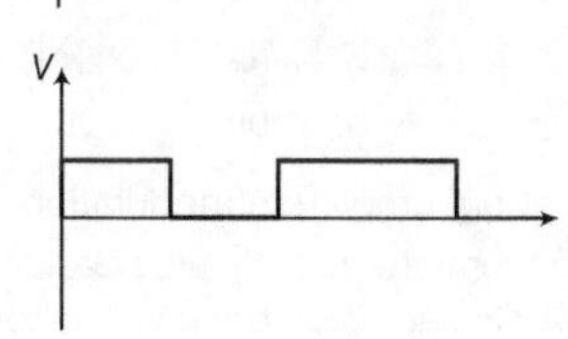

4 A digital signal

I. do not provides a continuous set of values
II. represents values as discrete steps
III. can utilize binary system
IV. can utilize decimal as well as binary systems. correct statements are
Then, mark the correct option.

a. Both I and II
b. Both I and III
c. I, II and III
d. All of the above

5 In which of the following cases, there is a successful transmission of a TV signal ?

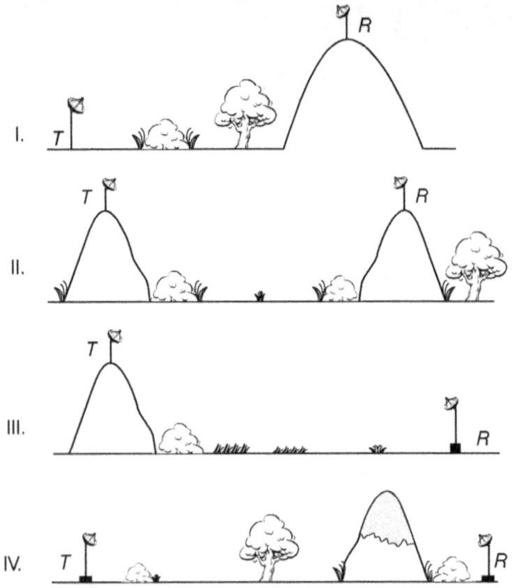

Your choice is

a. Only II
b. I, II and III
c. Only III
d. All of these

6 In a particular case of amplitude modulation, carrier signal used is

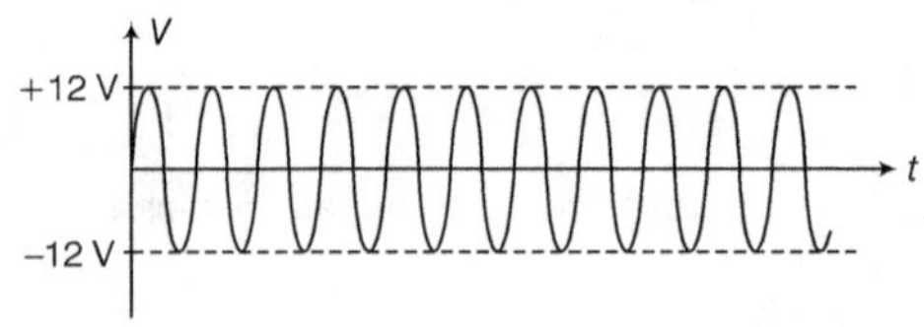

and the modulating signal is

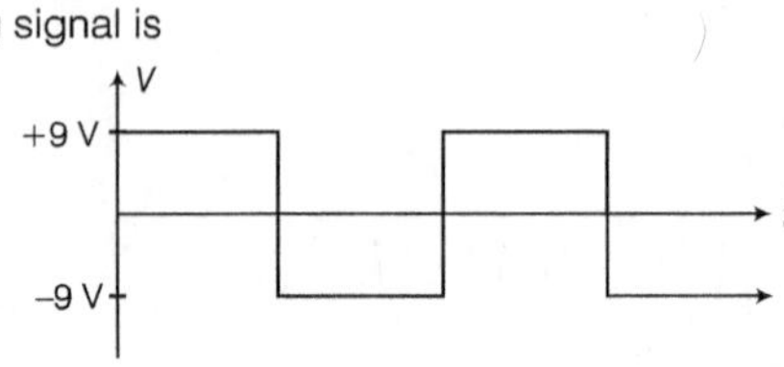

The modulation index will be

a. 0.5
b. 0.25
c. 0.75
d. 1.00

7 Consider the following statements about amplitude modulation

I. When modulation index = 1, then there are multiple side bands causing distortion.
II. When modulation index = 1, power in each side band is one fourth of carrier power.
III. Carrier power is wasted power as carrier itself conveys no information.
IV. Received signal gets weak, if modulation index drops to 0.7 from initial value of 0.8.

In your opinion correct choices are

a. Only I
b. I, II and III
c. II, III and IV
d. Both III and IV

8 An antenna has an impedance of 40 Ω. An unmodulated AM signal produces a current of 4.8 A. Let the modulation is 90%. The power contained in each side band will be

a. 921.6 W
b. 460.8 W
c. 647.5 W
d. 186.7 W

Working Space

9 A standard AM broadcast station uses upto 5 kHz frequency for modulating a carrier of frequency 980 kHz. The total bandwidth of AM station is

a. 10 kHz
b. 1960 kHz
c. 985 kHz
d. 975 kHz

10 Your local radio station relays news signal as $V(t) = (2.0 + 4\sin 500\pi t) \cdot \cos(2\pi \times 10^5 t)$ volts. The modulation index of the signal is

a. 20
b. 2
c. 0.2
d. 0.02

11 A spy has to relay three coded massages using frequencies of 1600 kHz, 5 MHz and 60 MHz to his headquarter. Then, mark the correct option.

a. *A* is transmitted *via* space wave while *B* and *C* are transmitted *via* sky wave
b. *A* is transmitted *via* ground wave, *B via* sky wave and *C via* space wave
c. *B* and *C* are transmitted *via* ground wave while *A* is transmitted *via* sky wave
d. *B* is transmitted *via* ground wave while *A* and *C* are transmitted *via* space wave

12 A transmitter puts out a total power of 25 W of 30% modulated AM signal. Percentage of total power carried by carrier wave is

a. 50%
b. 80%
c. less than 35%
d. more than 95%

13 Consider the carrier wave and modulating signal given

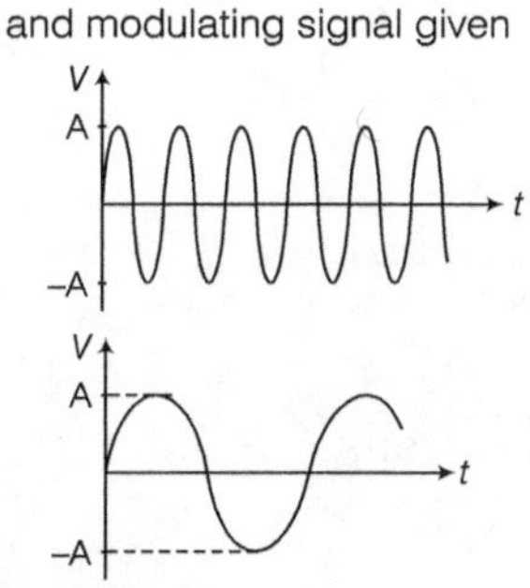

Then, mark the correct options.

a. Modulated signal is of constant amplitude *A*
b. Modulated signal is of constant amplitude 2*A*
c. Amplitude of modulated signal varies from *A* to 2*A*
d. Amplitude of modulated signal varies from 0 to 2*A*

14 The new 4G technology service provider is allotted a bandwidth of 2×75 MHz for mobile phone communication. Let one GSM channel has a bandwidth of 200 kHz. The total number of GSM users which can be accomodated by this service provider is nearly

a. 750 users if we assume that gap between two users is included in 200 kHz bandwidth
b. 7500 users
c. 750 users if we assume that gap between two users is included in 2×75 MHz band width
d. 75000 users

15 A signal consisting of only two wavelengths (λ_A and λ_B) are reflected off the F_2 layer of ionosphere as shown in the figure.

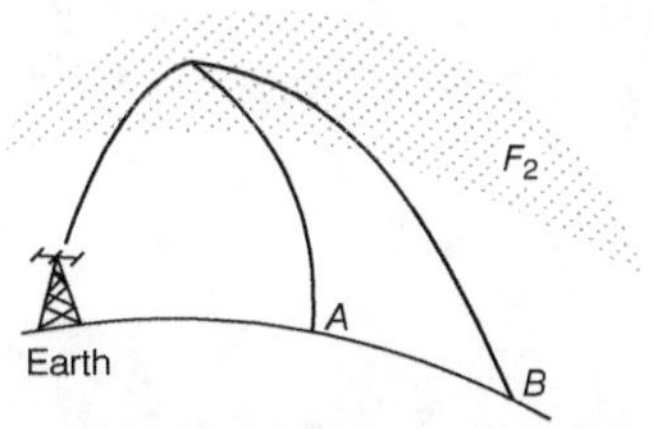

Then, the relation between λ_A and λ_B will be

a. $\lambda_A = \lambda_B$
b. $\lambda_A > \lambda_B$
c. $\lambda_A < \lambda_B$
d. Data insufficient

Working Space

16 An amplitude modulated wave is as shown in the figure.

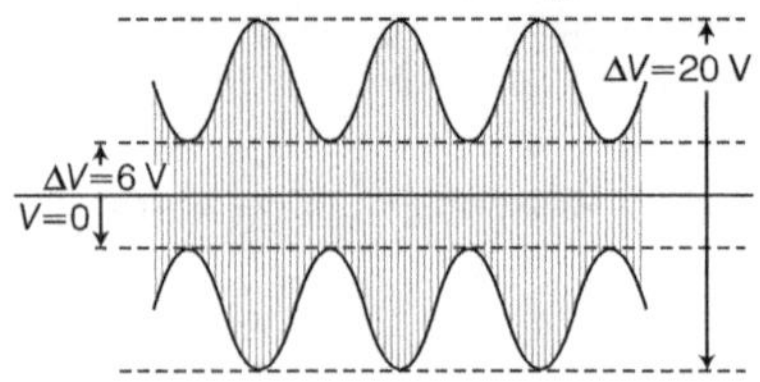

Modulation for this amplitude modulation will be

a. around 100%
b. around 20%
c. around 50%
d. around 80%

17 Minimum number of transducers required to assemble an optical communication link for transmitting voice signals is

a. 1
b. 2
c. 3
d. 4

18 To the given circuit, a time varying electrical signal is fed.

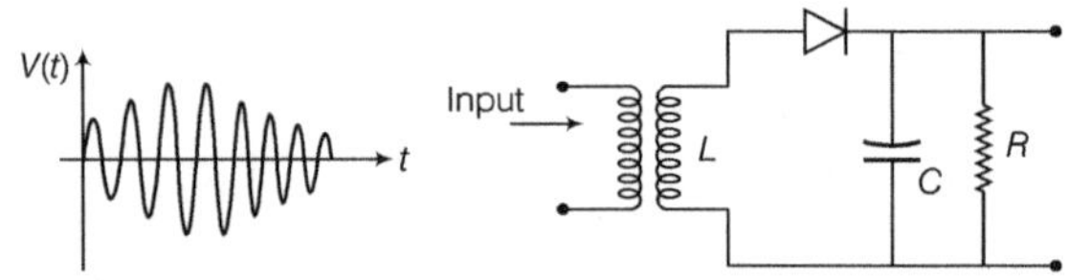

The output across resistor R, when f(wave) << RC, is

a. 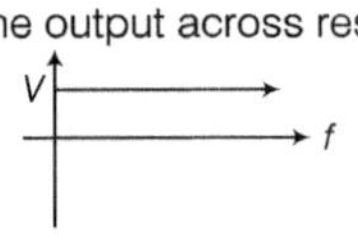
b. 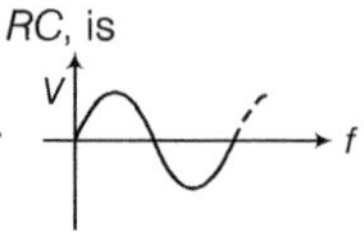
c. 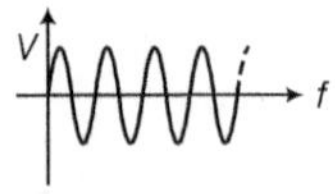
d. 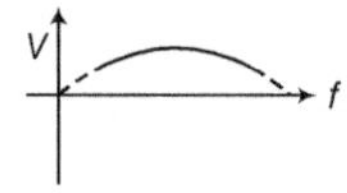

19 Except for the upper layer of atmosphere called ionosphere, rest of the atmosphere is composed mainly of neutral molecules. This is because

a. change in lower layers is conducted to ground
b. lower layers consists of gases with higher work functions/ionisation potentials
c. most of the ionising radiation is already reflected off upper layers and does not reach lower layers
d. atmospheric pressure decreases with height above earth's surface

20 Consider the following circuit.

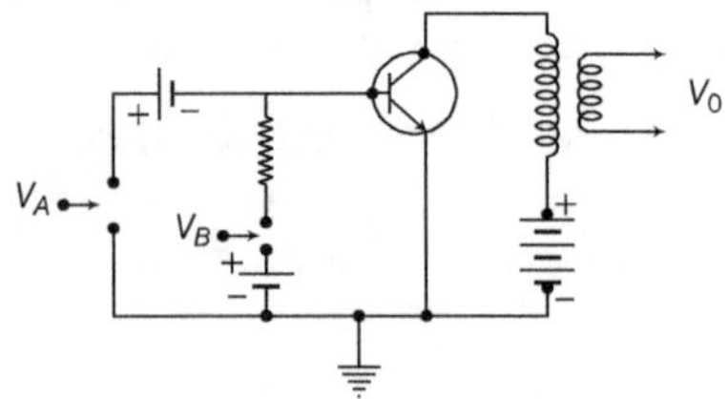

Input voltages at A and B are,

and $V_A = 2\sin(8\pi t)$V,

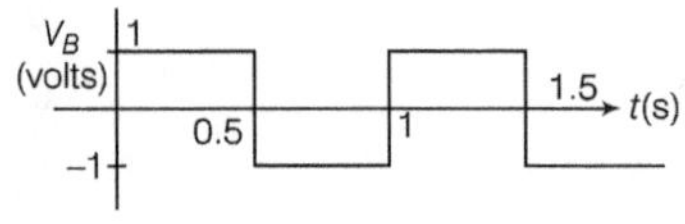

Then, output V_0 is

a. $V_0 = \begin{cases} 3\sin 8\pi t, t = 0 \text{ to } t = 0.5\text{s} \\ 2\sin 8\pi t, t = 0.5 \text{ to } t = 1\text{s} \end{cases}$

b. $V_0 = 2\sin 8\pi t \pm 1$ V

c. $V_0 = \begin{cases} 3\sin 8\pi t, t = 0 \text{ to } t = 0.5\text{s} \\ 0, t = 0.5 \text{ to } t = 1\text{s} \end{cases}$

d. $V_0 = \begin{cases} 3\sin 8\pi t, t = 0 \text{ to } t = 0.5\text{s} \\ \sin 8\pi t, t = 0.5 \text{ to } t = 1\text{s} \end{cases}$

Working Space

Statement Type

Direction *(Q. No. 21) This section is based on Statement I and Statement II. Select the correct answer from the codes given below.*

a. Both Statement I and Statement II are correct and Statement II is the correct explanation of Statement I
b. Both Statement I and Statement II are correct but Statement II is not the correct explanation of Statement I
c. Statement I is correct but Statement II is incorrect
d. Statement II is correct but Statement I is incorrect

21 Statement I In day time a certain low frequency signal may not reach *F* layers of ionosphere.

Statement II For low frequency signal atmospheric attenuation is large.

Matching List Type

Direction *(Q. No. 22) Choices for the correct combination of elements from Column I and Column II are given as options (a), (b), (c) and (d), out of which one is correct.*

22 Match the different parts of electromagnetic spectrum (given in Column I) with the corresponding photon energies (in Column II) in joules and mark the correct option from the codes given below.

	Column I		Column II
i.	AM radio	p.	7×10^{-13}
ii.	FM radio	q.	7×10^{-16}
iii.	Microwaves	r.	4×10^{-19}
iv.	Visible light	s.	7×10^{-24}
v.	X-rays	t.	7×10^{-26}
vi.	Gamma rays	u.	7×10^{-28}

Codes

	i	ii	iii	iv	v	vi
a.	p	q	r	s	t	u
b.	p	q	s	r	u	t
c.	q	p	r	t	u	s
d.	u	t	s	r	q	p

JEE Main & AIEEE Archive

(Compilation of Last 13 Years' Questions)

1 Match Column I (Electromagnetic wave type) with Column II (Its association/application) and select the correct option from the choices given below the Columns. **(JEE Main 2014)**

	Column I		Column II
i.	Infrared waves	p.	To treat muscular strain
ii.	Radio waves	q.	For broadcasting
iii.	X-rays	r.	To detectfracture of bones
iv.	Ultraviolet	s.	Absorbed by the ozone layer of the atmosphere

Codes

	i	ii	iii	iv		i	ii	iii	iv
a.	s	r	q	p	b.	p	w	d	t
c.	r	q	p	s	d.	p	q	r	s

2 A diode detector is used to detect an amplitude modulated wave of 60% modulation by using a condenser of capacity 250 pF in parallel with a load resistance 100 kΩ. Find the maximum modulated frequency which could be detected by it. **(JEE Main 2013)**

a. 10.62 MHz
b. 10.61 kHz
c. 5.31 MHz
d. 5.31 kHz

3 A radar has a power of 1 kW and is operating at a frequency of 10 GHz. It is located on a mountain top of height 500 m. The maximum distance upto which it can detect object located on the surface of the earth is
(radius of the earth $= 6.4 \times 10^6$ m) **(AIEEE 2012)**

a. 80 km
b. 16 km
c. 40 km
d. 64 km

4 Which of the following four alternatives is not correct? We need modulation. **(AIEEE 2011)**

a. To increase the selectivity
b. To reduce the time lag between transmission and reception of the information signal
c. To reduce the size of antenna
d. To reduce the fractional bandwidth, i.e. the ratio of the signal bandwidth to the centre frequency

5 Statement I Sky wave signals are used for long distance radio communication. These signals are in general, less stable than ground wave signals.

Statement II The state of ionosphere varies from hour to hour, day to day and season to season. **(AIEEE 2011)**

a. Both Statement I and Statement II are correct and Statement II is the correct explanation of Statement I
b. Both Statement I and Statement II are correct but Statement II is not the correct explanation of Statement I
c. Statement I is correct but Statement II is incorrect
d. Statement II is correct but Statement I is incorrect

6 Consider the telecommunication through optical fibres. Which of the following statements is not correct? **(AIEEE 2003)**

a. Optical fibres can be of graded refractive index
b. Optical fibres are subjected to electromagnetic interference from outside
c. Optical fibres have extremely low transmission loss
d. Optical fibres may have homogeneous core with a suitable cladding

Answers *with* Explanations

Answers

1. Dual Nature of Matter and Radiation

DPP-1 Dual Nature of Electromagnetic Waves

1. (i) $-3\times10^{8}\,ms^{-1}$ (iii) $86\,MHz$
(v) 10 nT (nano-tesla) (vi) $10\cos[(1.8)y+(5.4\times10^{8})t]\hat{k}$
(vii) $\approx 42p\,Jm^{-3}$ (viii) $12.6\,mWm^{-2}$
(ix) (a) 84 pico-pascals (b) 42 pico-pascals (c) 63 pico-pascals

2. (i) 35A (iii) $4\times10^{14}\,Vm^{-1}s^{-1}$ (iv) $1.2\times10^{-4}\,T$

3. $\approx1000\times10^{-15}\,m\approx1pm$ 4. *c* 5. *c* 6. *b* 7. *d* 8. *d* 9. *c* 10. *b*
11. *b* 12. *a,d* 13. *a,c,d* 14. *a,d* 15. *a,c,d* 16. *a,b,c,d* 17. *b,c* 18. *b,d* 19. *a,b* 20. *d*
21. *c* 22. *a* 23. *d* 24. *d* 25. *a* 26. *b* 27. *d* 28. *a*

DPP-2 Matter Waves

1. (i) $\frac{0.101}{\sqrt{V}}$ Å (ii) $\frac{h}{\sqrt{3mk_BT}}$ 2. *c* 3. *c* 4. *c* 5. *a* 6. *b* 7. *c* 8. *c*
9. *d* 10. *d* 11. *d* 12. *a,d* 13. *b,d* 14. *c,d* 15. *b,d* 16. *a,c* 17. *a* 18. *d*
19. *d* 20. *a* 21. *b* 22. *d* 23. *d* 24. 5

DPP-3 Emission of Electrons and Photoelectric Effect

1. (i) $6\times10^{9}\,m^{-2}s^{-1}$ (ii) $1.92\times10^{-9}\,J$
(iii) $1.76\times10^{-19}\,J$ (iv) 0 to 1.1 eV

2. (i) 4133 Å (ii) 2%
(iii) 13 μA

3. *c* 4. *d* 5. *d* 6. *d* 7. *c* 8. *c* 9. *d* 10. *a* 11. *c* 12. *d*
13. *a,c* 14. *a,b,c,d* 15. *a,c,d* 16. *a,b,c,d* 17. *a,b,c* 18. *b,d* 19. *c* 20. *d* 21. *c* 22. *d*
23. *d* 24. *a* 25. *d* 26. *d*

Revisal Problems (JEE Main)

1. *c* 2. *d* 3. *a* 4. *a* 5. *a* 6. *c* 7. *c* 8. *a* 9. *a* 10. *c*
11. *b* 12. *d* 13. *c* 14. *c* 15. *d* 16. *a* 17. *b* 18. *b* 19. *c* 20. *d*
21. *a* 22. *d* 23. *d*

Revisal Problems(JEE Advanced)

1. *d* 2. *d* 3. *c* 4. *a* 5. *d* 6. *c* 7. *c* 8. *c* 9. *c* 10. *a*
11. *c* 12. *d* 13. *b* 14. *c* 15. *c* 16. *c* 17. *d* 18. *a,c,d* 19. *a,b,c* 20. *b,d*
21. *a* 22. *d* 23. *d* 24. 2

JEE Main & AIEEE Archive

1. *b* 2. *d* 3. *c* 4. *c* 5. *d* 6. *a* 7. *b* 8. *b* 9. *b* 10. *d*
11. *d* 12. *a* 13. *b* 14. *d* 15. *a* 16. *c* 17. *b* 18. *d* 19. *c* 20. *a*
21. *a* 22. *c*

JEE Advanced & IIT JEE Archive

1. *a,c* 2. 2 3. 1 4. 7 5. 3 6. *a* 7. *b* 8. *d* 9. *a*
10. *a* 11. *a,d* 12. *c* 13. 24

2. Atoms

DPP-1 Early Atomic Models

1. (i) 180° (ii) Angle of deflection decreases
(iii) 1.1×10^{-14} m (v) Scattering angle decreases

2. *d* 3. *c* 4. *c* 5. *c* 6. *d* 7. *d* 8. *d* 9. *a* 10. *b*
11. *a* 12. *a* 13. *a* 14. *b* 15. *d* 16. *a,b,c* 17. *b,c,d* 18. *c,d* 19. *a,b* 20. *a*
21. *d* 22. *d*

DPP-2 Bohr's Theory for Hydrogen and Hydrogen-like Atoms

3. *b* 4. *b* 5. *a* 6. *a* 7. *b* 8. *a* 9. *d* 10. *c*
11. *a* 12. *a,b,c,d* 13. *a,c,d* 14. *a,b,d* 15. *a,d* 16. *a,b* 17. *d* 18. 4 19. 5 20. *a*
21. *d* 22. *d* 23. *b*

DPP-3 Electronic Transitions in Hydrogen and Hydrogen-like Atoms

2. *d* 3. *c* 4. *d* 5. *a* 6. *c* 7. *d* 8. *c* 9. *c* 10. *a*
11. *a* 12. *d* 13. *a,c,d* 14. *a,c,d* 15. *a,b* 16. *b,c* 17. *b,c* 18. *b* 19. *a* 20. *d*
21. *d* 22. *d* 23. *b* 24. *d* 25. *a* 26. *a* 27. *a* 28. 3

DPP-4 X-Rays

1. *a* 2. *d* 3. *d* 4. *a* 5. *c* 6. *b* 7. *a* 8. *d* 9. *d* 10. *d*
11. *b,c,d* 12. *a,c,d* 13. *a,d* 14. *b,c* 15. *c,d* 16. *c* 17. *b* 18. *d* 19. *c* 20. *d*
21. *c* 22. *c* 23. *c*

Revisal Problems (JEE Main)

1. *d* 2. *d* 3. *a* 4. *c* 5. *d* 6. *c* 7. *c* 8. *b* 9. *d* 10. *b*
11. *d* 12. *a* 13. *a* 14. *b* 15. *a* 16. *c* 17. *d* 18. *c* 19. *d* 20. *d*

Revisal Problems(JEE Advanced)

1. *c* 2. *c* 3. *d* 4. *c* 5. *a* 6. *b,c,d* 7. *b,d* 8. *a,b,c,d* 9. *a,b,c,d* 10. *d*
11. *c* 12. *d* 13. *c* 14. *b* 15. *d* 16. *c* 17. *a* 18. *d* 19. *d*
20. *c* 21. *a* 22. *a*

JEE Main & AIEEE Archive

1. *c* 2. *d* 3. *d* 4. *d* 5. *b* 6. *d* 7. *b* 8. *c* 9. *a* 10. *a*

JEE Advanced & IIT JEE Archive

1. *b* 2. *a* 3. *d* 4. *b* 5. *c* 6. *a,c* 7. *b* 8. *c* 9. *c* 10. *a*
11. *b* 12. *b* 13. *a* 14. *a*

3. Nuclei

DPP-1 Atomic Nucleus (Composition, Size, Atomic Masses, Isotopes, Isobars, Isotones and General Stability Trends of Nucleus)

1. (i) 10^9 eV (ii) 2.4×10^{17} kgm^{-3} (iii) 40 N

2. *c* 3. *d* 4. *c* 5. *a* 6. *d* 7. *c* 8. *d* 9. *d* 10. *d* 11. *d*
12. *d* 13. *a,b,c* 14. *a,b,c,d* 15. *b,c,d* 16. *b,c,d* 17. *a,b,d* 18. *a* 19. *a* 20. *b* 21. *c*

DPP-2 Mass Defect, Binding Energy, Nuclear Stability, Criteria for Fission and Fusion Processes

1. (i) $m(^2H) = 2.0141u$, $m(^1H) = 1.0078u$, $m(^{16}O) = 15.9949u$ (ii) 207.16 Mev

2. *d* **3.** *d* **4.** *d* **5.** *a* **6.** *c* **7.** *a* **8.** *a* **9.** *c* **10.** *d*
11. *d* **12.** *b,c* **13.** *a,b* **14.** *a,c,d* **15.** *a,b* **16.** *a,b,c,d* **17.** *d* **18.** *a* **19.** *a* **20.** *d*
21. 5

DPP-3 Radioactivity and Law of Radioactive Disintegration

1. (i) 4224 yr (ii) No **2.** *c* **3.** *d* **4.** *b* **5.** *a* **6.** *b* **7.** *b* **8.** *c* **9.** *c*
10. *a* **11.** *d* **12.** *a,b,c,d* **13.** *a,c* **14.** *a,d* **15.** *a,c* **16.** *a,c,d* **17.** *d* **18.** *a* **19.** *a*
20. *c* **21.** *b* **22.** *c* **23.** *d* **24.** *d* **25.** *a*

Revisal Problems (JEE Main)

1. *a* **2.** *a* **3.** *c* **4.** *c* **5.** *c* **6.** *b* **7.** *d* **8.** *c* **9.** *d* **10.** *d*
11. *c* **12.** *c* **13.** *d* **14.** *a* **15.** *b* **16.** *c* **17.** *a* **18.** *a* **19.** *a* **20.** *b*

Revisal Problems (JEE Advanced)

1. *d* **2.** *b* **3.** *a* **4.** *a,c,d* **5.** *a,b,c,d* **6.** *b,c* **7.** *a,d* **8.** *a* **9.** *a* **10.** *c*
11. *c* **12.** *a* **13.** *d* **14.** *d* **15.** *c* **16.** *c* **17.** *b* **18.** *b* **19.** *a* **20.** *d*
21. *d*

JEE Main & AIEEE Archive

1. *a* **2.** *c* **3.** *b* **4.** *c* **5.** *b* **6.** *b* **7.** *a* **8.** *b* **9.** *b* **10.** *b*
11. *b* **12.** *b* **13.** *c* **14.** *c* **15.** *c* **16.** *a* **17.** *b* **18.** *d* **19.** *c* **20.** *c*
21. *c* **22.** *a* **23.** *b* **24.** *a* **25.** *d* **26.** *a*

JEE Advanced & IIT JEE Archive

1. *c* **2.** *a* **3.** 4 **4.** *d* **5.** *c* **6.** 1 **7.** 8 **8.** *d* **9.** *a* **10.** *b*
11. *a* **12.** *c* **13.** *b,d* **14.** *a* **15.** *b* **16.** *a* **17.** *c*

4. Electronic Devices

DPP-1 Semiconductors

1. *a* **2.** *b* **3.** *d* **4.** *d* **5.** *d* **6.** *d* **7.** *a* **8.** *d* **9.** *c* **10.** *c*
11. *a,b* **12.** *a,b,c,d* **13.** *b,d* **14.** *a,b,c,d* **15.** *a,b,c* **16.** *d* **17.** *d* **18.** *b* **19.** *c* **20.** *d*

DPP-2 Semiconductor Diodes

1. *b* **2.** *a* **3.** *d* **4.** *a* **5.** *a* **6.** *c* **7.** *a* **8.** *a* **9.** *d*
10. *d* **11.** *a,b,c,d,* **12.** *a,b,c* **13.** *a,b,d* **14.** *b,d* **15.** *b,d* **16.** *a* **17.** *b* **18.** *d* **19.** *d*
20. *d* **21.** *c* **22.** *d* **23.** *c*

DPP-3 Junction Transistors

1. *b* **2.** *c* **3.** *a* **4.** *d* **5.** *d* **6.** *c* **7.** *c* **8.** *d* **9.** *b* **10.** *a*
11. *b,c,d* **12.** *a,b,d* **13.** *c,d* **14.** *b,d* **15.** *b,d* **16.** *d* **17.** *d* **18.** *c* **19.** *a* **20.** *c*
21. *d* **22.** *c*

DPP-4 Logic Gates

2. *c*	**3.** *b*	**4.** *b*	**5.** *a*	**6.** *d*	**7.** *b*	**8.** *c*	**9.** *b*	**10.** *c*	**11.** *d*
12. *c*	**13.** *b,c*	**14.** *a,b,c*	**15.** *b,c,d*	**16.** *c,d*	**17.** *c*	**18.** *b*	**19.** *a*	**20.** *a*	**21.** *a*

Revisal Problems (JEE Main)

1. *b*	**2.** *c*	**3.** *c*	**4.** *c*	**5.** *b*	**6.** *d*	**7.** *a*	**8.** *a*	**9.** *c*	**10.** *a*
11. *c*	**12.** *b*	**13.** *a*	**14.** *c*	**15.** *d*	**16.** *a*	**17.** *c*	**18.** *a*	**19.** *c*	**20.** *b*
21. *d*									

JEE Main & AIEEE Archive

1. *a*	**2.** *a*	**3.** *a*	**4.** *c*	**5.** *a*	**6.** *c*	**7.** *a*	**8.** *d*	**9.** *d*	**10.** *d*
11. *a*	**12.** *c*	**13.** *b*	**14.** *c*	**15.** *b*	**16.** *d*	**17.** *c*	**18.** *c*	**19.** *c*	**20.** *d*
21. *d*	**22.** *c*	**23.** *c*	**24.** *b*	**25.** *a*	**26.** *c*	**27.** *a*	**28.** *c*	**29.** *c*	

5. Communication System

DPP-1 Propagation of Electromagnetic Waves

1. *b*	**2.** *b*	**3.** *a*	**4.** *d*	**5.** *c*	**6.** *b*	**7.** *c*	**8.** *d*	**9.** *a*	**10.** *a*
11. *a,c,d*	**12.** *b,c,d*	**13.** *a,b,c,d*	**14.** *a,b,c,d*	**15.** *a,b,c*	**16.** *d*	**17.** 5	**18.** *d*	**19.** *d*	**20.** *c*

DPP-2 Elements of Communication System

1. *c*	**2.** *a*	**3.** *c*	**4.** *c*	**5.** *a*	**6.** *d*	**7.** *c*	**8.** *d*	**9.** *b*	**10.** *d*
11. *a,b,c*	**12.** *a,b,d*	**13.** *b,c,d*	**14.** *a,b,c*	**15.** *a,b,c*	**16.** *c*	**17.** *a*	**18.** *b*	**19.** *a*	**20.** *d*

DPP-3 Modulation : AM and FM

1. *d*	**2.** *c*	**3.** *d*	**4.** *a*	**5.** *b*	**6.** *d*	**7.** *d*	**8.** *d*	**9.** *c*	**10.** *b*
11. *a*	**12.** *b*	**13.** *d*	**14.** *c*	**15.** *a*	**16.** *c*	**17.** *c*	**18.** *c*	**19.** *a,b*	**20.** *a,b*
21. *a,b,c*	**22.** *d*								

Revisal Problems (JEE Main)

1. *b*	**2.** *d*	**3.** *d*	**4.** *c*	**5.** *b*	**6.** *c*	**7.** *c*	**8.** *d*	**9.** *a*	**10.** *c*
11. *b*	**12.** *d*	**13.** *d*	**14.** *a*	**15.** *b*	**16.** *c*	**17.** *d*	**18.** *d*	**19.** *d*	**20.** *d*
21. *b*	**22.** *d*								

JEE Main & AIEEE Archive

1. *d*	**2.** *b*	**3.** *a*	**4.** *b*	**5.** *b*	**6.** *b*

1. Dual Nature of Matter and Radiation

DPP-1 Dual Nature of Electromagnetic Waves

1. (i) Comparing with a travelling wave equation, we have $1.8y + 5.4\times10^8 t$ = phase of wave. As, phase remains constant.

$$\therefore \quad \frac{d\phi}{dt} = \frac{d}{dt}(1.8y + 5.4\times10^8 t)$$
$$= \frac{d}{dt}(\text{constant})$$
$$\Rightarrow \quad 1.8\frac{dy}{dt} + 5.4\times10^8 = 0$$
$$\Rightarrow \quad \frac{dy}{dt} = \frac{-5.4\times10^8}{1.8}$$
$$= -3\times10^8\ \text{ms}^{-1}$$

Now $\frac{dy}{dt}$ = speed of wave $= -3\times10^8\ \text{ms}^{-1}$

which is in negative y-direction.

(ii) As phase of wave is given by $ky + \omega t$,

$$\therefore \quad k = \text{wave number} = \frac{2\pi}{\lambda}$$
$$= 1.8\left(\frac{\text{rad}}{\text{m}}\right)$$

So, $$\frac{2\pi}{\lambda} = 1.8 \text{ or } \lambda = \frac{2\pi}{1.8}(\text{m})$$
$$\Rightarrow \quad \lambda = 3.5\ \text{m}$$

When this wave passes through a medium of refractive index n_2. It is travelling in a medium of refractive index n_1. Then,

n_{21} = refractive index of medium 2 w.r.t. 1

$$= \frac{n_2}{n_1} = \frac{v_1}{v_2} = \frac{f_1\lambda_1}{f_1\lambda_2}$$
$$= \frac{\lambda_1}{\lambda_2} \quad \text{(As } f \text{ remains constant)}$$
$$\Rightarrow \quad n_2\lambda_2 = n_1\lambda_1$$
$$\Rightarrow \quad \frac{3}{2}\times\lambda_2 = n_1\lambda_1 \quad \ldots\text{(i)}$$

If $n_1 = 1$

Then, $$\lambda_2 = \frac{2}{3}\times n_1\times\lambda_1 = \frac{2}{3}\times1\times3.5$$

$\Rightarrow \lambda_2$ = Wavelength of wave in glass = 2.33 m

(iii) Frequency, $f = \frac{1}{T} = \frac{\omega}{2\pi} = \frac{5.4\times10^8}{2\pi}$

$$= 0.86\times10^8 = 86\times10^6\ \text{Hz}$$
$$= 86\ \text{MHz}$$

(iv) Frequency of light does not change in reflection, refraction, diffraction, interference or in travel through a medium. It is the characteristic of source.

(v) $\frac{E}{B} = c$

$$\Rightarrow \quad B = \frac{E}{c} = \frac{3.1}{3\times10^8}$$
$$\Rightarrow \quad B \approx 1\times10^{-8} = 10\times10^{-9}$$
$$\Rightarrow \quad B \approx 10\ \text{nT (nano-tesla)}$$

(vi) As electric oscillations are in (+) $\hat{\mathbf{i}}$-direction, wave propagates in (−) $\hat{\mathbf{j}}$-direction. So, magnetic field is perpendicular to both in (+) $\hat{\mathbf{k}}$-direction.

$$\therefore \quad B = B_{max}\,[(\cos(ky + \omega t))]\hat{\mathbf{k}}$$
$$= 10\cos[(1.8)y + (5.4\times10^8)t]\,\hat{\mathbf{k}}\ (\text{nT})$$

(vii) Energy is equally distributed among electric and magnetic fields.

$$\therefore \quad u = u_E + u_B$$
$$= \frac{1}{2}\varepsilon_0 E_{rms}^2 + \frac{1}{2}\cdot\frac{B_{rms}^2}{\mu_0}$$
$$= \frac{1}{2}\varepsilon_0 E_{rms}^2 + \frac{1}{2\mu_0}\cdot\frac{E_{rms}^2}{c^2}$$
$$= \frac{1}{2}\varepsilon_0 E_{rms}^2 + \frac{\varepsilon_0\mu_0}{2\mu_0}E_{rms}^2$$
$$\left(\text{Using } \frac{E}{B} = c \text{ and } c^2 = \frac{1}{\mu_0\varepsilon_0}\right)$$
$$\therefore \quad u = \varepsilon_0 E_{rms}^2 = \varepsilon_0\left(\frac{E_{max}}{\sqrt{2}}\right)^2$$
$$= 8.85\times10^{-12}\times\frac{(3.1)^2}{2}$$
$$= 42.5\times10^{-12}\ \text{Jm}^{-3} \approx 42\ \text{pJm}^{-3}$$

(viii) For normal incidence, energy delivered by wave per square metre per second

$$= \text{Intensity} = I = u\times c$$
$$= 42\times10^{-12}\times3\times10^8$$
$$= 126\times10^{-4}$$
$$= 12.6\ \text{mWm}^{-2}\ \text{(milliwatts per square metre)}$$

(ix) (a) Radiation pressure in case of total reflection is

$$p = \frac{2I}{c} = 2u$$
$$= 2\times42\times10^{-12} = 84\ (\text{Nm}^{-2})\times10^{-12}$$
$$= 84\ \text{(pico-pascals)}$$

(b) Radiation pressure in case of total absorption

$$= \frac{I}{c} = u = 42\times10^{-12}\ \text{Nm}^{-2}$$
$$= 42\ \text{pico-pascals}$$

(c) With 50% reflection and 50% absorption, half of intensity is reflected and half of it is absorbed.

$$\therefore \quad p = \underbrace{2\left(\frac{I/2}{c}\right)}_{\text{reflection}} + \underbrace{\left(\frac{I/2}{c}\right)}_{\text{absorption}},$$

(As $I/2$ is available for each mode)

$$= \frac{I}{c} + \frac{I}{2c} = \frac{I}{c}\left(1+\frac{1}{2}\right)$$
$$= \frac{3}{2}\times\frac{I}{c} = \frac{3}{2}\times u = \frac{3}{2}\times42\ \text{pico-pascals}$$
$$= 63\ \text{pico-pascals}$$

2. At $t = 0$, when switch is closed, $V_C + V_R = V_0$ – Kirchhoff's rule.

$$\Rightarrow \quad \frac{q}{C} + IR = V_0 = \frac{q_{max}}{C}, \qquad \left(\because i = \frac{dq}{dt}\right)$$
$$\Rightarrow \quad \frac{dq}{(q - q_{max})} = -\frac{dt}{RC}$$

$\Rightarrow$ Integrating, $\quad q = q_{max}(1 - e^{-t/RC})$

Current at any instant is

$$I = \frac{dq}{dt} = q_{max}\left(-\frac{d}{dt}e^{-t/RC}\right)$$
$$= q_{max}\left(\frac{e^{-t/RC}}{RC}\right), \text{ where } q_{max} = \frac{V_0}{C}$$

(i) At $t = 0$, current through the circuit is

$$I_C = \text{conduction current}$$
$$= \frac{dq}{dt} = \frac{q_{max} \cdot e^{-t/RC}}{RC}\Bigg|_{t=0}$$
$$= \frac{CV_0}{RC} \cdot e^0 = \frac{V_0}{R} = \frac{70\text{ V}}{2\ \Omega} = 35\text{ A}$$

(ii) Value of conduction and displacement current is always same at every instant.

$\therefore$ I_D = displacement current = I_C = 35 A, at $t = 0$

(iii) As $E = \frac{\sigma}{\varepsilon_0} = \frac{Q}{A\,\varepsilon_0}$

So, rate of change of field between plates,

$$\frac{dE}{dt} = \frac{1}{A\,\varepsilon_0} \cdot \frac{dQ}{dt} = \frac{I_C}{A\,\varepsilon_0}$$
$$= \frac{35}{8.85 \times 10^{-12} \times 1 \times 10^{-2}}$$
$$= 4 \times 10^{14}\ \text{Vm}^{-1}\text{s}^{-1}$$

(iv) By Ampere's law,

$$\oint \mathbf{B} \cdot d\mathbf{l} = \mu_0 I_C = \mu_0 \varepsilon_0 \frac{d\phi_E}{dt}$$

$\Rightarrow$ $$B \cdot 2\pi r = \mu_0 \varepsilon_0 \frac{d}{dt}(\pi r^2 E)$$

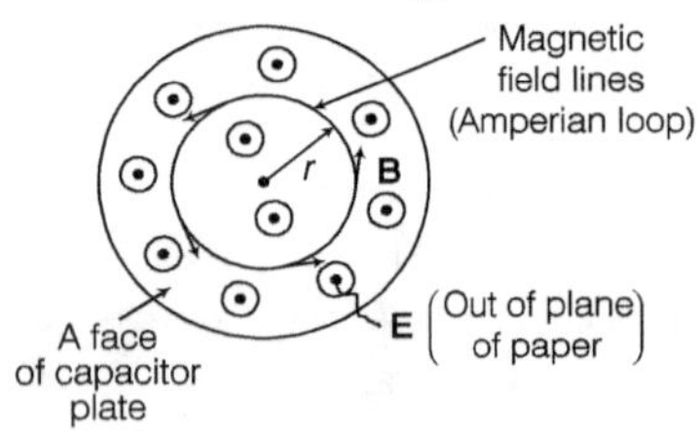

$\Rightarrow$ $$B = \frac{\mu_0 \varepsilon_0}{2} \cdot r \cdot \frac{dE}{dt}$$

As $E = 0$ for $r > r_0$ (radius of plate) and $\phi_E = E(\pi r_0^2)$

$\therefore$ $$B(2\pi r) = \mu_0 \varepsilon_0 \cdot \frac{d}{dt}(\pi r_0^2 E)$$

$\Rightarrow$ $$B = \frac{\mu_0 \varepsilon_0 r_0^2}{2r} \cdot \frac{dE}{dt}$$

B is maximum at $r = r_0$

$\therefore$ $$B\ (\text{at } r = r_0) = \frac{\mu_0 \varepsilon_0 r_0}{2} \cdot \frac{dE}{dt}$$
$$= \frac{\mu_0 \varepsilon_0}{2}(5.6 \times 10^{-2}) \times (4 \times 10^{14})$$
$$= 1.2 \times 10^{-4}\ \text{T}$$

3. A photon has no rest mass (because a light beam is never at rest in any reference frame).

Inertial mass of a photon is

$$m = \frac{p}{v} = \frac{hf}{c^2} \qquad \left(\text{As } p = \frac{h}{\lambda} = \frac{hf}{c} \text{ and } v = c\right)$$

Now let a photon is emitted from a star with frequency f_1, then PE associated with photon at surface of star

$$= \frac{-GMm}{R} = \frac{-GM\left(\frac{hf}{c^2}\right)}{R} = -\frac{GMhf_1}{c^2R}$$

Total energy of photon is its gravitational PE and its quantum energy (hf).

$\therefore$ $$\text{Total energy, } E = hf_1 + \left(\frac{-GMhf_1}{c^2R}\right)$$
$$= hf_1\left(1 - \frac{GM}{c^2R}\right)$$

When this photon is away from the star's surface the photon is free from gravitational field of Sun and its energy then will be

$E = hf_2$, where f_2 = current frequency of photon

From law of conservation of energy,

$$hf_2 = hf_1\left(1 - \frac{GM}{c^2R}\right)$$

$\Rightarrow$ $$\frac{f_2}{f_1} = \left(1 - \frac{GM}{c^2R}\right)$$

So, gravitational red shift (fractional shift)

$$\frac{\Delta f}{f_1} = \frac{f_1 - f_2}{f_1} = 1 - \frac{f_2}{f_1} = \frac{GM}{c^2R}$$

Frequency of a photon decreases at it moves away from the star.

Now in given case,

$$\text{Red shift} = \Delta f = f \cdot \frac{GM}{c^2R}$$
$$= 500\text{ nm} \times \left(\frac{6.67 \times 10^{-11} \times 2 \times 10^{30}}{(3 \times 10^8)^2 \times 7 \times 10^8}\right)$$
$$\approx 500 \times 10^{-9} \times 2 \times 10^{-6}\text{ m}$$
$$\approx 1000 \times 10^{-15}\text{ m} \approx 1\text{ pm}$$

4. (c) Average power transferred per unit area = magnitude of poynting vector = S

$$= \frac{E_{rms} \cdot B_{rms}}{\mu_0}, \quad \frac{E}{B} = c$$

and $$E_{rms} = \frac{E_{max}}{\sqrt{2}} = \frac{E_0}{\sqrt{2}}$$

So, E_0 = maximum value of electric field

$$= \sqrt{\frac{2S}{\varepsilon_0 c}} = \sqrt{\frac{2 \times 1350}{8.85 \times 10^{-12} \times 3 \times 10^8}}$$
$$= 1.01 \times 10^3\ \text{Vm}^{-1}$$

and $$B_0 = \frac{E_0}{c} = \frac{1.01 \times 10^3}{3 \times 10^8} = 3.37 \times 10^{-6}\ \text{T}$$

5. (c) To bring light's particle at rest means to get it absorb it by an atom.

Now, initial momentum of a photon just striking the surface,

$p_i = \frac{h}{\lambda}$ – de-Broglie's hypothesis

When it is completely absorbed, $p_f = 0$

So, change in momentum (opposite to direction of initial velocity)

$= \Delta p = \frac{h}{\lambda}$.

Energy incident per unit time = IA

$\therefore$ Number of photons incident per unit time

$$= n = \frac{IA}{hf} = \frac{IA\lambda}{hc}$$

So, total change in momentum per unit time

$$= n\Delta p = \frac{IA\lambda}{hc} \times \frac{h}{\lambda} = \frac{IA}{c}$$

So, force on the surface

= Total change of momentum per unit time

$$F = \frac{IA}{c}$$
$$= \frac{\text{Energy falling per unit time}}{\text{Speed of light}}$$
$$= \frac{\Delta U}{c}$$

6. (b) I_D = displacement current at that instant = $\varepsilon_0 A \frac{dE}{dt}$

$$= 8.85 \times 10^{-12} \times (0.058)^2 \times (2 \times 10^6)$$
$$= 6 \times 10^{-8} \text{A}$$

or $I_D = I_C$

$\therefore$ $I_C = 6 \times 10^{-8}$ A

7. (d) Charge on plates of capacitor at any instant is $Q = CV$

$\therefore$ Current that flows instanty is $I_C = \frac{dQ}{dt}$

$\Rightarrow$ $$I_C = \frac{d}{dt}CV = C\left(\frac{d}{dt}V\right)$$
$$= 6 \times 10^{-6} \times 12$$
$$= 72 \times 10^{-6} \text{ A}$$

$\therefore$ $I_C = I_D = 72 \times 10^{-6}$ A

8. (d) Using definition for Poynting vector,

$$S = \frac{P_0}{A} = \frac{P_0}{4\pi r^2} = \frac{1}{2} c \varepsilon_0 E_0^2$$

$\Rightarrow$ $$E_0 = \sqrt{\frac{\mu_0 c P_0}{2\pi r^2}}$$

9. (c) Intensity at a distance r from the transmitting tower

$$= \frac{p}{A} = \frac{1}{2} c \, \varepsilon_0 E_0^2$$

$\Rightarrow$ $$p = 4\pi r^2 \times \frac{1}{2} c \, \varepsilon_0 E_0^2$$
$$= 2\pi r^2 c \, \varepsilon_0 E_0^2$$
$$= 2 \times \pi \times (0.5)^2 \times (3 \times 10^8) \times (8.85 \times 10^{-12}) \times (3 \times 10^6)^2$$
$$\approx 4 \times 10^{10} \text{ W}$$

10. (b) Assuming a 100% absorption, radiation pressure

$$= p = \frac{I}{c} = \frac{\left(\frac{75}{4\pi \times (8 \times 10^{-2})^2}\right)}{(3 \times 10^8)}$$
$$= 3.108 \times 10^{-6} \text{ Nm}^{-2}$$

Now, area of tip is around 1 cm^2.

So, force on finger-tip

$$= p \times A = 3.108 \times 10^{-6} \times 1 \times 10^{-4}$$
$$= 3.1 \times 10^{-10} = 3 \times 10^{-10} \text{ N}$$

11. (b) As power, $P = \frac{dU}{dt} = \frac{\text{Energy emitted}}{\text{Time}}$

Hence, $E = P\Delta t$

Also, E = number of photons $\times$ energy of 1 photon

So, number of photons

$$N = \frac{E}{hf} = \frac{P\Delta t}{hf}$$

$\therefore$ Number of photons emitted by source per unit time

$$= n = \frac{N}{\Delta t} = \frac{P}{hf}$$

As $f_{\text{blue}} > f_{\text{red}}$ and $\frac{n_r}{n_b} = \frac{f_b}{f_r} > 1$

$\Rightarrow$ $n_r > n_b$

12. (a,d) $\frac{E}{B} = c \Rightarrow \mathbf{B} = \frac{1}{c} \cdot (\mathbf{E})$

So, $$\mathbf{B} = \frac{1}{c}(E_1 \hat{\mathbf{i}} + E_2 \hat{\mathbf{j}}) \cos(kz - \omega t)$$

As vibrations are confined to only XOY-plane.

$\therefore$ Wave is plane polarised.

13. (a,c,d) $f \cdot \lambda = c = 3 \times 10^8 \text{ ms}^{-1}$

$\Rightarrow$ $$\lambda = \frac{3 \times 10^8}{10^9} = 0.3 \text{ m}$$

This lies in region of radio waves.

An oscillating particle emits electromagnetic wave of same frequency.

14. (a,d) Only an accelerating charge can produce an electromagnetic wave.

A circulating charge which is moving with constant speed and a charge which is travelling in region of electric field are examples of accelerating charge.

15. (a,c,d) In case of reflection, radiation pressure $p = \frac{2I}{c}$ and in case of absorption it is $p = \frac{I}{c}$

For any real surface, part of energy is reflected (let a fraction r of total energy) and part of it is absorbed.

$\therefore$ Pressure, $p = \frac{I}{c}(1 + r)$, which is more than $\frac{I}{c}$.

16. (a,b,c,d) In microwave ovens used for cooling purposes, microwaves of frequency 2450 MHz are incident over food to be warmed or cooked, this frequency produces a photon energy of 1×10^{-5} eV and matches frequency of dipole water molecules.

17. (b,c) As $X_C = \frac{1}{C\omega}$, so when ω decreases, X_C increases and hence, current decreases. Also $I_C = I_D$, then both simultaneously decrease.

18. (b,d) Average energy density is time average value.

$\therefore$ $$u_{av} = \frac{1}{2}\varepsilon_0\left(\frac{E_0^2}{2}\right) + \frac{1}{2}\left(\frac{B_0^2}{2\mu_0}\right)$$
$$= \frac{1}{4}\varepsilon_0 E_0^2 + \frac{1}{4}\frac{B_0^2}{\mu_0}$$

Also, $$I_{av} = \frac{1}{4} c \, \varepsilon_0 E_0^2$$
$$E_0 = E_{max}, B_0 = B_{max}$$

19. (a,b) For a perfect reflecting surface, pressure

$$p = \frac{2I_{av}}{c}$$

Force, $$F = pA = 2\frac{I_{av}}{c} \times A$$
$$= \frac{2 \times 100 \times 10^{-3} \times 1}{3 \times 10^8}$$
$$= 6.67 \times 10^{-10} \text{ N}$$

For a perfect absorber,

$$F = \frac{I_{av}}{c} \times A = \frac{100 \times 1}{3 \times 10^8} = 3.33 \times 10^{-10} \text{ N}$$

20. (d) Young's double slit experiment shows interference of waves.

Electron microscope uses electrons as waves.

Photoemission shows particle nature of light.

Hence, (i) $\rightarrow$ (p), (ii) $\rightarrow$ (q), (iii) $\rightarrow$ (r), (iv) $\rightarrow$ (s).

21. (c) 2 MHz $= 2 \times 10^6$ Hz $\rightarrow$ Radio wave

2 GHz $= 2 \times 10^9$ Hz $\rightarrow$ Microwave

2 THz $= 2 \times 10^{12}$ Hz $\rightarrow$ Infrared ray

2 EHz $= 2 \times 10^{18}$ Hz $\rightarrow$ X-ray

2 ZHz $= 2 \times 10^{21}$ Hz $\rightarrow$ Gamma ray

Hence, (i) $\rightarrow$ (t), (ii) $\rightarrow$ (r), (iii) $\rightarrow$ (s), (iv) $\rightarrow$ (q), (v) $\rightarrow$ (p).

22. (a) Using relation between temperature and wavelength of a black body radiation, we have,

λ_m = wavelength corresponding to maximum intensity of emitted radiation $= \frac{0.29}{T}$ cm - K

We have, $T = \frac{0.29}{\lambda_m} = \frac{0.29}{1 \times 10^2} = 0.0029 \approx 0$ kelvin

$\therefore$ For a wavelength upto 1 cm, temperature is around zero kelvin.

When $\lambda = 1\,\mu m = 1 \times 10^{-6}\,m = 1 \times 10^{-4}$ cm

$$T = \frac{0.29 \text{ cm-K}}{1 \times 10^{-4}}$$
$$= 0.29 \times 10^4 \text{ K}$$
$$= 2900 \text{ K} \approx 3000 \text{ K}$$

and for $\lambda = 500 \text{ nm} = 500 \times 10^{-9}$ m

$$= 500 \times 10^{-9} \times 10^2 \text{ cm} = 5 \times 10^{-5} \text{ cm}$$
$$T = \frac{0.29}{5 \times 10^{-5}} = \frac{0.29}{5} \times 10^5$$
$$= 5800 \text{ K} \approx 6000 \text{ K}$$

Hence, (i) → (r), (ii) → (r), (iii) → (p), (iv) → (q).

23. (d) Total energy falling over the surface in 1s

$$= \text{Intensity} \times \text{Area}$$
$$= 750\,(\text{Wm}^{-2}) \times (50 \times 10^{-2} \times 100 \times 10^{-2})\,\text{m}^2$$
$$= 750 \times 50 \times 100 \times 10^{-4} \text{ W}$$

Total energy falling per minute

$$= 750 \times 50 \times 100 \times 10^{-4} \times 60 \text{ J}$$
$$= 22500 \text{ J} = 22.5 \text{ kJ}$$

$$\text{Energy absorbed} = \frac{1}{2} \times 22.5 = 11.25 \text{ kJ}$$

24. (d) Total momentum transferred to the surface

$$\Delta p = \underset{\text{(absorption)}}{\frac{\Delta U}{c}} + \underset{\text{(reflection)}}{2\left(\frac{\Delta U}{c}\right)}$$
$$= \frac{11.25 \times 10^3}{c} + \frac{2 \times 11.25 \times 10^3}{c}$$
$$= 1.13 \times 10^{-4} \text{ kg-ms}^{-1}$$

25. (a) Radiation pressure of the flashlight on reflector and on the astronaut,

$$\therefore \quad p = \frac{2 \times \left(\frac{1000}{0.5}\right) \times \left(\frac{1}{2}\right)}{3 \times 10^8}$$
$$= 0.67 \times 10^{-5} \text{ Nm}^{-2}$$

As only $\frac{1}{2}$ of power is pushing the astronaut.

$\therefore$ There is a factor $\frac{1}{2}$ in above calculation.

$$\therefore \quad p = 0.67 \times 10^{-5} \text{ Nm}^{-2}$$

26. (b) Force, $F = p \times A$

$$= 0.67 \times 10^{-5} \times 0.5$$
$$= 0.34 \times 10^{-5} \text{N}$$

$$\text{Acceleration, } a = \frac{F}{m} = \frac{0.35 \times 10^{-5}}{150}$$
$$= 0.22 \times 10^{-7} \text{ ms}^{-2}$$

$$\text{Time, } t = \sqrt{\frac{2d}{a}}$$
$$= \sqrt{\frac{2 \times 100 \times 10^3}{0.22 \times 10^{-7}}}$$

$$\therefore \quad t \approx \sqrt{10} \times \sqrt{10^{12}} \approx 3 \times 10^6$$
$$\approx \frac{3000}{3600} \times 10^3 \text{ h} \approx 0.8 \times 10^3 \text{ h}$$
$$\approx 0.8 \times 42 \text{ days}$$
$$\approx 33 \text{ days} / 34 \text{ days}$$

27. (d) Statement I is incorrect and Statement II is correct.
An accelerating charge produces an electromagnetic wave whose frequency depends on the acceleration of charge. Frequency of radiation changes with acceleration not with speed.

28. (a) Statement I is correct and Statement II correctly explains Statement I.
When an ultroviolet photon strikes phosphor coating, its atom absorbs this photon and becomes excited. On deexcitation a lower frequency photon in visible region is emitted.

DPP-2 **Matter Waves**

1. (i) The de-Broglie wavelength associated with the charged particles is $\lambda = \frac{h}{p}$, where p = momentum of the particle

When particle is accelerated by V volts, then energy of particle is

$$E = \frac{1}{2}mv^2 = qV$$

$\Rightarrow$ Momentum, $p = \sqrt{2mE} = \sqrt{2mqV}$

When charged particle is an electron

where, $m = 9.1 \times 10^{-31}$kg, $q = 1.6 \times 10^{-19}$C,

$h = 6.62 \times 10^{-34}$ J-s

$$\therefore \quad \lambda = \frac{12.27}{\sqrt{V}} \text{ Å} \approx \sqrt{\left(\frac{150}{V}\right)} \text{ Å}$$

For a proton, $m_p = 1.67 \times 10^{-27}$kg

$$\lambda = \frac{0.286 \times 10^{-10}}{\sqrt{V}} \text{m} = \frac{0.286}{\sqrt{V}} \text{ Å}$$

For a deutron, $q = 1.6 \times 10^{-19}$C

$m_d = 2 \times 1.67 \times 10^{-27}$kg

$$\therefore \quad \lambda = \frac{0.202}{\sqrt{V}} \text{ Å}$$

For an α-particle,

$q = 2 \times 1.6 \times 10^{-19}$C

$m = 4 \times 1.67 \times 10^{-27}$kg

$$\therefore \quad \lambda = \frac{0.101}{\sqrt{V}} \text{ Å}$$

(ii) Wavelength associated with a moving uncharged particle of energy E,

$$\lambda = \frac{h}{\sqrt{2mE}}$$

For a thermal neutron,

$m = 1.67 \times 10^{-27}$kg

$E = k_BT$ where k_B = Boltzmann's constant $= 1.38 \times 10^{-23}$ JK^{-1}

$\therefore$ λ = wavelength associated with a thermal neutron

$$= \frac{30.835}{\sqrt{T}} \text{ Å}$$

For a gas molecule at temperature T, energy $E = \frac{3}{2}k_BT$

$$\therefore \quad \lambda = \frac{h}{\sqrt{3mk_BT}}$$

2. (c) $\lambda = \frac{h}{p}$,

So, $$\frac{\lambda_2 - \lambda_1}{\lambda_1} \times 100 = \frac{\frac{h}{p_2} - \frac{h}{p_1}}{\frac{h}{p_1}} \times 100$$

$$\Rightarrow \quad \frac{\lambda_2 - \lambda_1}{\lambda_1} = \frac{p_1 - p_2}{p_2}$$

$$\Rightarrow \quad \frac{\lambda_2 - \lambda_1}{\lambda_1} = \frac{p_1}{p_2} - 1$$

$$\Rightarrow \quad \frac{\left(\frac{0.25}{100}\right)\lambda_1}{\lambda_1} = \frac{p_1}{p_2} - 1$$

$$\Rightarrow \quad \frac{1}{400} + 1 = \frac{p_1}{p_2}$$

$$\Rightarrow \quad \frac{p_2}{p_1} = \frac{400}{401}$$

$$\Rightarrow \quad \frac{p_2 - p_1}{p_1} = \frac{400 - 401}{401} = \frac{-1}{401}$$

So, $$\frac{\Delta p}{p_1} = \frac{-1}{401}$$

$$\Rightarrow \quad 401\,\Delta p = p_1$$

$$\Rightarrow \quad p_1 = 401\,p_0 \qquad \text{(Given, } \Delta p = p_0\text{)}$$

3. (c) $$2\pi r_1 n = \lambda$$

$$\Rightarrow \quad n = \frac{\lambda}{2\pi r_1}$$

$$\Rightarrow \quad n = \frac{10^{-9}}{2 \times 3.14 \times 5.13 \times 10^{-11}}$$

$$\Rightarrow \quad n = 3$$

4. (c) For an electron microscope, wavelength associated with electrons must be less than 1.22 Å ≈ 1Å.

Now, using $\lambda = \sqrt{\frac{150}{V}}$ we have, $V = 150\,\text{V}$

when $\lambda = 1\text{Å}$

5. (a) For a particle of rest mass m_0,

$$\lambda = \frac{h}{m_0 v}\sqrt{\left(1 - \frac{v^2}{c^2}\right)}$$

When $v = c$, $\quad \lambda = 0$

6. (b) $\lambda = \frac{h}{\sqrt{2mE}}$

$E = -13.6\,\text{eV} \Rightarrow \text{KE} = 13.6\,\text{eV}$

$$\therefore \quad \lambda = \frac{6.63 \times 10^{-34}}{\sqrt{2 \times 9.1 \times 10^{-31} \times 13.6 \times 1.6 \times 10^{-19}}}$$

$$= 3.3 \times 10^{-10}\text{m} = 3.3\text{Å}$$

7. (c) When an incident electron enters the metal surface, its kinetic energy increases by an amount equals to work function of metal and hence, its speed increases. As no account of this increase in speed is made hence, a slight lower value of wavelength is observed. As $\lambda = \frac{h}{p}$, and when v increases, p also increases, causing a decrease in λ.

8. (c) When a metal is heated, aggregates consisting of many crystals fuses to form larger or sometimes a single crystal.

In Davisson-Germer experiment, a single crystal diffraction is required. Nickel target is not kept at higher temperature but it is baked, "Annealing" is done to grow crystal size. Experiment is performed at room temperature only. Also, observations are not temperature dependent.

9. (d) Longest wavelength is specified by $\lambda = 2L$, next by $\lambda = L$, then $\lambda = L \times \frac{2}{3}$, etc...

In general $\lambda_n = \frac{2L}{n}$, $n = 1, 2, 3, \ldots$

For (a) $\lambda = \frac{2L}{3}$

For (b) $\lambda = L$

For (c) $\lambda = 2L$. Hence, all are possible.

10. (d) Net force on electron at any instant is

$$F = q(\mathbf{E} + \mathbf{v} \times \mathbf{B})$$
$$= q(-E_0\hat{\mathbf{i}} + v_0\hat{\mathbf{i}} \times B_0\hat{\mathbf{j}})$$
$$= q(-E_0\hat{\mathbf{i}} + v_0 B_0\hat{\mathbf{k}})$$

Acceleration of electron,

$$a = \frac{F}{m} = \frac{q}{m}(-E_0\hat{\mathbf{i}} + v_0 B_0\hat{\mathbf{k}})$$

Velocity after t seconds $= u + at$

$$= 0 + \frac{q}{m}(-E_0\hat{\mathbf{i}} + v_0 B_0\hat{\mathbf{k}})\cdot t$$

So, momentum of electron after t seconds is

$$\mathbf{p} = m\mathbf{v} = q(-E_0\hat{\mathbf{i}} + v_0 B_0\hat{\mathbf{k}})t$$

or $$|\mathbf{p}| = \sqrt{E_0^2 q^2 t^2 + v_0^2 B_0^2 q^2 t^2}$$

or $$\mathbf{p} = qt\sqrt{E_0^2 + (v_0 B_0)^2}$$

Thus, $\lambda =$ wavelength associated with moving electron

$$= \frac{h}{p} = \frac{h}{qt(\sqrt{E_0^2 + (v_0 B_0)^2}}$$

11. (d) Let v_i be initial speed of α-particle and v_α and v_p are final speeds of α-particle and proton.

Then, $4m_p \cdot v_i = 4m_p v_\alpha + m_p v_p \Rightarrow$ momentum is conserved.

Also, coefficient of restitution = 1 ⇒ Collision being elastic.

$$\Rightarrow \quad v_i - 0 = v_p - v_\alpha$$

Eliminating v_i from there, we get

$$4v_p - 4v_\alpha = 4v_\alpha + v_p$$

$$\Rightarrow \quad 3v_p = 8v_\alpha$$

or $$v_\alpha = \frac{3}{8}v_p$$

Now, de-Broglie wavelengths of proton and α-particle are

$$\lambda_p = \frac{h}{m_p v_p}$$

and $$\lambda_\alpha = \frac{h}{m_\alpha v_\alpha}$$

$$\Rightarrow \quad \frac{\lambda_\alpha}{\lambda_p} = \frac{h}{m_\alpha v_\alpha} \times \frac{m_p v_p}{h}$$

$$= \left(\frac{m_p}{m_\alpha}\right)\cdot\left(\frac{v_p}{v_\alpha}\right) = \frac{1}{4} \times \frac{8}{3} = \frac{2}{3}$$

12. (a,d) Net force is Lorentz force.

$$\therefore \quad F_{net} = q(\mathbf{E} + \mathbf{v} \times \mathbf{B})$$

$$\text{Acceleration} = a_{net} = \frac{F_{net}}{m_\alpha} = \frac{q_\alpha}{m_\alpha}(\mathbf{E} + \mathbf{v} \times \mathbf{B})$$

So, $$a_{net} = \frac{q_\alpha}{m_\alpha}(13\hat{\mathbf{i}} - 5\hat{\mathbf{j}})$$

so, $$\mathbf{a} \cdot \hat{\mathbf{k}} = 0$$

Hence, acceleration of α-particle is perpendiucular to Z-axis.

And de-Broglie's wavelength associated with α-particle at any instant t is

$$\lambda = \frac{h}{p} = \frac{h}{mv} = \frac{h}{m(v_0 + at)}$$

So, wavelength keeps on decreasing with time.

13. (b,d) Neutron beam scatters and form a wave like diffraction pattern only when spacing of atoms is of order of wavelength of neutron beam.

As, $$d\sin\theta = n\lambda$$

$$\Rightarrow \quad 0.1\times\sin 30^\circ = 1\times\lambda$$

$$\Rightarrow \quad 0.1\times\frac{1}{2} = 1\times\lambda \Rightarrow \lambda = \frac{0.1}{2} = 0.05\text{ nm}$$

$$\Rightarrow \quad \lambda = 0.5\times10^{-10}\text{m}$$

Also, momentum of neutrons is

$$p = \frac{h}{\lambda} = \frac{6.6\times10^{-34}}{0.5\times10^{-10}}$$

$$\therefore \quad p = 2\times6.6\times10^{-24}\text{ kg-ms}^{-1}$$

$$E = \frac{p^2}{2m} = \frac{(2\times6.6\times10^{-24})^2}{2\times1.6\times10^{-27}} = 54.45\times10^{-21}\text{ J}$$

$$= \frac{54.45\times10^{-21}}{1.6\times10^{-19}} = 34.04\times10^{-2}\text{ eV}$$

$$= 0.34\text{ eV}$$

14. (a,d) Relativistic corrections are required, when speed of particle is not very far from speed of light.

For $\lambda = 0.1$ nm, momentum, $p = \frac{h}{\lambda}$

$$\Rightarrow \quad p = mv = \frac{h}{\lambda} \Rightarrow v = \frac{h}{m\lambda}$$

$$\Rightarrow \quad v = \frac{6.6\times10^{-34}}{0.1\times10^{-9}\times9.11\times10^{-31}} \approx 7\times10^{6}\text{ms}^{-1}$$

which is nearly 43 times lesser than speed of light. But for $\lambda = 10^{-4}$ nm and $\lambda = 10^{-6}$nm, corrections are required.

15. (b,d) As there are two different values of wavelengths, particles momentum is changing. This is possible only when particle is following an elliptical path.

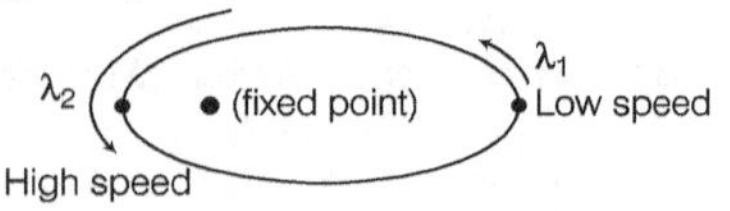

A high speed and high momentum correspond to a lower wavelength as $\lambda = \frac{h}{p}$.

So, particle is near to that fixed point when emitted the wavelength is λ_2.

16. (a,c) Momentum is conserved

so, $$p_C = p_A + p_B \qquad \ldots\text{(i)}$$

$$\Rightarrow \quad \frac{h}{\lambda_C} = \frac{h}{\lambda_A} + \frac{h}{\lambda_B} \qquad (\text{when } p_A, p_B > 0 \text{ or } p_A, p_B < 0)$$

$$\Rightarrow \quad \lambda_C = \frac{\lambda_A\lambda_B}{\lambda_A + \lambda_B}$$

$$\Rightarrow \quad \lambda_C = \frac{\lambda_1\lambda_2}{\lambda_1+\lambda_2} = \frac{\lambda_1\cdot2\lambda_1}{\lambda_1+2\lambda_1} = \frac{2\lambda_1^2}{3\lambda_1} = \frac{2}{3}\lambda_1$$

When $p_A > 0$, $p_B < 0$

or $p_A < 0$, $p_B > 0$

then, Eq. (i) gives

$$\lambda_C = \frac{\lambda_B\lambda_A}{|\lambda_A - \lambda_B|} = \frac{2\lambda_1^2}{\lambda_1} = 2\lambda_1$$

17. (a) ω = Angular frequency of de-Broglie wave

$$= 2\pi f = 2\pi\frac{v}{\lambda} = \frac{2\pi v}{(h/p)} = \frac{2\pi m'v^2}{h} = \frac{2\pi mv^2}{h\sqrt{1-v^2/c^2}}$$

Note *A relativistic correction in mass is required.*

k = wave number of de-Broglie's wave

$$= \frac{2\pi}{\lambda} = \frac{2\pi m'v}{h} = \frac{2\pi mv}{h\sqrt{1-v^2/c^2}}$$

Now, $$\frac{d\omega}{dv} = \frac{2\pi mv}{h(1-v^2/c^2)^{3/2}}$$

and $$\frac{dk}{dv} = \frac{2\pi m}{h(1-v^2/c^2)^{3/2}}$$

So, v_g = group velocity of de-Broglie wave

$$= \frac{d\omega}{dk} = \frac{d\omega/dv}{dk/dv} = \frac{\dfrac{2\pi mv}{h(1-v^2/c^2)^{3/2}}}{\dfrac{2\pi m}{h(1-v^2/c^2)^{3/2}}} = v$$

$$\therefore \quad v_g = v$$

18. (d) v_p has no physical significance as the motion of the wave group not the motion of an individual wave that make up the group corresponds to the motion of the particle and $v_p > c$ for de-Broglie waves does not violate special relativity.

19. (d) KE for electron = eV = 100 eV

$\therefore$ λ = Wavelength of 100 eV electron

$$= \frac{hc}{\sqrt{2mc^2K}} = \frac{1240\text{ eV-nm}}{\sqrt{2\times(5.11\times10^5\text{ eV})\times100\text{eV}}}$$

Note $hc = 1240$ eV-nm *and* $mc^2 = 5.11\times10^5$eV

$$\Rightarrow \quad \lambda = 0.123\text{ nm}$$

For first order maxima,

$$d\sin\theta = \lambda = 0.123, \text{ where } d = 0.091\text{ nm}$$

$\Rightarrow \sin\theta = 1.35$ which is not possible. So, no peak is observed.

20. (a) Diffraction is caused when electrons are scattered from surface layers of atoms (electrons are of low energy).

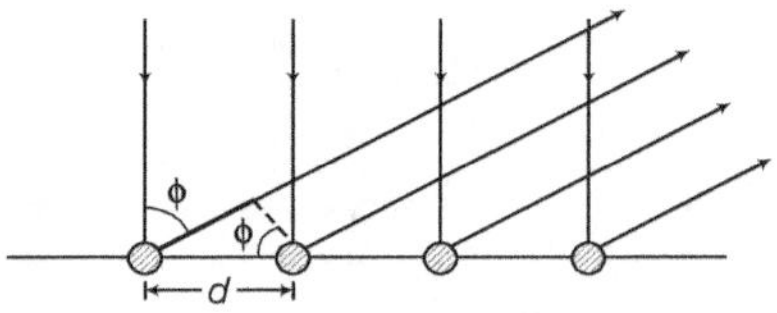

Constructive interference (maxima) occurs, when path length difference between two adjacent rays is an integral multiple of wavelengths or,

$$d\sin\phi = n\lambda = \frac{nh}{p}$$

21. (b) Backing of Nickel sample the "Annealing" process, causes large single crystal regions to develop in the polycrystalline sample.

No maxima and minima occurs in polycrystalline scattering, which is the initial case.

22. (d) Statement I is incorrect but Statement II is correct. A particle behaves like a wave not become a wave.

23. (d) Statement I is incorrect and Statement II is correct.

In Davisson-Germer experiment, electron are not scattered from two different layers. They are low energy electrons ($V = 54$ V) and so they are scattered from the surface of metal slab.

24. (5) $$d\sin\theta = n\cdot\lambda$$

n = order of diffraction maxima.

$$\Rightarrow \quad \lambda = \frac{d\sin\theta}{n}$$

Also, $$\lambda = \frac{h}{p} = \frac{h}{mv}$$

$$\Rightarrow \quad v = \frac{h}{m\lambda} = \frac{nh}{md\sin\theta}$$

$$\Rightarrow \quad v = \frac{6.63\times10^{-34}\times2}{9.11\times10^{-31}\times3\times10^{-6}\times\sin55^\circ}$$

$$= 0.59168\times10^3\text{ ms}^{-1} = 590\text{ ms}^{-1} = 118\times5\text{ ms}^{-1}$$

$$\therefore \quad K = 5$$

DPP-3 Emission of Electrons and Photoelectric Effect

1. Energy of an light photon, $E = hf = \frac{hc}{\lambda} = \frac{1240 \text{ eV-nm}}{400 \text{ nm}}$,

$\therefore \quad hc \approx 1240 \text{ eV-nm}$

$= \frac{1240}{400} \text{ eV}$

$= 3.1 \text{ eV} = 3.1 \times 1.6 \times 10^{-19} \text{ J}$

$= 4.96 \times 10^{-19} \text{ J}$

$(\because 1 \text{ eV} = 1.6 \times 10^{-19} \text{ J})$

$\approx 5 \times 10^{-19} \text{ J}$

Intensity of light over the surface $= 3 \times 10^{-9} \text{ Wm}^{-2}$

$\therefore$ Number of photons striking the surface per square metre per second

$$= \frac{I}{E} = \frac{3 \times 10^{-9}}{5 \times 10^{-19}} \frac{\text{Wm}^{-2}}{\text{J}}$$

$$= 6 \times 10^{9} \text{m}^{-2}\text{s}^{-1}$$

(i) Assuming 100% absorption, (as photoemission efficiency is not given in question)

For each incident photon an electron is emitted, when $hf > \phi_0$.

As, photon energy is 3.1 eV and work function is 2 eV so, photon is absorbed and electron is emitted.

So, number of electrons emitted per square metre in 1s and number of photons incident per m^2 per second.

$= 6 \times 10^{9} (\text{m}^{-2}\text{s}^{-1})$

(ii) Energy is absorbed by metal plate and electrons. Energy absorbed by metal plate is work function.

So, for each emitted electron metal plate absorbs 2 eV of energy.

$\therefore$ Total energy absorbed

$= \left(2 \text{eV} \times 1.6 \times 10^{-19} \frac{\text{J}}{\text{eV}}\right) \times (6 \times 10^{9}) \text{m}^{-2}\text{s}^{-1}$

$= 19.2 \times 10^{-10} \text{ Jm}^{-2}\text{s}^{-1}$

$= 1.92 \times 10^{-9}$ joules of energy is absorbed for per square metre per second.

(iii) Maximum KE of emitted electron

= KE associated with most energised electron

= energy of photon − work function

$= (3.1 - 2) \text{ eV} = 1.1 \text{ eV}$

$= 1.1 \times 1.6 \times 10^{-19} \text{ J}$

$= 1.76 \times 10^{-19} \text{ J}$

(iv) Range of KE for emitted electrons is from 0 to 1.1 eV.

Few electrons have least 0 energy.

2. (i) Let W = Work function of cathode for a surface electron, then

***Case* I** K_1 = Maximum KE of emitted photoelectron.

$$= \frac{hc}{\lambda} - W = \frac{12400}{3000} - W = 4.13 - W$$

***Case* II** K_2 = Maximum KE of emitted photoelectron.

$$= \frac{12400}{1650} - W = 7.51 - W$$

Since, $v_2 = 2v_1$

$\Rightarrow \quad K_2 = 4K_1$

$\Rightarrow \quad \frac{12400}{3000} - W = 4\left(\frac{12400}{1650} - W\right)$

$\Rightarrow \quad W = 3 \text{ eV}$

So, threshold wavelength, $\lambda_0 = \frac{hc}{W}$ $\quad \left(\because \frac{hc}{\lambda_0} = W\right)$

$= \frac{12400}{3} = 4133 \text{Å}$

(ii) Energy of a photon of wavelength 3000Å is

$E_1 = \frac{hc}{\lambda} = hf$

$= \frac{12400}{3000} = 4.13 \text{ eV}$

$= 4.13 \times 1.6 \times 10^{-19} \text{ J}$

$\Rightarrow \quad E_1 = 6.6 \times 10^{-19} \text{ J}$

Rate of photon incident

$= \frac{P_1}{E_1} = \frac{10^{-3}}{6.6 \times 10^{-19}}$

$= 1.52 \times 10^{15} \text{s}^{-1}$

Number of electrons ejected per second,

$n = \frac{(q/t)}{e} = \frac{I}{e}$

$= \frac{4.8 \times 10^{-6}}{1.6 \times 10^{-19}} = 3 \times 10^{13} \text{s}^{-1}$

Efficiency of electron emission from cathode

$= \frac{(\text{Number of emitted } e^-)}{(\text{Number of photons incident})} \times 100$

$= \frac{3 \times 10^{13}}{1.52 \times 10^{15}} \times 100 \approx 2\%$

(iii) Energy of photon of wavelength 1650 Å is E_2,

$E_2 = \frac{hc}{\lambda}$

$= \frac{12400}{1650} = 7.51 \text{ eV}$

$\approx 12 \times 10^{-19} \text{ J}$

Number of photons falling per second

$= \frac{5 \times 10^{-3}}{12 \times 10^{-19}}$

$= 4.17 \times 10^{15} \text{s}^{-1}$

$\therefore$ Number of electrons emitted per second

$= \frac{2}{100} \times 4.17 \times 10^{15}$

$= 8.34 \times 10^{13}$ electrons per second

$= 8.34 \times 10^{13} \times 1.6 \times 10^{-19} \text{C/s}$

$= 13.34 \times 10^{-6} \text{ A}$

$\approx 13 \,\mu\text{A}$

3. (c) According to Einstien's photoelectric equation,

$eV_0 = hf - \phi_0$

$\Rightarrow \quad V_0 = \frac{h}{e} \cdot f - \frac{\phi_0}{e}$

So, slope of V_0 *versus* f graph is

$\frac{h}{e} = 4.12 \times 10^{-15}$

$\Rightarrow \quad h = 4.12 \times 10^{-15} \times 1.6 \times 10^{-19}$

$= 6.6 \times 10^{-34}$ J-s

So, students observations are correct but the graph plotted is between stopping potential and frequency.

4. (d) As, $hf = E$

$\Rightarrow \quad \frac{hc}{\lambda} = E$

$\Rightarrow \quad \frac{1}{\lambda} = \frac{E}{hc}$

Now, $hc \approx 1240$ eV-nm

$\Rightarrow \quad hc = 1240 \times 10^{-7}$ eV-cm

So, $\frac{1}{\lambda} = \frac{1\text{eV}}{1240 \times 10^{-7}\text{ eV-cm}}$ (when $E = 1$ eV)

$= 8065.8 \text{ cm}^{-1}$

So, a wavelength reciprocal $\frac{1}{\lambda}$ of 8065.8 cm^{-1} corresponds to an energy of 1eV.

5. (d) Maximum kinetic energies are given by

$$K = \frac{hc}{\lambda} - \phi_0 \quad \ldots(i)$$

and $$K' = \frac{4hc}{3\lambda} - \phi_0 \quad \ldots(ii)$$

From Eqs. (i) and (ii), we get

$\Rightarrow \quad K' - K = \frac{4hc}{3\lambda} - \frac{hc}{\lambda}$

$K' - K = \frac{hc}{3\lambda}$

But from Eq. (i) $\frac{hc}{\lambda} = K + \phi_0$

$\therefore \quad K' - K = \frac{K}{3} + \frac{\phi_0}{3}$

$\Rightarrow \quad K' = \frac{4K}{3} + \frac{\phi_0}{3}$

or $K' > \frac{4K}{3}$

6. (d) As light is generated due to de excitation of a hydrogen electron so, maximum possible energy of photon is 13.6 eV. Work function of metal is 4 eV.

$\therefore \quad K_{max}$ = Maximum KE of emitted photoelectrons

$= 13.6 - 4$

$= 9.6$ eV

So, stopping potential = 9.6 V

So, for anode potential of –10V is able to make photocurrent zero.

7. (c) $K_{max} = hf_1 - \phi_0 = hf_1 - hf_0 = h(f_1 - f_0)$

8. (c) $eV_1 = \frac{hc}{\lambda_1} - \phi_0, \quad eV_2 = \frac{hc}{\lambda_2} - \phi_0, \; eV_3 = \frac{hc}{\lambda_3} - \phi_0$

$\Rightarrow \quad V_1 = \frac{hc}{e\lambda_1} - \frac{\phi_0}{e}$

$V_2 = \frac{hc}{e\lambda_2} - \frac{\phi_0}{e}$

$V_3 = \frac{hc}{e\lambda_3} - \frac{\phi_0}{e}$

$\because V_1, V_2$ and V_3 are in AP.

$\Rightarrow \quad 2V_2 = V_1 + V_3$

$\Rightarrow \quad \frac{2hc}{e\lambda_2} - \frac{2\phi_0}{e} = \frac{hc}{e\lambda_1} + \frac{hc}{e\lambda_3} - \frac{2\phi_0}{e}$

$\Rightarrow \quad \frac{2}{\lambda_2} = \frac{1}{\lambda_1} + \frac{1}{\lambda_3}$

$\Rightarrow \quad \lambda_1, \lambda_2$ and λ_3 are in harmonic progression.

9. (d)

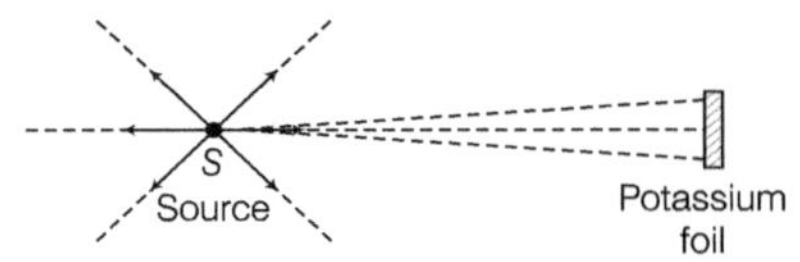

As source radiates energy in all directions, amount of energy reaching the potassium foil per square metre in 1s and is

$$I = \frac{P_0}{4\pi r^2} = \frac{1.0}{4\pi \times (0.5)^2}$$

$= 0.32 \text{ Wm}^{-2}$

Area over which energy is absorbed $= \pi R^2$

where, R = radius of atom.

Rate at which energy falls over the potassium atom is

$$P = IA = \frac{P_0}{4\pi r^2} \cdot \pi R^2$$

$\Rightarrow \quad P = IA$

$= 0.32 \times 5.3 \times 10^{-20}$

$= 1.7 \times 10^{-20}$ W or Js^{-1}

If the incoming energy is fully absorbed, the time required to accumulate sufficient energy (1.8 eV) for the electron to escape is

$$t = \frac{1.8 \text{ eV}}{1.7 \times 10^{-20} \text{ Js}^{-1}} \times \frac{1.6 \times 10^{-19} \text{ J}}{1\text{eV}} = 17 \text{ s}$$

10. (a) $E_1 = \frac{12400}{4000} = 3.1\text{eV}$

$E_2 = \frac{12400}{4800} = 2.58$ eV

$E_3 = \frac{12400}{6000} = 2.06$ eV

$E_4 = \frac{12400}{7000} = 1.77$ eV

$E_5 = \frac{12400}{7800} = 1.58$ eV

Clearly 4th and 5th wavelengths are not emitting any electron.

Now, number of photoelectrons emitted per second

= Number of photons incident per second

$= \frac{I_1A_1}{E_1} + \frac{I_2A_2}{E_2} + \frac{I_3A_3}{E_3}$

$= IA\left(\frac{1}{E_1} + \frac{1}{E_2} + \frac{1}{E_3}\right)$

$= \frac{7.5 \times 10^{-3}}{5} \times 10^{-4} \times \frac{1}{1.6 \times 10^{-19}} \times \left(\frac{1}{3.1} + \frac{1}{2.58} + \frac{1}{2.06}\right)$

$= 1.12 \times 10^{12}$ electrons per second

11. (c) Higher frequency causes photoemission of electron with largest kinetic energy

$$f = \frac{\omega}{2\pi} = \frac{8 \times 10^{15}}{2\pi} \text{ s}^{-1}$$

Maximum kinetic energy of photoelectrons is,

$K_{max} = hf - \phi_0$

$= \frac{4.14 \times 10^{-15} \times 8 \times 10^{15}}{2\pi} - 2.0$

$= 5.27 - 2.0 = 3.27\text{eV}$

12. (d) Only 4% of energy is absorbed and only 3% of this can cause photoemission, part of intensity available for photoemission is

$$I' = 0.03 \times 0.04\, I_0 = 1.2 \text{ nWcm}^{-2}$$

$\therefore$ Number of electrons/second

$$= \frac{1.2 \times 10^{-9} \text{W}}{hf} = 1.5 \times 10^{9} \text{ electrons per second}$$

13. (a,c) Energy of emitted photons

$$= 5 \text{ eV} = 8 \times 10^{-19} \text{ J}$$

Power of source $= 3.2 \times 10^{-3}$ W

$\therefore$ Number of photons emitted by source per second

$$= n_1 = \frac{3.2 \times 10^{-3}}{8 \times 10^{-19}} = 4 \times 10^{15} \text{ photons per second}$$

Number of photons reaching the metal sphere

$$= \frac{n_1}{4\pi r^2} = \frac{4 \times 10^{15}}{4\pi (0.8)^2} = \frac{4 \times 10^{15}}{4\pi \times 0.64}$$

$$\approx 5 \times 10^{14} \text{ photons per unit area per unit time}$$

$$\approx 5 \times 10^{14} \text{ m}^{-2}\text{s}^{-1}$$

Area (effective area) over which photons fall, $A = \pi r^2$

$$= \pi (8 \times 10^{-3})^2 = 2.01 \times 10^{-4} \text{ m}^2$$

$\therefore$ Number of photons incident on sphere in 1s

$$= 5 \times 10^{14} \times 2.01 \times 10^{-4} \approx 10^{11} \text{s}^{-1}$$

As each photon can make emission of an electron, so number of electrons emitted from sphere each second

$$= 10^{11} \text{s}^{-1}$$

And $K_{max} = 2 \text{eV} = (5 - 3) \text{eV}$

So, stopping potential is 2V. When potential of sphere becomes 2 V, photoemission stops.

$$\therefore \quad 2 = \frac{1}{4\pi\varepsilon_0} \cdot \frac{q}{r}$$

$$\Rightarrow \quad q = \frac{2 \times r}{1/4\pi\varepsilon_0} = \frac{2 \times 8 \times 10^{-3}}{9 \times 10^{9}}$$

$$= 1.78 \times 10^{-12} \text{C}$$

So, maximum number of electrons ejected $= \dfrac{q}{e}$

$$= \frac{1.78 \times 10^{-12}}{1.6 \times 10^{-19}}$$

$$= 1.1125 \times 10^{7}$$

So, photoemission stops with in 1s.

14. (a,b,c,d) Concave lens forms image of source S at S'.

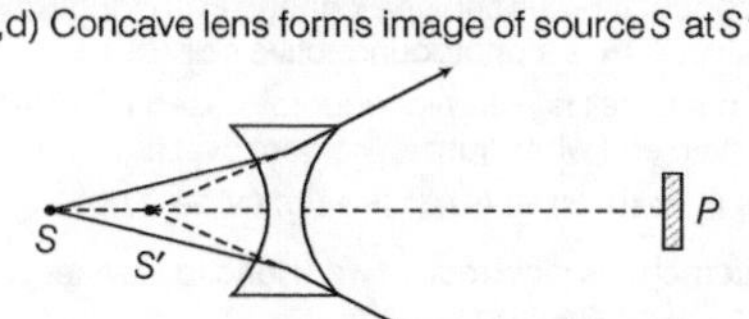

So, S' now acts like a source and as its distance is less, photoemission rate increases.

With a convex lens in place of aperture, it delivers all photons reaching upto the lens to metal surface.

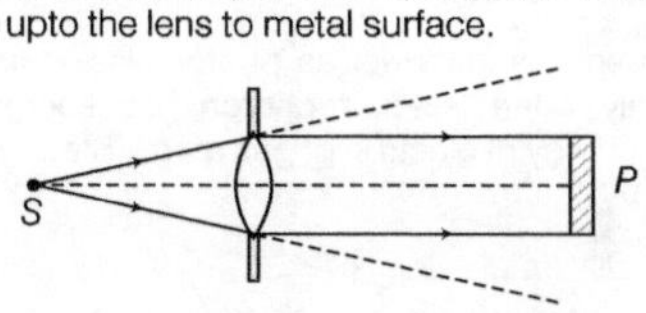

$\therefore$ Photoemission rate increases when a glass slab is placed in place of aperture, It absorbs a part of light and delivered remaining part reducing photoemission rate.

When a prism is placed in place of aperture, it may deviates light beam (monochromatic).

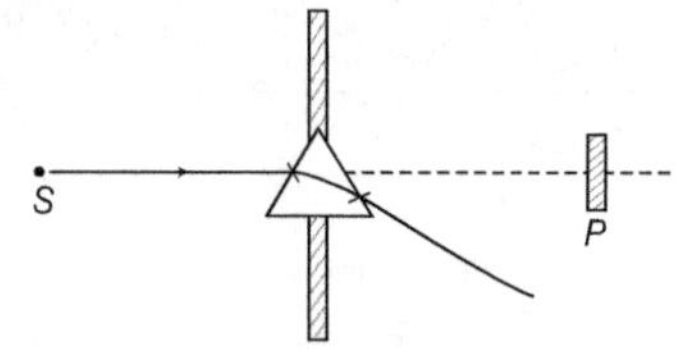

Upto such an extent that light does not reach P. So, all options are correct.

15. (a,c,d) Stopping potential is more negative for a metal, whose threshold frequency is less.

As $I_e \propto I_\varepsilon$ and photocurrent does not increase, when frequency is increased. So, option (b) is incorrect.

As, $V_0 = \dfrac{h}{e} \cdot f - \dfrac{\phi_0}{e}$, V *versus* f graph of every metal has slope $\dfrac{h}{e}$.

Also, stopping potential does not depend on intensity.

Also, V_s remains constant with f, so graph is a straight line.

16. (a,b,c,d) When a wave is incident over a particle, particle is set into vibrations. Amplitude of vibration is proportional to amplitude of impending wave. Also, KE $\propto A_e^2$.

17. (a,b,c) Photoelectric effect defines a definite energy and definite momentum associated with light (a wave), which is the property of a moving particle.

18. (b,d) Stopping potential depends on energy of incident light and work function of photosensitive surface.

Both of these remains same.

Hence, V_s remains constant.

As, intensity $I_\varepsilon \propto \dfrac{1}{r^2}$, r = distance and photocurrent $\propto$ Intensity

So, photocurrent $I \propto \dfrac{1}{r^2} \Rightarrow \dfrac{I_1}{I_2} = \left(\dfrac{r_2}{r_1}\right)^2$

$$\Rightarrow \quad I_2 = I_1 \left(\frac{r_1}{r_2}\right)^2 = 18 \left(\frac{0.2}{0.6}\right)^2 = 0.2 \text{ mA}$$

19. (c) Area of lens through which light passes initially, $A_1 = \dfrac{\pi d^2}{4}$

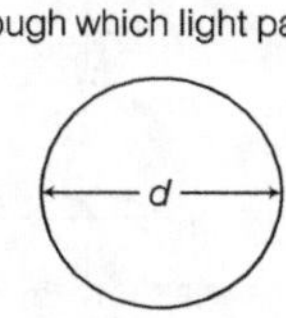

Area of lens through which light passes after covering central portion is $A_2 = \left[\pi \dfrac{d^2}{4} - \dfrac{\pi (d^2/4)}{4} = \dfrac{3\pi}{4}\left(\dfrac{d^2}{4}\right)\right]$

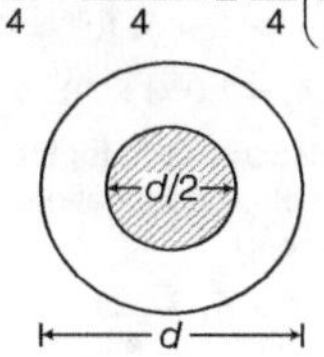

Ratio of number of photons in both cases $= \dfrac{A_1}{A_2} = \dfrac{I_1}{I_2}$

$$\Rightarrow \quad I_2 = I_1 \cdot \left(\frac{A_2}{A_1}\right) = I \cdot \frac{\frac{3\pi}{4}\left(\frac{d^2}{4}\right)}{\pi\left(\frac{d^2}{4}\right)}$$

$$\Rightarrow \quad I_2 = \frac{3}{4} I$$

20. (d)

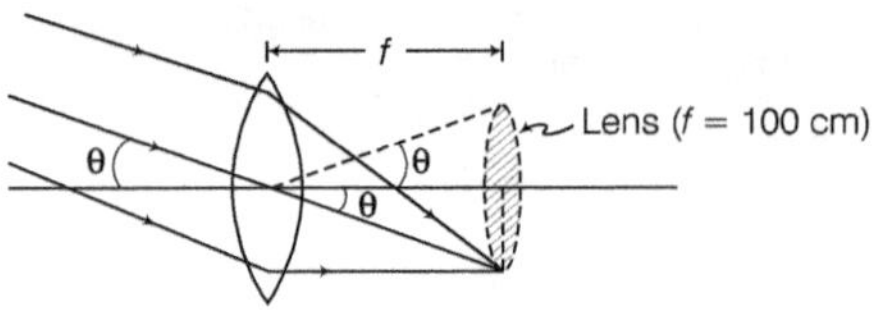

$$\theta = 30 \text{ min} = \frac{1}{2} \times 1^\circ$$
$$= \frac{1}{2} \times \frac{\pi}{180^\circ}$$

Image is formed in focal plane.

∴ Radius of image, $r = f \cdot \theta \Rightarrow$ diameter of image is

$$2r = 2 \times f \times \theta$$
$$= 2 \times 100 \times \frac{1}{2} \times \frac{\pi}{180}$$
$$= \frac{100 \times 3.14}{180}$$
$$= \frac{314}{180} = 1.74 \text{ cm}$$
$$= 17.4 \text{ mm} \approx 17 \text{mm}$$

21. (c) Energy of 1 photon $= hf = \dfrac{hc}{\lambda}$

$$= \frac{12400 \text{ eV-Å}}{6000 \text{Å}} = 2.066 \text{ eV}$$
$$= 2.07 \times 1.6 \times 10^{-19} \text{ J}$$
$$= 3.3 \times 10^{-19} \text{ J}$$

Power of source = 2W

∴ Number of photons emitted from source (per second)

$$n_1 = \frac{2}{3.3 \times 10^{-19}} = 6.06 \times 10^{18} \text{ s}^{-1}$$

Number of photons incident over aperture per square metre per second,

$$n_2 = \frac{n_1}{4\pi r^2}$$
$$= \frac{6.06 \times 10^{18}}{4\pi \times (0.6)^2}$$
$$= 1.34 \times 10^{18} \text{ m}^{-2}\text{s}^{-1}$$

$$\text{Area of aperture} = \pi r^2 = \frac{\pi d^2}{4}$$
$$= A_1 = \frac{\pi \times (0.1)^2}{4}$$
$$= 7.85 \times 10^{-3} \text{m}^2$$

Photons passing through the aperture

$$= n_3 = n_2 \times A_1$$
$$= 1.34 \times 10^{18} \times 7.85 \times 10^{-3}$$
$$\Rightarrow \quad n_3 = 1.052 \times 10^{16} \text{s}^{-1}$$

Now, aperture acts like a source for the metal surface number of photons reaching at the location of metal surface per squaremetre, per second

$$= n_4 = \frac{n_3}{4\pi r^2}, r = 5.4 \text{ m}$$
$$\Rightarrow \quad n_4 = \frac{1.052 \times 10^{16}}{4\pi \times (5.4)^2} = 2.87 \times 10^{13} \text{ m}^{-2}\text{s}^{-1}$$

$$\text{Area of metal plate} = 0.5 \text{ cm}^2$$
$$= 0.5 \times 10^{-4} \text{ m}^2$$

So, number of photons incident over metal surface

$$= n_5 = n_4 \times \text{area of metal plate}$$
$$= 2.87 \times 10^{13} \times 0.5 \times 10^{-4}$$
$$= 1.435 \times 10^9 \text{s}^{-1}$$

22. (d) When a concave lens is placed in the aperture opening, lens focus an image S' of source S which acts like source (with lens).

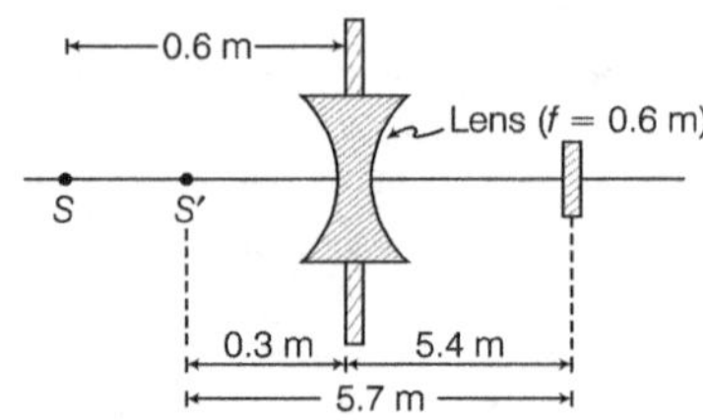

By lens formula,

$$\frac{1}{v} - \frac{1}{-0.6} = \frac{1}{-0.6}$$
$$\Rightarrow \quad v = -0.3 \text{ m}$$

Now, number of photons emitted from this new source S',

$$n_6 = 80\% \text{ of } n_3$$
$$\Rightarrow \quad n_6 = \frac{80}{100} \times 1.052 \times 10^{16} \text{s}^{-1}$$

Due to this source S', photons reaching metal surface per unit area per unit time

$$= n_7 = \frac{n_6}{4\pi r^2}, \; r = 5.7 \text{ m}$$
$$\Rightarrow \quad n_7 = \frac{\frac{80}{100} \times 1.052 \times 10^{16}}{4\pi \times (5.7)^2}$$
$$\Rightarrow \quad n_7 = 2.06 \times 10^{3} \text{m}^{-2}\text{s}^{-1}$$

This is the photon flux at the metal surface with lens.

23. (d) Energy of incident photons in both case is same and work function for metal also remains same. So, kinetic energy of most energetic electrons will also be same.

∴ Ratio of stopping potentials of both cases is 1.

24. (a) A photocell converts light energy into electrical energy and it can be used as a converting device.

In photoemissive cell, electrons are emitted from cathode which are attracted towards positive anode.

Photoconductive cell shows a change in conductivity in presence of light. An LDR is a photoconductive cell.

Photovoltaic cell is a semiconductor based p-n junction used to develop an emf when light is incident over it.

Hence, (i) → (t), (ii) → (q), (iii) → (s), (iv) → (t,r,p).

25. (d) Statement I is incorrect as work function is energy, required for emission of an electron

Work function is more for an electron deep inside metal surface whereas it is least for a surface electron.

Also, work function increases with number of electrons emitted. Statement II is correct.

26. (d) Statement I is incorrect as photoemission in a gas sample occurs only when incident photon has energy greater than ionisation energy of electron of gas atom. Statement II is correct.

Revisal Problems (JEE Main)

1. (c) $E = hf$

$\Rightarrow \quad f = \frac{E}{h} = \frac{14.4 \times 10^3 \times 1.6 \times 10^{-19}}{6.6 \times 10^{-34}}$

$\Rightarrow \quad f = 3.490 \times 10^{18}$

$\Rightarrow \quad f \approx 3.5 \times 10^{18}$ Hz

So, $\lambda = \frac{c}{f} = \frac{3 \times 10^8}{3.5 \times 10^{18}} = 0.86 \times 10^{-10}$m

This lies in region of X-rays.

2. (d) Intensity at a distance of 3.5 m from the source

= Energy falling per unit area per unit time

$= \frac{800\text{W}}{4\pi(3.5)^2\text{m}^2} = 5.194 \text{ Wm}^{-2}$

or $I \approx 5.2 \text{ Wm}^{-2}$

Half of this intensity is provided by electric field and half is by magnetic field.

So, $\frac{1}{2} \cdot I = \frac{1}{2}(uc)$

$u = \text{energy density} = \frac{1}{2}\varepsilon_0 c E_{rms}^2$

$\Rightarrow \quad \frac{1}{2} \times 8.85 \times 10^{-12} \times 3 \times 10^8 \times E_{rms}^2 = \frac{1}{2} \times 5.2$

$\Rightarrow \quad E_{rms}^2 = \frac{5.2}{8.85 \times 10^{-2} \times 3 \times 10^8}$

$= \frac{5.2 \times 10^4}{8.85 \times 3} \approx 1959$

$\Rightarrow \quad E_{rms} = \sqrt{1959} \approx 44 \text{ Vm}^{-1}$

3. (a) Each photon has energy E and momentum p,

$E = hf$

and $p = \frac{hf}{c}$

From Einstein's relativistic energy-momentum relation

$E = \sqrt{c^2p^2 + m_0c^4}$

where, m_0 = rest mass of a photon = 0. As, photon moves with speed c relative to all reference frames.

$\therefore \quad E = cp$

$\Rightarrow \quad E = \frac{ch}{\lambda_1}$ ...(i)

$\left(\because \text{de-Broglie wavelength, } \lambda = h/p\right)$

and $E = hf = \frac{hc}{\lambda_2}$ ...(ii)

Form Eqs. (i) and (ii), we get

$\Rightarrow \quad \frac{\lambda_1}{\lambda_2} = 1$

$\Rightarrow \quad \frac{\lambda_1}{\lambda_2} = E^0 \qquad (E^0 = 1)$

4. (a) For curves b and c, stopping potential is same so, frequencies for curve a and b are same.

$\Rightarrow \quad f_a = f_b$

Also, $f_c > f_a$

As saturation current is same for curves b and c. So, intensity is same for curves b and c.

$\Rightarrow \quad I_b = I_c$

Also, $I_a < I_b$

5. (a) Tungsten is preferred as its melting point is high (3650 K). Its disadvantages are its high work function (4.52 eV) and high operating temperature (2500K). Tungsten also have an advantage of to withstand high voltages. Inspite of option (a) all other options are correct.

6. (c) Radius of circular path is

$r = \frac{\sqrt{2mK}}{qB}$

$\Rightarrow$ KE of emitted electron is

$K = \frac{r^2q^2B^2}{2m} = \frac{1}{2}\left(\frac{e}{m}\right) \times er^2B^2$

or $K = \frac{1}{2} \times 1.7 \times 10^{11} \times 1.6 \times 10^{-19} \times (100 \times 10^{-2})^2 \times \left(\frac{1}{\sqrt{17}} \times 10^{-5}\right)^2$

$\Rightarrow \quad K = 8 \times 10^{-20}\text{J} = 0.5 \text{ eV}$

Energy of photon,

$E = \frac{hc}{\lambda} = \left(\frac{12375}{2475}\right) = 5 \text{ eV}$

Now, work function of metal is

$\phi_0 = E - K = 5 - 0.5 = 4.5 \text{ eV}$

7. (c) As, $\lambda = \frac{h}{p}$

$\Rightarrow \quad \lambda - \Delta\lambda = \frac{h}{p + \Delta p}$

$\Rightarrow \quad \frac{199\lambda}{200} = \frac{h}{p + \Delta p}$

$\Rightarrow \quad p + \Delta p = \frac{200}{199}p$

$\Rightarrow \quad p = 199\,\Delta p$

8. (a) From momentum conservation, initial momentum = 0. Hence, after the explosion, each particle must have equal and opposite momentum.

So, $\lambda = \frac{h}{p}$

$\Rightarrow \quad \frac{\lambda_1}{\lambda_2} = \frac{p_2}{p_1} = 1$

9. (a) $E = n \cdot \frac{hc}{\lambda}$

$\Rightarrow n = \frac{E\lambda}{hc} = \frac{IA\cos\theta\lambda}{hc}$

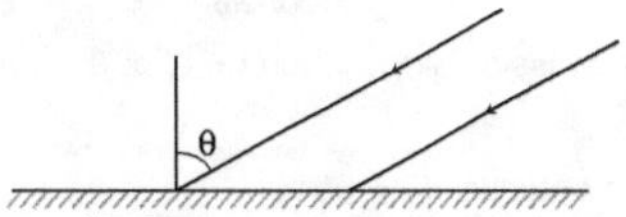

Surface

Case I When $\theta = 0°$,

$\therefore \quad n_1 = \frac{IA\cos 0°\,\lambda}{hc} = \frac{IA\lambda}{hc}$

Case II When, $\theta = 60°$

$n_2 = \frac{IA\cos 60°\,\lambda}{hc}$

$= \frac{1}{2} \cdot \frac{IA\lambda}{hc}$

$\therefore \quad \text{Ratio} = \frac{n_2}{n_1} = \frac{1}{2}$

10. (c) Energy of incident photon

$$= hf = K_{max} + W$$
$$= 0.73 + 1.82 = 2.55 \text{ eV}$$

Momentum of emitted photon

$$= \frac{E}{c} = \frac{hf}{c}$$
$$= \frac{2.55 \times 1.6 \times 10^{-19}}{3 \times 10^8}$$
$$= 1.36 \times 10^{-27} \text{kg-ms}^{-1}$$

As momentum is conserved.

$\therefore$ Momentum of recoiling atom $= 1.36 \times 10^{-27}$ kg-ms^{-1} in opposite direction.

Hence, recoil speed of the atom is

$$v = \frac{\text{Momentum}}{\text{Mass}} = \frac{1.36 \times 10^{-27}}{1.67 \times 10^{-27}}$$
$$= 0.814 \text{ ms}^{-1}$$

11. (b) $E = a(1 + \cos\omega t)\cdot\cos\omega_0 t$

$$= a\cos\omega_0 t + a\cos\omega t\cdot\cos\omega_0 t$$
$$= a\cos\omega_0 t + \frac{1}{2}a\cos(\omega + \omega_0)t + \frac{1}{2}a\cos(\omega - \omega_0)t$$

So, radiation consists of a mixture of 3 different frequencies out of which highest in $\omega + \omega_0$. This causes emission of electron with greatest KE.

Now, $K_{max} = hf - \phi_0$

$$= h\frac{(\omega + \omega_0)}{2\pi} - \phi_0$$
$$= \frac{h}{2\pi}(6 \times 10^{14} + 3.6 \times 10^{15}) \text{ J} - 2.39 \text{ eV}$$
$$= (4.41 \times 10^{-19} - 3.82 \times 10^{-19})\text{J}$$
$$= 0.37 \text{ eV}$$

12. (d)

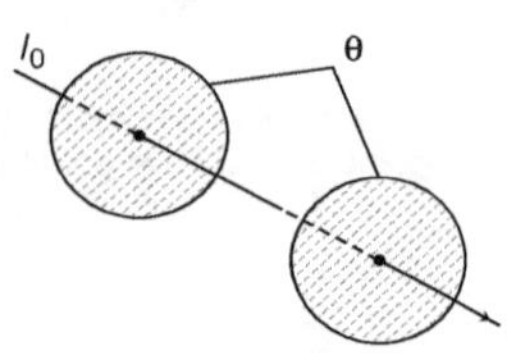

If θ = angle between polarisation planes of polaroids, then resultant intensity of polarised light obtained from Malus law is

$$I = \frac{I_0}{2}\cos^2\theta$$

But frequency of radiation remains constant.

$\therefore$ f_{max}^2 = constant

And photoemission stops when $I = 0$, at $\theta = \pi/2$ so, $f_{max}^2 = 0$ at $\theta = \pi/2$.

13. (c) Stoping potential, $\phi_0 = hf_0$

$$= 6.6 \times 10^{-34} \times 0.49 \times 10^{15} \text{ J}$$
$$= \frac{6.6 \times 10^{-34} \times 0.49 \times 10^{15}}{1.6 \times 10^{-19}} \text{ eV}$$
$$= \frac{6.6 \times 0.49}{1.6}$$
$$= 2.02 \text{ eV} \approx 2\text{eV}$$

14. (c) KE of a particle at high speeds is

$$K = \frac{mc^2}{\sqrt{1 - v^2/c^2}} - mc^2$$

Because it is equal to its rest energy and kinetic energy

$\Rightarrow$ $E = K + mc^2$

where, $E = \frac{mc^2}{\sqrt{1 - v^2/c^2}}$

For a particle at rest in a given reference frame $K = 0$.

$\therefore$ Total rest energy, $E = mc^2$

Now, $E = K + mc^2$ and $E = \frac{mc^2}{\sqrt{1 - v^2/c^2}}$

$$\Rightarrow \quad E^2 = \frac{m^2c^2(v^2 - v^2 + c^2)}{1 - v^2/c^2}$$

(We added and subtracted v^2)

$$= p^2c^2 + \frac{m^2c^4(1 - v^2/c^2)}{1 - v^2/c^2}$$
$$= p^2c^2 + m^2c^4$$

or $E = p^2c^2 + m^2c^4$

Now, $E^2 = (K + mc^2)^2$

$$\Rightarrow \quad p^2c^2 + m^2c^4 = (K + mc^2)^2$$
$$\Rightarrow \quad K^2 + 2mc^2K = p^2c^2 = \frac{h^2c^2}{\lambda^2} \quad \text{(by de-Broglie's } \lambda = h/p)$$
$$\Rightarrow \quad \lambda^2 = \frac{h^2c^2}{K^2 + 2mc^2K}$$
$$\Rightarrow \quad \lambda = \frac{hc}{\sqrt{K^2 + 2mc^2K}}$$

15. (d) For low speed when mass and energy do not interchange, we do not require more complex relativistic formulae and a much simpler classical view is used.

The v-factor, $r = \frac{1}{1 - \sqrt{1 - v^2/c^2}}$ which appears in relativistic formulae has a value of 1.005 at $r = \frac{1}{10}$c. So, for speeds as high as 0.1 c $= 3 \times 10$ ms^{-1}, the factor $\sqrt{1 - v^2/c^2}$ gives a numerical correction of less than 1%. If we have mass and KE of a particle, then we use

$$\frac{K}{mc^2} = r - 1$$
$$= \frac{1}{\sqrt{1 - v^2/c^2}} - 1$$

If $\frac{K}{mc^2} < 0.01$, then $v \le 1.01$ and accuracy is less than 1%.

Now, for $K = 100$ eV and $mc^2 = 0.511$ MeV

We have, $\frac{K}{mc^2} \approx \frac{1000}{0.5 \times 10^8} << 1\%$

no correction is required.

For $\frac{K}{mc^2} = \frac{10^5}{0.5 \times 10^8} << 1\%$

no correction is required.

16. (a) KE of electron = | Total Energy | = 3.4 eV

$$\lambda = \frac{h}{p} = \frac{h}{\sqrt{2Km}}$$
$$= \frac{6.6 \times 10^{-34}}{\sqrt{(2 \times 3.4 \times 1.6 \times 10^{-19} \times 9.1 \times 10^{-31})}}$$
$$= 6.63 \times 10^{-10}\text{m} = 6.63 \text{ Å}$$

Alternate Method

(a) Total energy of electron in a stationary state = − 3.4 eV

As $|E| = K = \frac{-U}{2}$

KE of electron = |− 3.4| = 3.4eV

Using $\lambda = \frac{h}{p} = \frac{h}{\sqrt{2mK}}$

We have, $\lambda = \frac{6.6 \times 10^{-34}}{\sqrt{2 \times 9.1 \times 10^{-31} \times 3.4 \times 1.6 \times 10^{-19}}}$

$= \frac{6.6 \times 10^{-34}}{\sqrt{99.01 \times 10^{-50}}} \approx \frac{6.6 \times 10^{-34}}{10 \times 10^{-25}} = 6.6\ \text{Å}$

17. (b) For a standing wave with nodes at its ends, possibilities are

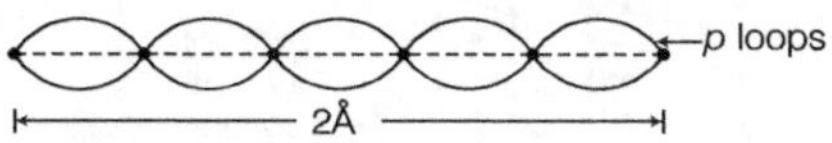

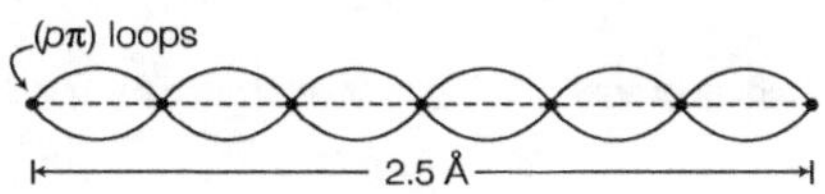

So, $p \times \frac{\lambda}{2} = 2\ \text{Å}$

and $(p+1)\frac{\lambda}{2} = 2.5\text{Å}$

$\Rightarrow (p + 1 - p)\frac{\lambda}{2} = 2.5 - 2$

$\Rightarrow \lambda = 2 \times 0.5 = 1\text{Å} = 10^{-10}\text{m}$

As, $\lambda = \frac{h}{p}$ and $p = \sqrt{2mK}$

$\therefore K = \frac{h^2}{2m\lambda^2} = \frac{(6.6 \times 10^{-34})}{2 \times 9.1 \times 10^{-31} \times (10^{-10})^2} = 150.8\ \text{eV}$

18. (b) When lens is moved away, initially it is focused over cathode forming an intense spot of less area.

So, intensity over a smaller area increases. This results in lowering of emission rate as area over which light is focused is reduced.

19. (c) When zero of potential energy is changed, change occurs only in potential energy of electron and in total energy of electron. Kinetic energy does not depend on choice of zero of potential energy as kinetic energy obtained by both is same.

$\therefore \frac{\lambda_1}{\lambda_2} = 1$

20. (d) Only Statement IV is incorrect as

$\phi_0 = hf - K$

and $K \neq 0 \Rightarrow \phi_0 \neq hf$

21. (a) Both Statement I and Statement II are correct and Statement II is a correct explanation of Statement I.

In a metal, valence band and conduction band overlap. So, during excitation, an electron can occupy all possible values of energies from 1.8 eV to 2.8 eV, which are photon energies in visible region.

22. (d) Statement I is incorrect but Statement II is correct.
A surface cannot strike an electron like a bat to a ball. Electron absorbs and re-emit photons.

23. (d) Statement I is incorrect but Statement II is correct.

As, $d\sin\theta = n\lambda$ and $\sin\theta = n \cdot \frac{\lambda}{d}$

Also, $\frac{\lambda}{d} \approx 1$ for a clear and prominent diffraction pattern.

$\Rightarrow \sin\theta \approx n$

So, $n \neq 2$

No second order maxima observed in Davisson-Germer experiment.

Revisal Problems (JEE Advanced)

1. (d) According to classical theory, energy is uniformly distributed uniformly continuous over the entire area over which light is falling.

Now, we assume here

(i) 100% absorption.

(ii) Absorption occurs in top atomic layers.

(iii) Energy absorbed by atom ∝ cross-section area of atom.

(iv) Energy absorbed is given to only one of electron.

So, kinetic energy of emitted electron

= Absorbed energy − Work function

As, energy absorbed = area × energy per unit area unit time × time

= area × intensity × time.

$K_{max} = AIt - \phi_0$

which increases with time.

2. (d) As, $K_{max} = AIt - \phi_0$ at start of photoemission an electron accumulate just sufficient energy to overcome work function.

With $K_{max} = 0$, we have

$t = \frac{\phi_0}{AI} = \frac{2.28 \times 1.6 \times 10^{-16}\text{mJ}}{\left(10^{-7}\frac{\text{mJ}}{\text{cm}^2\text{s}} \times \pi \times 10^{-16}\text{cm}^2\right)}$

≈ 130 days

3. (c) As 96% of light is reflected, thus only 4% of light is absorbed. Violet region is only 3%. So, intensity available for photoemission is

$I' = 3\%$ of (4% of I_0)

$= 0.030 \times 0.040 \times I_0 = 1.2\ \text{nWcm}^{-2}$

4. (a) Number of electrons = Number of photons absorbed

$= \frac{\text{Energy falling per second}}{\text{Energy of 1 photon}}$

$= \frac{1.2 \times 10^{-9}}{hf} = \frac{\lambda(1.2 \times 10^{-9})}{h \times c}$

$= \frac{250 \times 10^{-9} \times 1.2 \times 10^{-9}}{6.6 \times 10^{-34} \times 3 \times 10^{8}}$

$= 1.5 \times 10^9$ (per second)

5. (d) Work function for metal = $\phi_0 = hf_0$

$= 4.14 \times 10^{-15}\ \text{eV-s} \times 1.1 \times 10^{15}\ \text{s}^{-1}$

$= 4.5\ \text{eV}$

So, stopping potential using

$eV_0 = hf - \phi_0$

$\Rightarrow V_0 = 0.46\ \text{V}$

6. (c) Plants appears green/yellow in bright sunlight. So, a plant leaf reflects most of green/yellow region.

It is violet region which causes the photosynthesis.

7. (c) From the given SPD graph intensity maxima is in visible region near violet.

8. (c) λ_{min} = 200 nm = 2×10^{-7}m, which is for UV. So, no X-rays are observed near earth's surface, most of them are absorbed in ionosphere.

9. (c) $\lambda' - \lambda = \frac{h}{mc}(1 - \cos\phi)$

$$\Rightarrow \quad \lambda' = \lambda + \frac{h}{mc}(1 - \cos\phi)$$

$$= 10 + 0.293 \times \frac{h}{mc}$$

As, $\frac{h}{mc}$ = 2.426 pm (called Compton wavelength)

We have,

$$\lambda' = 10.7\,\text{pm}$$

Now, $\lambda' - \lambda$ is maximum, when $1 - \cos\phi$ is maximum or when $1 - \cos\phi = 2$.

Which gives,

$$\lambda' = \lambda + 2\frac{h}{mc}$$

$$= 10 + 4.9 = 14.9\,\text{pm}$$

10. (a) Maximum recoil KE = Change in energy of incident photon

$$= h(f - f')$$

$$= hc\left(\frac{1}{\lambda} - \frac{1}{\lambda'}\right)$$

$$= hc\left(\frac{1}{10} - \frac{1}{14.9}\right)$$

$$= 40.8\,\text{keV}$$

11. (c) Photoemission stops, when potential on the surface of sphere is equal to stopping potential.

Now, $$\frac{hc}{\lambda} = eV_0 + \phi_0$$

$$\Rightarrow \quad eV_0 = \frac{hc}{\lambda} - \phi_0 \quad \ldots(i)$$

Also, V_0 = surface potential of sphere = $\frac{kq}{r}$

$$\Rightarrow \quad V_0 = \frac{1}{4\pi\varepsilon_0} \cdot \frac{q}{r} \quad \ldots(ii)$$

On equating Eqs. (i) and (ii), we get

$$\frac{hc}{\lambda} - \phi_0 = eV_0 = e\left(\frac{1}{4\pi\varepsilon_0} \cdot \frac{q}{r}\right)$$

On substituting the values, we have

$$\left(\frac{1240}{1200} - 4.76\right) = 1.44\,\text{eV}$$

$$= e\left(9 \times 10^9 \times \frac{q}{10^{-2}}\,\text{V}\right)$$

$$\Rightarrow \quad q = \frac{1.44 \times 10^{-2}}{9 \times 10^9}$$

$$\therefore \quad q = 1.6 \times 10^{-12}\,\text{C}$$

As, $$q = ne$$

n = number of electrons emitted = $\frac{q}{e} = \frac{1.6 \times 10^{-12}}{1.6 \times 10^{-19}}$

$$= 10^7 \text{ electrons}$$

So, mass lost by sphere due to photoemission is

$$m' = n \times m_e = 10^7 m_e$$

12. (d) Graphs for only one sample, Intensity in case A is more as current I is more.

13. (b) Size of atom is around 10^{-4}m and speed of photon

$$= 3 \times 10^{-8}\,\text{ms}^{-1}$$

So, absorption time is

$$t = \frac{\text{distance}}{\text{speed}}$$

$$= \frac{10^{-4}\,\text{m}}{3 \times 10^8\,\text{ms}^{-1}} = \frac{1}{3} \times 10^{-22}\,\text{s}$$

14. (c) Let there is one free electron per atom in metal.

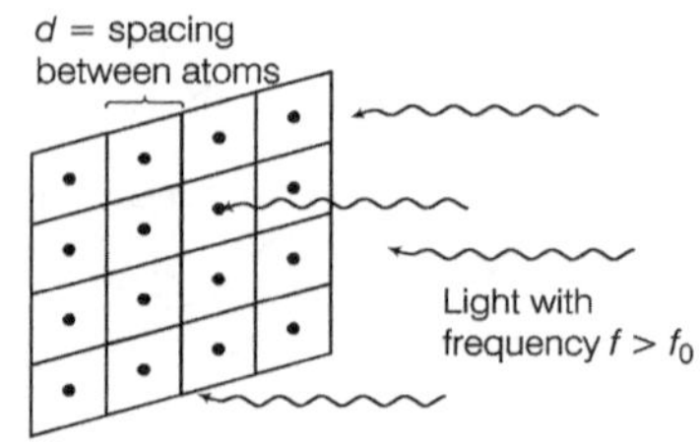

Total amount of energy falling over each atom per unit time

$$= \frac{dE}{dt} = Id^2 \qquad \begin{pmatrix} d^2 = \text{Area} \\ I = \text{Intensity} \end{pmatrix}$$

Energy absorbed in time, $t = Id^2t$

As volume of a cube of side d is d^3 and we assume that there is one atom in d^3 volume.

So, number of atoms per unit volume is

$$N = \frac{1}{d^3}$$

But $$N = \frac{\rho}{A}$$

So, $$d = \left(\frac{A}{\rho}\right)^{1/3}$$

For photoemission, absorbed energy = work function

$$\therefore \quad Id^2t = \phi_0$$

or $$t = \frac{\phi_0}{Id^2}$$

$$\Rightarrow \quad t = \text{time lag} = \frac{\phi_0}{I}\left(\frac{\rho}{A}\right)^{2/3}$$

15. (c) According to classical physics, calculation of momentum and energy shows that scattered electron will emerge in all possible directions without any dependence on KE of incoming electrons.

16. (c) Group speed of de-Broglie wave is $v_g = \frac{\omega}{k} = \frac{p}{m}$, this quantity has a physical meaning. But the frequency associated with de-Broglie wave does not have any physical significance.

$\therefore$ Product $f\lambda$ = Phase speed has no physical significance.

17. (d) $\frac{h}{e}$ = Slope of V_0-f graph.

From first set of readings, we have

$$h = 1.6 \times 10^{-19}\left(\frac{2 \times 10^{-14}}{5}\right)$$

$$= 6.4 \times 10^{-34}\,\text{Js}$$

From second set of readings, we have

$$h = \frac{1.6 \times 10^{-19} \times 2.5 \times 10^{-14}}{5}$$

$$= 8 \times 10^{-34}\,\text{Js}$$

As, h obtained from both set of readings is not same. So, experiment is not done properly.

18. (a,c,d) As, $p = \frac{h}{\lambda}$, they have same linear momentum.

Photon's energy is $E_{ph} = hf = \frac{hc}{\lambda} = pc$ and the particle's total energy $= E_P = \sqrt{p^2c^2 + E_0^2}$

As, $\sqrt{p^2c^2 + E_0^2} > pc$

$$E_P > E_{ph}$$

Also, KE of particle, $K = E_P - E_0 = \sqrt{p^2c^2 + E_0^2} - E_0$

So, $K < pc$

19. (a,b,c) $\lambda = \frac{h}{p} = \frac{hc}{pc}$

For a particle with kinetic energy K,

$$K = E - E_0 = \sqrt{p^2c^2 + E_0} - E_0$$

$$= \sqrt{p^2c^2 + m_0^2c^4} - m_0c^2$$

$\therefore$ $pc = \sqrt{K(K + 2m_0c^2)}$

So, $\lambda = \frac{hc}{\sqrt{K(K + 2m_0c^2)}}$

For $v << c$, we have

$$K = m_0c^2\left(\frac{1}{\sqrt{1 - v^2/c^2}} - 1\right) \leq m_0c^2$$

$\therefore$ $\lambda = \frac{h}{\sqrt{2m_0K}}$

Also, for a particle $E = \sqrt{p^2c^2 + E_0^2}$

$$pc = \sqrt{E^2 - E_0^2}$$

When $E >> E_0$, we have $pc \cong E$. In this case the de-Broglie wavelength of the particle is

$$\lambda_p = \frac{h}{p} \text{ or } \lambda_p \cong \frac{hc}{E} \quad \ldots\text{(i)}$$

And for a photon, $E = pc$ and

$\therefore$ $\lambda_{ph} = \frac{h}{p} = \frac{hc}{E}$ ...(ii)

Hence, Eqs. (i) and (ii) are equal or, $\lambda_p \cong \lambda_{ph}$

20. (b,d) Scattering of electron is from a single crystal which is a set of many planes of atoms called the Bragg planes.

Kinetic energy of electron actually increases (as it undergoes a reverse process of photoemission) by an amount equal to work function of target crystal.

21. (a) Graph between V_0 and f is

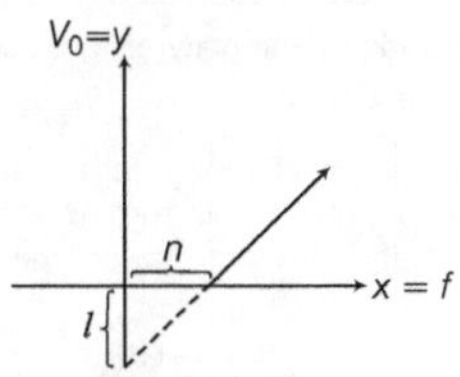

As, $V_0 = \frac{h}{e} \cdot f - \frac{\phi_0}{e}$

$\therefore$ Slope $= \frac{h}{e} = \frac{l}{n} \Rightarrow h = \frac{le}{n}$

Graph between K_{max} and f is

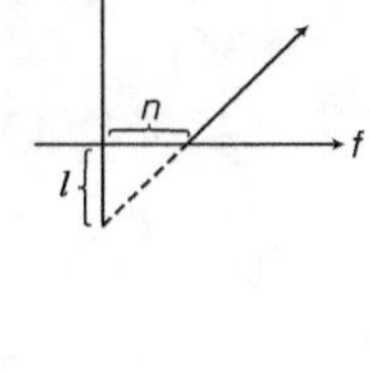

$$K_{max} = hf - \phi_0$$

$\therefore$ Slope $= h = \frac{l}{n}$

Graph between v_{max}^2 and f is

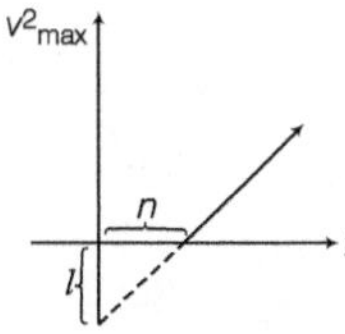

$$v^2 = \frac{2h}{m}f - \frac{2\phi_0}{m}$$

$\Rightarrow$ $\tan\theta = \frac{l}{n} = \frac{2h}{m} \Rightarrow \frac{lm}{2n}$

Hence, (i) $\rightarrow$ (p), (ii) $\rightarrow$ (q), (iii) $\rightarrow$ (r).

22. (d) Statement I is incorrect but Statement II is correct.
An isolated electron cannot absorb a photon (only a bound electron in crystal lattice can absorb a photon).
Total energy of an electron, $E = (m_0^2c^4 + p^2c^2)^{1/2}$

$$= m_0c^2 + hf$$

On squaring both sides, we get

$$p^2c^2 + 2m_0c^2pc = p^2c^2$$

or $2m_0c^2pc = 0$

but this is impossible. So, a free electron cannot absorb a photon.

23. (d) Statement I is incorrect. In reflection, frequency remains constant.

Thus, energy of reflected photon is same as the energy of incident photon but number of photons is not conserved. Thus, Statement II is correct.

24. (2) Number of photoelectrons emitted upto 10 s

= Number of photons falling over plate per unit area $\times$ area $\times$ time $\times \frac{1}{10^6}$

$\Rightarrow$ $\frac{1}{10^6} \times 10^{16} \times 5 \times 10^{-4} \times 10 = 5 \times 10^7$

At time $t = 10$ s,

Change on plate A,

$$q_A = ne$$

$$= 5 \times 10^7 \times 1.6 \times 10^{-19}$$

$$= 8 \times 10^{-12} \text{C}$$

Charge on plate B,

$$q_B = 33.7 \times 10^{-12} - 8 \times 10^{-12}$$

$$= 25.7 \times 10^{-12} \text{C}$$

So, electric field between plates

$$E = \frac{q_B - q_A}{2A\varepsilon_0}$$

$$= \frac{(25.7 - 8) \times 10^{-2}}{2 \times 5 \times 10^{-4} \times 8.85 \times 10^{-12}}$$

$$= 2 \times 10^3 \frac{\text{N}}{\text{C}}$$

So, increase in energy of photoelectrons

$$= qV = eEd$$

$$= (Ed)\text{eV}$$

$$= 2 \times 10^3 \times 10^{-2} \text{ eV}$$

$$= 20 \text{ eV}$$

$\therefore$ $k = 2$

JEE Main & AIEEE Archive

1. (b) When an electron moves in a circular path, then

Radius, $r = \dfrac{mv}{eB}$

$$\Rightarrow \quad \frac{r^2e^2B^2}{2} = \frac{m^2v^2}{2}$$

$$\text{KE}_{max} = \frac{(mv)^2}{2m}$$

$$\Rightarrow \quad \frac{r^2e^2B^2}{2m} = (\text{KE})_{max}$$

Work function of the metal (W),

i.e. $\quad W = h\nu - \text{KE}_{max}$

$$1.89 - \phi = \frac{r^2e^2B^2}{2m}\cdot\frac{1}{2}\text{eV}$$

$$= \frac{r^2eB^2}{2m}\text{eV}$$

[$h\nu \to 1.89$ eV, for the transition from third to second orbit of H-atom]

$$= \frac{100\times10^{-6}\times1.6\times10^{-19}\times9\times10^{-8}}{2\times9.1\times10^{-31}}$$

$$\phi = 1.89 - \frac{1.6\times9}{2\times9.1} = 1.89 - 0.79 = 1.1\text{ eV}$$

2. (d) As λ is increased, there will be a value of λ above which photoelectron will be cease to come out, so photocurrent will becomes zero.

3. (c) Davisson and Germer experimentally established wave nature of electron by observing diffraction pattern while bombarding electrons on Ni-crystal.

4. (c) $\quad K_{max} = eV_0 = h(\nu - \nu_0)$

If $\nu' = 2\nu$

$\therefore \quad K'_{max} = eV'_0 = h(2\nu - \nu_0) = 2K_{max} + h\nu_0$

$\therefore \quad K'_{max} > 2K_{max}$ and $\nu'_0 > 2\nu_0$

Alternate Method

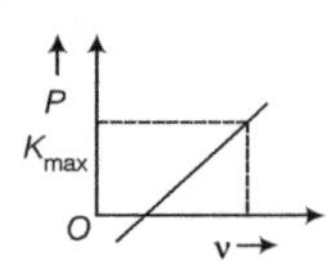

From graph, we can predict that K_{max} and ν do not correspond linear graph from origin.

5. (d) de-Broglie wavelength,

$$\lambda = \frac{h}{mv} = \frac{h}{p}$$

where, p = momentum

By conservation of momentum,

$$p_1 + p_2 = 0 \text{ or } p_1 = p_2$$

$$\Rightarrow \quad \frac{h}{p_1} = \frac{h}{p_2}$$

$$\therefore \quad \lambda_1 = \lambda_2 = \lambda \qquad \left(\because \lambda = \frac{h}{p}\right)$$

6. (a) As, power = Energy × time

or $\quad 4\times10^3 = hf\times10^{20} \qquad \left(\because \dfrac{\text{energy}}{\text{time}} = \text{power}\right)$

$$f = \frac{4\times10^3}{10^{20}\times6.023\times10^{-34}}$$

$$f = 6.64\times10^{16}\text{ Hz}$$

The obtained frequency lies in the band of X-rays.

7. (b) $\because \quad \text{KE}_{max} = eV_0$

$$\Rightarrow \quad \frac{1}{2}mv^2 = eV_0 = 1.68\text{ eV}$$

$$\Rightarrow \quad h\nu = \frac{hc}{\lambda} = \frac{1240\text{ eV-nm}}{400\text{ nm}} = 3.1\text{ eV}$$

$$\Rightarrow \quad 3.1\text{ eV} = W_0 + 1.6\text{ eV}$$

[from Einstein equation, $E = W_0 + K_{max}$]

$$W_0 = 1.42\text{ eV}$$

8. (b) For constructive interference,

$$2d\cos i = n\lambda = \frac{h}{\sqrt{2meV}}$$

On substituting values, we get

$$V \simeq 50\text{ V}$$

9. (b) Expression is given by $2d\cos i = n\lambda_{dB}$.

10. (d) As diffraction pattern has to be wider than slit width, so option (d) is the correct.

11. (d) The momentum of the photon

$$p = \frac{h}{\lambda} = \frac{h\nu}{c}$$

12. (a) $h\nu_0 = 6.2$ eV, $eV_0 = 5$ eV

From Einstein's photoelectric equation

$$h\nu = h\nu_0 + eV_0 = 6.2 + 5 = 11.2\text{ eV}$$

$$\Rightarrow \quad \frac{hc}{\lambda} = 11.2\text{ eV}$$

or $\quad \lambda = \dfrac{hc}{112} = 1108.9$ Å

which belongs to ultraviolet region.

13. (b) The photoelectric effect is an instantaneous phenomenon (experimentally proved). It takes approximate time of the order of 10^{-10} s.

14. (d) According to the photoelectric effect in a photocell, if a light of wavelength λ is incident on a cathode, then electrons are emitted, which constitute the photoelectric current.

Photocell is based on the principle of photoelectric effect. As the wavelength of light changes, there is no change in number of electrons emitted and hence, no change in current (plate current of photocell). Thus, the two wavelength of incident light and plate current are independent to each other.

Platecurrent depends on intensity of light used.

Hence, option (d) is correct.

15. (a) When **E**, **v** and **B** are all along same direction, then magnetic force experienced by electron is zero while electric force is acting opposite to velocity of electron, so velocity of electron will decrease.

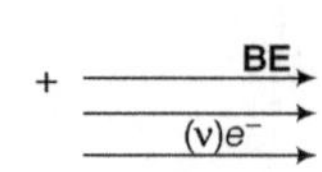

$\left(\begin{array}{l}\because \text{Lorentz force} = 0\\ \therefore \quad \theta = 0\\ \Rightarrow \quad qvB\sin\theta = 0\end{array}\right)$

16. (c) We know that, $\lambda = \dfrac{h}{mv}$

and $\quad K = \dfrac{1}{2}mv^2 = \dfrac{(mv)^2}{2m}$

$$\Rightarrow \quad mv = \sqrt{2mK}$$

Thus, $\quad \lambda = \dfrac{h}{\sqrt{2mK}} \Rightarrow \lambda \propto \dfrac{1}{\sqrt{K}}$

$$\Rightarrow \quad \frac{\lambda_2}{\lambda_1} = \frac{\sqrt{K_1}}{\sqrt{K_2}} = \frac{\sqrt{K_1}}{\sqrt{2K_1}} \qquad (\because K_2 = 2K_1)$$

or $$\frac{\lambda_2}{\lambda_1} = \frac{1}{\sqrt{2}}$$

or $$\lambda_2 = \frac{\lambda_1}{\sqrt{2}}$$

17. (b) We know intensity depends on source distance.

$\therefore$ $$\frac{I_2}{I_1} = \frac{(r_1)^2}{(r_2)^2} \qquad \left(\text{as } I \propto \frac{1}{r^2}\right)$$

$\Rightarrow$ $$\frac{I_2}{I_1} = \frac{(1)^2}{(1/2)^2}$$

$$I_2 = 4I_1$$

Now, since number of electrons emitted per second is directly proportional to intensity, so number of electrons emitted by photocathode would increase by a factor of 4.

18. (d) Einstein's photoelectric equation is

$$KE_{max} = h\nu - \phi \qquad \ldots(i)$$

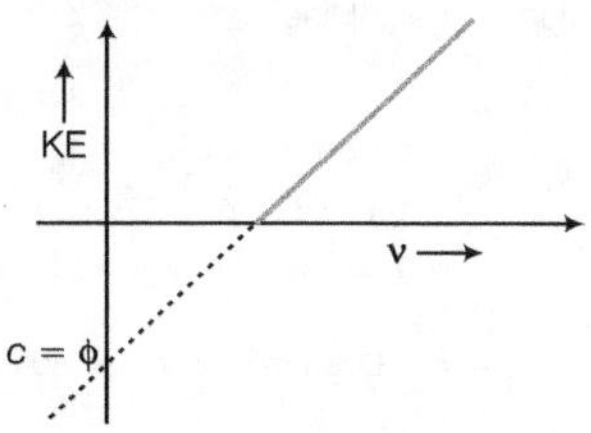

The equation of line is

$$y = mx + c \qquad \ldots(ii)$$

Comparing above two equations, we get

$$m = h, \quad c = -\phi$$

Hence, slope of graph is equal to Planck's constant (non-variable) and does not depend on intensity of radiation. But interception c depends on work function ϕ.

19. (c) $\frac{hc}{\lambda_0} = \phi_0 \Rightarrow \lambda_{max} = \frac{hc}{\phi_0}$

$$= \frac{6.6 \times 10^{-34} \times 3 \times 10^8}{4 \times 1.6 \times 10^{-19}}$$

$$= 310 \text{ nm}$$

20. (a) We know, Einstein's equations

$$hf = hf_0 + \frac{1}{2}mv^2$$

(hf: Energy of incident light; hf_0: Work function; $\frac{1}{2}mv^2$: Kinetic energy)

$\Rightarrow$ $$v_1^2 = \frac{2hf_1}{m} - \frac{2hf_0}{m}$$

and $$v_2^2 = \frac{2hf_2}{m} - \frac{2hf_0}{m}$$

$\therefore$ $$v_1^2 - v_2^2 = \frac{2h}{m}[f_1 - f_2]$$

21. (a) Formation of covalent bonds due to the wave nature of electrons.

22. (c) Work function, $W = \frac{hc}{\lambda}$

(here, they are interested in asking threshold wavelength)

where, h = Planck's constant,

c = velocity of light.

Therefore, $$\frac{W_{Na}}{W_{Cu}} = \frac{\lambda_{Cu}}{\lambda_{Na}}$$

or $$\frac{\lambda_{Na}}{\lambda_{Cu}} = \frac{W_{Cu}}{W_{Na}} = \frac{4.5}{2.3} = 2 \text{ (nearly)}$$

JEE Advanced & IIT JEE Archive

1. (a,c) **Key concept** It is based on photoelectric effect, a part of the energy of the photon is used in liberating the electron from the metal surface which is equal to the work function W_0 of the metal while remaining energy is used in imparting kinetic energy to the ejected electron.

For photoelectric effect with incident photon having wavelength λ. Stopping potential is V_0.

Such that, work function (ϕ) by a photon is given by

$$\phi = h\nu - eV_0$$

$\Rightarrow$ $$h\nu = \phi + eV_0$$

$\Rightarrow$ $$\frac{hc}{\lambda} = \phi + eV_0$$

$\Rightarrow$ $$V_0 = \frac{ch}{e\lambda} - \frac{\phi}{e}$$

So, V_0 *versus* λ curve will be hyperbola and V_0 *versus* $1/\lambda$ curve will be straight line with positive slope.

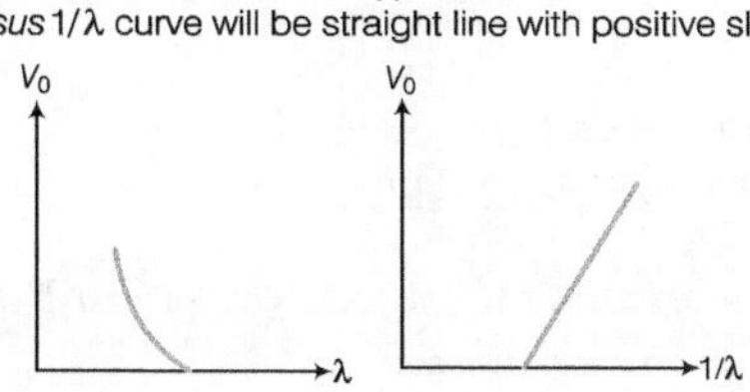

2. (2) According to Bohr's second postulate, we have

$$mvr = \frac{nh}{2\pi} = \frac{3h}{2\pi} \qquad \text{(given)}$$

So, value of n is 3.

So, angular momentum of an electron, $mvr = \frac{3h}{2\pi}$ i.e. $mv = \frac{3h}{2\pi r}$

Now, radius of nth shell, $r = \left(\frac{n^2}{z}\right)a_0$

i.e. $$r = \frac{(3)^2}{3} . a_0 \qquad (\because z_{Li} = 3)$$

$$r = 3a_0$$

According to de-Broglie wavelength of an electron, we get

$$\lambda = \frac{h}{mv}$$

$$\lambda = \frac{h}{\frac{3h}{2\pi r}} = \frac{2\pi r}{3}$$

i.e. $$\lambda = \frac{2\pi \times 3a_0}{3} = 2\pi a_0$$

$\Rightarrow$ $$\lambda = 2\pi a_0 \qquad \ldots(i)$$

Given that de-Broglie wavelength of the electron in Li^{2+} is $P\pi a_0$.

Comparing above relation with Eq. (i), we get

$$\lambda = 2\pi a_0 = P\pi a_0 \Rightarrow P = 2$$

3. (1) $eV_0 = hf - W$

$$\therefore \quad V_0 = \left(\frac{h}{e}\right) f - \frac{W}{e}$$

V_0 *versus* f graph is a straight line with slope $= \frac{h}{e}$ = a universal constant. Therefore, the ratio of two slopes should be 1.

4. (7) Photoemission will stop when potential on silver sphere becomes equal to the stopping potential.

$$\therefore \quad \frac{hc}{\lambda} - W = eV_0$$

Here, $$V_0 = \frac{1}{4\pi\varepsilon_0} \frac{ne}{r}$$

$$\therefore \quad \left(\frac{1240}{1200} \text{eV}\right) - (4.7 \text{ eV}) = \frac{9 \times 10^9 \times n \times 1.6 \times 10^{-19}}{10^{-2}}$$

$$(6.2 - 4.7) = \frac{9 \times 10^9 \times n \times 1.6 \times 10^{-19}}{10^{-2}}$$

or $$n = \frac{1.5 \times 10^{-2}}{9 \times 1.6 \times 10^{-10}} = 1.04 \times 10^7$$

$$\therefore \quad Z = 7$$

5. (3) $\lambda = \frac{h}{p} = \frac{h}{\sqrt{2qqVm}}$ or $\lambda \propto \frac{1}{\sqrt{qm}}$

$$\frac{\lambda_p}{\lambda_\alpha} = \sqrt{\frac{q_\alpha}{q_p} \cdot \frac{m_\alpha}{m_p}} = \sqrt{\frac{(2)(4)}{(1)(1)}} = 2.828$$

The nearest integer is 3.

6. (a)

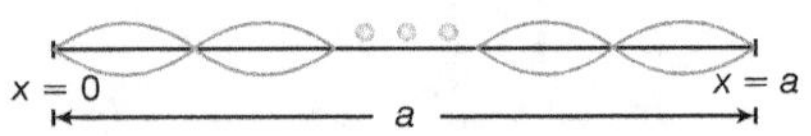

$$a = \frac{n\lambda}{2}$$

$$\therefore \quad \lambda = \frac{2a}{n} = \frac{h}{p} = \frac{h}{\sqrt{2Em}} \quad \text{...(i)}$$

or $$\sqrt{E} \propto \frac{1}{a}$$

$$\therefore \quad E \propto \frac{1}{a^2}$$

7. (b) From Eq. (i), $E = \frac{n^2h^2}{8a^2m}$

In ground state, $n = 1$

$$\therefore \quad E_1 = \frac{h^2}{8ma^2}$$

Substituting the values, we get

$$E_1 = 8 \text{ MeV}$$

8. (d) From Eq. (i), $\quad p \propto n$

$$\therefore \quad mv \propto n \text{ or } v \propto n$$

9. (a) $$E_1 = \frac{1240}{550} = 2.25 \text{ eV}$$

$$E_2 = \frac{1240}{450} = 2.75 \text{ eV}$$

$$E_3 = \frac{1240}{350} = 3.54 \text{ eV}$$

E_1 cannot emit photoelectrons from q and r plates. E_2 cannot emit photoelectrons from r.

Further, work function of p is least and it can emit photoelectrons from all three wavelengths. Hence, magnitude of its stopping potential and saturation current both will be maximum.

10. (a) Momentum of striking electrons,

$$p = \frac{h}{\lambda}$$

$\therefore$ Kinetic energy of striking electrons

$$K = \frac{p^2}{2m} = \frac{h^2}{2m\lambda^2}$$

This is also, maximum energy of X-ray photons.

Therefore, $$\frac{hc}{\lambda_0} = \frac{h^2}{2m\lambda^2}$$

or $$\lambda_0 = \frac{2m\lambda^2 c}{h}$$

11. (a,d) From the relation, $eV = \frac{hc}{\lambda} - \phi$

or $$V = \left(\frac{hc}{e}\right)\left(\frac{1}{\lambda}\right) - \frac{\phi}{e}$$

This is equation of straight line.

Slope is $\tan\theta = \frac{hc}{e}$

$$\phi_1 : \phi_2 : \phi_3 = \frac{hc}{\lambda_{01}} : \frac{hc}{\lambda_{02}} : \frac{hc}{\lambda_{03}} = \frac{1}{\lambda_{01}} : \frac{1}{\lambda_{02}} : \frac{1}{\lambda_{03}} = 1 : 2 : 4$$

$$\frac{1}{\lambda_{01}} = 0.001 \text{ nm}^{-1} \text{ or } \lambda_{01} = 10000 \text{ Å}$$

$$\frac{1}{\lambda_{02}} = 0.002 \text{ nm}^{-1} \text{ or } \lambda_{02} = 5000 \text{ Å}$$

$$\frac{1}{\lambda_{03}} = 0.004 \text{ nm}^{-1} \text{ or } \lambda_{03} = 2500 \text{ Å}$$

Violet colour has wavelength 4000 Å.

So, violet colour can eject photoelectrons from metal-1 and metal-2.

12. (c) The first photon will excite the hydrogen atom (in ground state) to first excited state (as $E_2 - E_1 = 10.2$ eV). Hence, during de-excitation a photon of 10.2 eV will be released. The second photon of energy 15 eV can ionise the atom. Hence the balance energy i.e. $(15 - 13.6)$ eV $= 1.4$ eV is retained by the electron. Therefore, by the second photon an electron of energy 1.4 eV will be released.

13. (24) nth line of Lyman series means transition from $(n + 1)$th state to first state.

$$\frac{1}{\lambda} = RZ^2\left[1 - \frac{1}{(n+1)^2}\right] \quad \text{...(i)}$$

de-Broglie wavelength in $(n + 1)$th orbit :

$$\lambda = \frac{h}{mv} = \frac{hr}{mvr} = \frac{(2\pi)(hr)}{(n+1)h} = \frac{2\pi r}{(n+1)}$$

or $$\frac{1}{\lambda} = \frac{(n+1)}{2\pi r} \quad \text{...(ii)}$$

Equating Eqs. (i) and (ii), we get

$$\left(\frac{n+1}{2\pi r}\right) = RZ^2\left[\frac{n(n+2)}{(n+1)^2}\right] \quad \text{...(iii)}$$

Now, as $$r \propto \frac{n^2}{Z}$$

$$\therefore \quad r = \frac{(n+1)^2}{11} r_0$$

Substituting in Eq. (iii), we get

$$\frac{11}{2\pi r_0} = \frac{R(11)^2 (n)(n+2)}{(n+1)}$$

or $\quad (n + 1) = (1.09 \times 10^7)(11)(2\pi) \times (0.529 \times 10^{-10})(n^2 + 2n)$

Solving this equation, we get

$$n = 24$$

2. Atoms

DPP-1 Early Atomic Models

1. (i) For a particle with scattering angle θ and impact parameter b,

$$b = \frac{Ze^2 \cot(\theta/2)}{4\pi\varepsilon_0 K}, \; K = \text{Kinetic energy of } \alpha\text{-particle}$$

For $b = 0$, α-particle is colliding 'head-on' with nucleus.

With $b = 0$, above formula gives

$\cot\theta/2 = 0 \Rightarrow \theta/2 = 90^\circ$ or $\theta = 180^\circ$

So, particle is deflected back on its path.

(ii) For a constant (given) b, increase in energy K gives an increase in value of $\cot(\theta/2)$. So, θ decreases.

(iii) For $Z = 79$, $K = 10 \times 10^6$eV and $\theta = 90^\circ$, we have,

$$\cot\theta/2 = \cot 90^\circ/2 = \cot 45^\circ = 1$$

So,
$$b = \frac{Ze^2(\cot\theta/2)}{4\pi\varepsilon_0 K}$$
$$= \frac{9 \times 10^9 \times 79 \times (1.6 \times 10^{-19})^2 \times 1}{10 \times 10^6 \times 1.6 \times 10^{-19}}$$
$$= 1.1 \times 10^{-14}\,\text{m}$$

(iv) As deflection is due to electrostatic force only (at $Z = 0$, $\theta = 0$)

∴ Mass of nucleus does not comes into the formula as recoil of nucleus is neglected.

(v) For a given energy as b decreases, θ increases.

(vi)

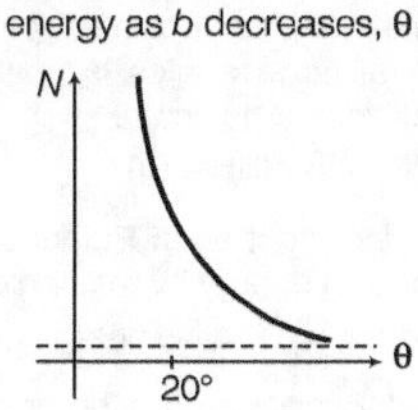

As, $N \propto \frac{1}{(\sin\theta)^4}$

So, graph is very steep.

2. (d) Dalton assumed atom to be smallest particle of matter, which is indivisible further.

3. (c) Atom's size is same in both Thomson's and Rutherford's model. They differs only in arrangement of electrons.

4. (c) For backward scattering $\theta > 90^\circ$, a very strong repulsion is required which is possible only when positive charge is concentrated in a very small region.

5. (c) Scattering is due to a single collision (in moderate thick sheets) and probability of a single collision increases linearly with thickness. Otherwise, if sheet is very thin then particle go straight without any collision (zero deflection).

6. (d) When particle passes through the region of electric field, it has two velocities.

∴ Path is parabolic.

But path of particle becomes again straight line when it is free of electric field's influence.

7. (d) By law of sines, $\frac{\Delta p}{\sin\theta} = \frac{mv}{\sin\left(\frac{\pi - \theta}{2}\right)}$

As,
$$\sin\left(\frac{\pi - \theta}{2}\right) = \cos\frac{\theta}{2}$$

and
$$\sin\theta = 2\sin\frac{\theta}{2}\cos\frac{\theta}{2}$$

We have
$$\frac{\Delta p}{2\sin\frac{\theta}{2}\cos\frac{\theta}{2}} = \frac{mv}{\cos\frac{\theta}{2}}$$

$\Rightarrow$
$$\Delta p = 2mv \cdot \sin\frac{\theta}{2}$$

8. (d) Relation between scattering angle and impact parameter is

$$\cot\frac{\theta}{2} = \frac{2\pi\varepsilon_0 mv^2}{Ze^2} \cdot b$$

or
$$\cot\frac{\theta}{2} \propto b$$

9. (a)

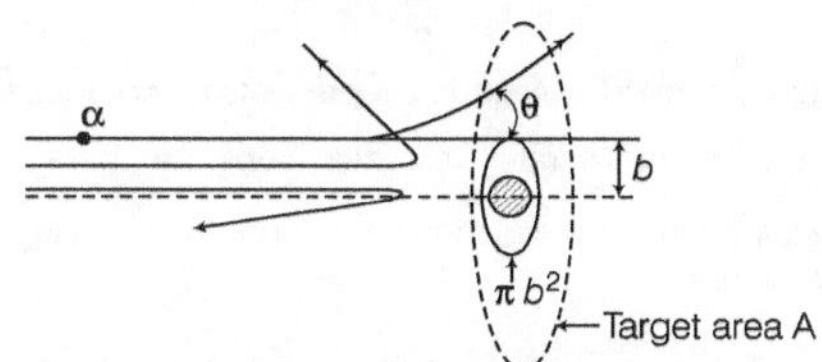

α-particle is scattered before it reaches near to nucleus and hence, it does not pass with in a distance b of it.

Now, fraction of α-particles, which are scattered by angle θ or more is

$$f = \frac{\alpha\text{-particles scattered by angle } \theta \text{ or more}}{\text{Total member of incident } \alpha\text{-particles}}$$
$$= \frac{\text{Aggregate cross-section}}{\text{Target area}}$$

Now, when α-particle beam is targetted over area A, it encounters ntA nuclei. But an α-particle which is initially directed anywhere in area πb^2 is scattered, through θ or more.

where,
$$\cot\frac{\theta}{2} = \frac{2\pi\varepsilon_0 \, mv^2}{Ze^2} \cdot b$$

So,
$$f = \frac{\text{Aggregate cross-section}}{\text{Area}} = \frac{ntA \times \pi b^2}{A}$$
$$= nt\pi b^2$$

10. (b) For no deflection, $\frac{e}{m} = \frac{E^2}{2VB^2}$

When m is increased by 208 times, then keeping other things constant, B^2 must be increased 208 times, or B is increased $\sqrt{208} = 14.4$ times.

11. (a) $\frac{1}{2}mv^2 = \text{eV}$ and $v = \frac{E}{B} = \frac{V}{Bd}$

where, d = Distance between capacitor plates.

$\Rightarrow$
$$\frac{1}{2}m\left(\frac{V^2}{B^2d^2}\right) = \text{eV}$$

$\Rightarrow$
$$V \propto B^2 \quad \text{or} \quad B = K\sqrt{V}$$

So, magnetic field must be increased by a factor of $\sqrt{2}$.

Explanations for Q Nos. (12 and 13)

Force due to electric field $F_e = qE$ in $+y$-direction

Acceleration of ion due to

$$\mathbf{E} = a_y = \frac{F_e}{m} = \frac{q}{m} \cdot E$$

Deflection in y-direction $= S_y = u_y t + \frac{1}{2}a_y t^2$

$$= 0 \cdot t + \frac{1}{2}\frac{qE}{m} \cdot \left(\frac{L}{v_x}\right)^2 = \frac{qEL^2}{2mv^2} \quad \text{...(i)}$$

Force due to magnetic field

$$F_m = q(\mathbf{v}_x \times \mathbf{B})$$
$$= q(v\hat{\mathbf{i}} \times B\hat{\mathbf{j}}) = Bqv(\hat{\mathbf{k}})$$

This causes a deflection in z-direction

$$= S_z = u_z t + \frac{1}{2}a_z t^2$$
$$= \frac{1}{2}\frac{Bqv}{m}\cdot\frac{L^2}{v^2} = \frac{BqL^2}{2mv} \quad \text{...(ii)}$$

Now, substituting v from Eqs. (ii) and (i), we get

$$S_y \cdot \frac{B^2q^2L^4}{4m^2S_z^2} = \frac{qEL^2}{2m}$$
$$\Rightarrow \quad S_y = 2\left(\frac{m}{q}\right)\left(\frac{E}{B^2L^2}\right)\cdot S_z^2$$
$$\Rightarrow \quad S_z^2 = \left(\frac{1}{2}\cdot\frac{q}{m}\cdot\frac{B^2L^2}{E}\right)\cdot S_y$$

which is of form $z^2 = 4ay$ and it represents a parabola. All positive ions with same $\frac{q}{m}$ ratio lies over same parabola. For lighter particles parabola is much broader and so have larger value of Latusrectum.

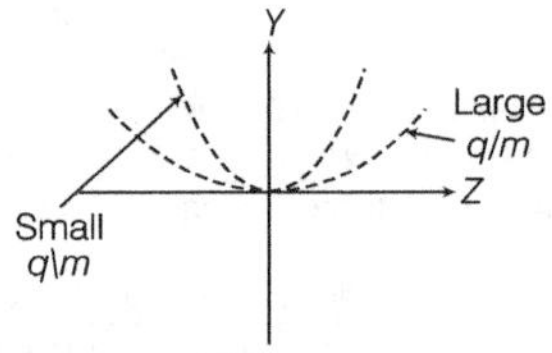

12. (a)

13. (a)

14. (b) Impact parameter is

$$b = \frac{Z_1Z_2e^2}{4\pi\varepsilon_0} \times \frac{1}{2K} \times (\cot\theta/2)$$

For a head on collision, distance of closest approach

$$d_0 = \frac{Z_1Z_2e^2}{4\pi\varepsilon_0} \times \frac{1}{K}$$

So, from above expressions we have,

$$b = d_0 \times \frac{1}{2\tan(\theta/2)}$$

Now, $$d_0 = \frac{1}{4\pi\varepsilon_0} \times Z_1Z_2 \times e^2 \times \frac{1}{K}$$
$$= \frac{9 \times 10^9 \times 2 \times 47 \times (1.6 \times 10^{-19})^2}{4.8 \times 10^6 \times 1.6 \times 10^{-19}}$$
$$= 2.8 \times 10^{-14}\,\text{m}$$

$\therefore$ $$b = \frac{2.8 \times 10^{-14}}{2 \times (\sqrt{2} - 1)} = 3.36 \times 10^{-14}\,\text{m}$$

15. (d) $\Delta p = 2p\sin\theta/2$

$$= \frac{2K}{m}\cdot\sin\frac{\theta}{2}$$
$$\approx 2.3 \times 10^{14}\,\text{kg-ms}^{-1}$$

16. (a,b,c) Except option (d) all are still correct. Isotopes are discovered after Dalton's atomic theory.

17. (b,c,d) Most of α-particles pass straight results that most of the matter is empty. Also, atoms have positively charged centre which causes deflection.

18. (c,d) A loss of energy by accelerating electron continuously leads to spiral motion of electron with a "continuous" emission spectrum.

19. (a,b) Till experimental confirmation of finding charged particle (electron) in atom, atom is assumed indivisible (Dalton's model).

20. (a) Most of α-particles pass straight as if gold foil is not there. Deflection as small angles indicates presence of a positive charge. Deflection at large angles is possible due to a highly concentrated positive charge centre.
Hence, (i) $\rightarrow$ (r), (ii) $\rightarrow$ (p), (iii) $\rightarrow$ (q)

21. (d) Statement I is incorrect as in Rutherford's model there is a continuous emission of energy by rotating electron.
Statement II is correct, a negative energy means a stable orbit.

22. (d) Statement I is incorrect as at high and moderate pressures gases are insulators and hence, no ionisation is possible. Statement II is correct.

DPP-2 **Bohr's Theory for Hydrogen and Hydrogen-like Atoms**

1. (i) When atom is bounded by gravitational force, then

$$\frac{m_e v^2}{r} = \frac{Gm_e m_p}{r^2} \quad \text{...(i)}$$

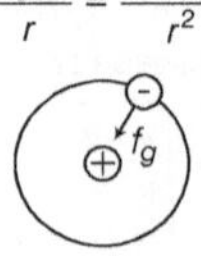

and $$m_e vr = \frac{nh}{2\pi}$$

For first Bohr's orbit, $n = 1$

$$\Rightarrow \quad m_e v_e r = \frac{h}{2\pi} \quad \text{...(ii)}$$

From Eqs. (i) and (ii), we get

$$r = \frac{(h/2\pi)^2}{Gm_p m_e^2} \approx 1.2 \times 10^{29}\,\text{m}$$

This is much greater than the estimated size of whole universe.

(ii) First we find orbital speed of the earth using data from gravitation chapter.
Then, orbital angular momentum associated with earth will be mvr assuming a circular orbit.
Angular momentum you find is relatively large in the order of $\left(10^{70} \times \frac{h}{2\pi}\right)$.
So $n \approx 10^{70}$; This is a very large value.

(iii) As for such large values of n, the difference in successive energies for $n = 10^{70}$ and $n = (10^{70}) + 1$, are so small compared to energy of any of level, so, for all practical purposes levels are considered continuous.

2. Ground state energy of each electron in He-atom

$$= -Z_{\text{eff}}^2 \times 13.6\,\text{eV}$$

So, for a 2-electron atom, total energy will be around

$$= 2 \times -Z_{\text{eff}}^2 \times 13.6$$
$$= -Z_{\text{eff}}^2 \times 27.2\,\text{eV}$$

Energy of He^+ (when it is ionised singly)

$= -(2)^2 \times 13.6$ eV

$= -54.4$ eV (now there is no screening, so $Z_{eff} = 2$).

Hence, ionisation energy $= (-54.4) - (-Z_{eff}^2 \times 27.2)$ eV

But ionisation energy measured experimentally = 24.46 eV

Equating both we get,

$$-54.4 + Z_{eff}^2 \times 27.2 = 24.46$$

$\Rightarrow \quad Z_{eff} = 1.70$

So, $\quad Z_{eff} \neq 2$ as we expected.

3. (b) When an atom that is in excited state E' and at rest in our frame of reference, emits a photon of energy hf, the photon also carries a momentum $p = \frac{hf}{c} = \frac{hc}{\lambda}$. Conservation of momentum requires the atom must recoil with a momentum p and so, will have a kinetic energy $\frac{p^2}{2m}$ where, m = mass of atom.

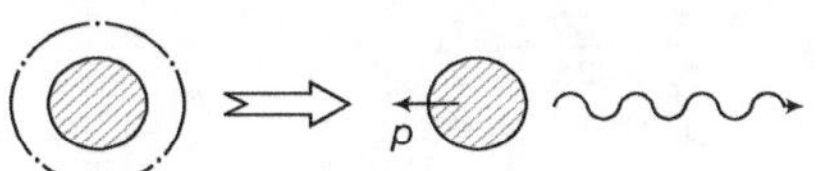

Excited atom Energy (E') | Recoiling atom in ground state Energy (E) | Photon of energy (hf)

So, conservation of energy gives :

$$\Delta E = E' - E$$
$$= hf + \text{recoil KE of atom}$$

$\Rightarrow \quad hf = \Delta E - \frac{(p)^2}{2m}$

$\Rightarrow \quad hf < \Delta E$

Note : *In Bohr's analysis recoil of atom is not considered.*

4. (b) Only momentum and energy are conserved. Mass is not conserved, a certain part of mass is usually converted into mass-energy $E = mc^2$.

5. (a) As energy of electrons is

$$E = 1.6 \times 10^{-19} \times 10^4$$
$$= 1.6 \times 10^{-15} \text{eV}$$

It is quite less for excitation of H_2 atom (which are in ground state). So, collisions are elastic.

6. (a) $\frac{hc}{\lambda} = 11$ eV

$$hc \approx 1240 \text{ eV-nm}$$

$\Rightarrow \quad \lambda = \frac{1240}{11}$ nm

$$= 112 \times 10^{-9} \text{ m}$$
$$= 1.12 \times 10^{-8} \text{m}$$

which is in ultraviolet region of spectrum.

7. (b) When two atoms collide inelastically, energy lost due to collision excites the colliding atoms.

Hence, there is energy lost (maximum)

$= 2 \times$ energy of excitation of atoms to their first excitation position

$= 2 \times 10.2$

$= 20.40$ eV

8. (a) In Bohr's atomic model only orbital angular momentum is considered. No account is taken for spin or orientation of electron and so, Bohr's model gives inaccurate value of angular momentum.

9. (c) In a multi-electron atom, electrons are not subjected to a central force as it is assumed in Bohr's atomic model.

10. (c) When an electron and a proton are infinitely apart, the energy of the system is zero.

$$\ominus \longleftarrow \infty\text{-distance} \longrightarrow \oplus$$

When they come together to form a hydrogen atom, the total energy of system = – 13.6 eV. This means + 13.6 eV of energy is released.

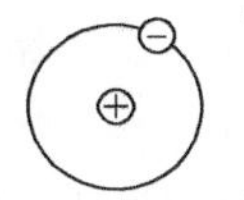

A stable H-atom is formed with release of 13.6 eV energy.

So, by Einstein's mass-energy equivalence, $E = mc^2$,

Mass of H-atom is

$$M = m_e + m_p - \frac{13.6 \times 1.6 \times 10^{-19}}{c^2}$$

$\Rightarrow \quad M = m_e + m_p - \frac{13.6 \times 1.6 \times 10^{-19}}{(3 \times 10^8)^2}$

$$= (m_e + m_p - 2.42 \times 10^{-35}) \text{ kg}$$

11. (a) Let the speed of electron = v

and Bohr's radius = a_0

Then, number of rounds made by electron in 1s = $\frac{2\pi a_0}{v}$

So, current constituted by electron

$$= \frac{2\pi a_0}{v} \cdot e$$
$$= \frac{2\pi \times 0.53 \times 10^{-10} \times 1.6 \times 10^{-19}}{2.2 \times 10^6}$$
$$\approx 2.42 \times 10^{-35} \text{ A}$$

12. (a,b,c,d) Bohr's has a definite radius and definite speed. So, an electron is known both in momentum and position which is violation of Heisenberg's principle.

Bohr's atom does not account for spin of electron and so, gives incorrect value of angular momentum of electron.

Bohr's atomic model is a classical model, which does not give explaination of relative intensities and is not applicable to atom which are not of Hydrogen-like atomic structure.

13. (a,c,d) Given, $E_3 - E_2 = 47.2$

$\Rightarrow \quad \left(-13.6\frac{Z^2}{9}\right) - \left(-13.6\left(\frac{Z^2}{4}\right)\right) = 47.2$

$\Rightarrow \quad 13.6\, Z^2\left(\frac{1}{4} - \frac{1}{9}\right) = 47.2$

$\Rightarrow \quad 13.6\, Z^2\left(\frac{5}{36}\right) = 47.2$

$\Rightarrow \quad Z^2 = \frac{47.2 \times 36}{13.6 \times 5} = 25$

$\Rightarrow \quad Z = 5$

In ground state,

$$\text{Total energy} = E_1 = \frac{-13.6Z^2}{n^2}$$
$$= -13.6 \times 5^2$$
$$= -340 \text{ eV}$$

So, $\quad KE = -E_1 = 340$ eV

and $\quad PE = 2E_1 = 2(-340)$

$$= -680 \text{ eV}$$

14. (a,b,d) In Bohr's atom, total energy of electron in ground state is –13.6 eV.

$\Rightarrow$ Potential energy in first orbit

$$= -2 \times 13.6$$
$$= -27.20 \text{ eV}$$

Now, total energy of electron in second orbit

$$= \frac{-13.6}{n^2} = \frac{-13.6}{4}$$
$$= -3.4 \text{ eV}$$

So, its PE in second orbit

$$= -2 \times 3.4 = -6.8 \text{ eV}$$

and its KE = 3.4 eV

If PE in first orbit is to make zero, then we have to increase potential energy by 27.20 eV.

This makes PE in second orbit

$$= (+27.20) + (-6.8) = 20.40 \text{ eV}$$

And total energy = PE + KE

$$= 20.40 + 3.4 = 23.80 \text{ eV}$$

15. (a,d) Loss of KE of collision will be used to excite the atom from one level to another.

So, ΔE = 0, 10.2 eV, 12.09 eV, 12.75 eV, 13.06 eV, ..., 13.6 eV

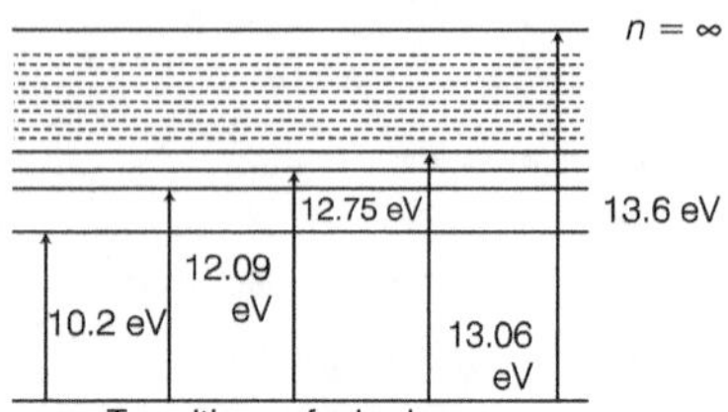

Transitions of a hydrogen atom

Let v_0 = initial speed of neutron and let an perfectly inelastic collision results. If v_f = final common speed, then by momentum conservation.

$$mv_0 = mv_f + mv_f$$

$$\Rightarrow \quad v_f = \frac{v_0}{2}$$

So, $$\text{final KE} = \frac{1}{2} \times 2m \times \frac{v_0^2}{4} = \frac{K_i}{2}$$

So, maximum loss of KE $= \Delta K = K_i - K_f$

$$= K - \frac{K}{2} = \frac{K}{2}$$

So, possible values of ΔK are 0 and $\frac{K}{2}$.

$$\Rightarrow \quad \Delta K = \left[0, \frac{K}{2}\right]$$

So, for $K < 20.4$, $\Delta K < 10.2$ eV, which is possible when collision is elastic.

Hence, collision is always elastic for $K = 20.4$, $\Delta K = 0$ or 10.2 eV.

$\Rightarrow$ Collision may perfectly inelastic or perfectly elastic.

For $K = 24.18$ eV,

$\Delta K = [0, 10.20, 12.09]$

When $\Delta K = 0$, collision is in elastic.

When $\Delta K = 10.20$, collision is inelastic.

When $\Delta K = 12.09$, collision is perfectly inelastic.

16. (a,b) Let both nucleus of charge (Ze and mass M), and electron of mass m revolves about their centre of mass.

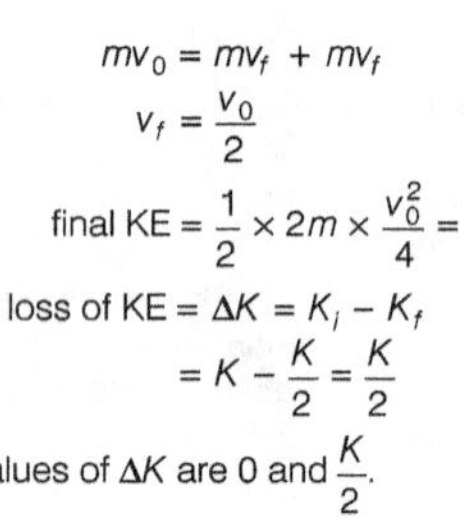

Now, $Mr_1 = mr_2$ and $r_1 + r_2 = r$

$$\Rightarrow \quad r_1 = \frac{mr}{M+m} \text{ and } r_2 = \frac{Mr}{M+m}$$

Centripetal force is due to electrostatic force so, CF on electron

$$= mr_2\omega^2 = \frac{1}{4\pi\varepsilon_0} \cdot \frac{Ze^2}{r^2}$$

$$\Rightarrow \quad m\left(\frac{Mr}{M+m}\right)\omega^2 = \frac{1}{4\pi\varepsilon_0} \cdot \frac{Ze^2}{r^2}$$

$$\text{or} \quad \left(\frac{Mm}{M+m}\right) r^3\omega^2 = \frac{e^2}{4\pi\varepsilon_0} \quad \text{...(i)}$$

Let $\frac{Mm}{M+m} = \mu$ = reduced mass $= \frac{2m}{3}$

By Bohr's postulate,

$$\frac{nh}{2\pi} = \text{Angular momentum}$$

$$\Rightarrow \quad \frac{nh}{2\pi} = I\omega = (Mr_1^2 + mr_2^2)\omega$$

$$\Rightarrow \quad \frac{nh}{2\pi} = \mu r^2 \omega \quad \text{...(ii)}$$

From Eqs. (i) and (ii), we get

$$r = \frac{\varepsilon_0 n^2 h^2}{\pi \mu e^2 Z} = \frac{mn^2}{\mu Z} \cdot a_0$$

$$= \frac{m(M+m)}{Mm} \cdot \frac{n^2}{Z} \cdot a_0$$

$$= \frac{3}{2} \cdot \frac{n^2}{Z} \cdot a_0 = 0.795 \frac{n^2}{Z} (\text{Å})$$

Also, $$\text{total energy} = -13.6 \times \frac{Z^2}{n^2} \times \frac{\mu}{m}$$

$$= -13.6 \times \frac{Z^2}{n^2} \times \frac{2}{3} = -9.1 \frac{Z^2}{n^2} \text{ (eV)}$$

17. (d) Magnetic moment,

$$M_n = i_n A = ef_n A$$

$$= \frac{mZ^2e^5}{4\varepsilon_0^2 n^3 h^3} \cdot \pi r_n^2$$

$$= \frac{mZ^2e^5}{4\varepsilon_0^2 n^3 h^3} \cdot \pi \cdot \left(\frac{n^2 h^2 \varepsilon_0}{\pi m Z e^2}\right)^2$$

$$\Rightarrow \quad M_n \propto n'$$

Also, $$\mathbf{B}_n = \frac{\pi m^2 Z^3 e^7 \mu_0}{8\varepsilon_0^3 n^5 h^5}$$

$$\Rightarrow \quad B_n \propto \frac{1}{n^5}$$

and $$\omega_n = \frac{n}{r_n} = \frac{\pi m Z^2 e^4}{2\varepsilon_0^2 n^3 h^3}$$

$$\Rightarrow \quad \omega_n \propto \frac{1}{n^3}$$

and $$i_n = e \cdot f_n = \frac{mZ^2e^5}{4\varepsilon_0^2 n^3 h^3}$$

$$\Rightarrow \quad i_n \propto \frac{1}{n^3}$$

Hence, (i) $\rightarrow$ (s), (ii) $\rightarrow$ (p), (iii) $\rightarrow$ (q), (iv) $\rightarrow$ (q).

18. (4) [Ionisation energy = 4*Ryd*]

Note *1 Rydberg* $= 2.2 \times 10^{-18}$ J $= Rhc$.

Radius of an atom $= \frac{r_H}{Z} = \frac{0.53\text{Å}}{Z}$

Now, $$\frac{0.53 \times 10^{-10}}{Z} = 0.265 \times 10^{-20}$$

$$\Rightarrow \quad Z = \frac{0.53}{0.265} = 2$$

Now, transition energy $\Delta E = E_2 - E_1 = \frac{E_1}{2^2} - E_1$

$$\Rightarrow \quad -\frac{3}{4}E_1 = \frac{hc}{\lambda} = \frac{12375}{300}$$

$$\Rightarrow \quad E_1 = \frac{-4 \times 12375}{3 \times 300} = -55 \text{ eV}$$

So, ionisation energy = 55 eV

$$= \frac{55 \times 1.6 \times 10^{-19}}{2.2 \times 10^{-18}} \text{ Ryd}$$

$$= 4 \text{ Ryd}$$

19. (5) In nth orbit, centripetal force = electrostatic attraction

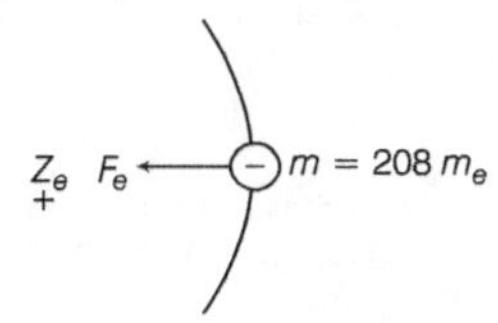

$$\Rightarrow \quad \frac{mv^2}{r} = \frac{1}{4\pi\varepsilon_0} \cdot \frac{Ze^2}{r^2} \quad \ldots\text{(i)}$$

Also, $$mvr = \frac{nh}{2\pi} \quad \ldots\text{(ii)}$$

On solving Eqs. (i) and (ii) we get,

$$r_n = \frac{n^2h^2\varepsilon_0}{Z\pi me^2}$$

With $Z = 3$ and $m = 208\, m_e$ we get

$$r_n = \frac{1}{624} \cdot \left(\frac{n^2h^2\varepsilon_0}{\pi m_e e^2}\right)$$

Now, Bohr's radius $a_0 = \frac{h^2\varepsilon_0}{\pi m_e e^2}$

$$n^2 = 624$$

$$\Rightarrow \quad n \approx 25$$

$$\Rightarrow \quad n/5 = 5$$

20. (a) Both Statement I and Statement II are correct and Statement II explains Statement I.

$$r_n \propto n^2$$

$$r_1 = \frac{0.53}{Z}$$

$$r_n = \frac{0.53}{Z} n^2 \Rightarrow \frac{r_n}{r_1} = n^2$$

$$\Rightarrow \quad \log_e\left(\frac{r_n}{r_1}\right) = 2\log_e n$$

This is a graph through origin with slope 2.

21. (d) Statement I is incorrect but Statement II is correct.

Using, $$Nx = N_0 \exp(-\Delta E/kT)$$

$$kT = 1.38 \times 10^{-23} \times 300$$

$$= 4.14 \times 10^{-21}$$

As $T = 300$ K = 27°C mean room temperature

$$\Delta E = E_x - E_0 = E_1 - E_0$$

$$= \left(\frac{-13.6}{4}\right) - \left(\frac{-13.6}{1}\right)$$

$$= 10.2 \text{ eV}$$

$$= 10.2 \times 1.6 \times 10^{-19} \text{ J}$$

Now, $$\frac{\Delta E}{kT} = \frac{10.2 \times 1.6 \times 10^{-19}}{4.14 \times 10^{-21}}$$

$$= 3.94 \times 10^2$$

So, number of atoms in excited state in given sample is

$$Nx \approx N_0 \cdot e^{-400} = \frac{N_0}{e^{400}}$$

As denominator is a very large number, practically we do not expect any of atom in excited state.

Moreover, due to intermolecular collisions, which are nearly elastic at room temperature, amount of energy available for excitation is very-very small to cross energy threshold for excitation of any of atom.

22. (d) First we find the location of COM

$$Mx = m(r - x) \quad \text{[Refer figure in question]}$$

$$\Rightarrow \quad x = \left(\frac{m}{M + m}\right)r \quad \ldots\text{(i)}$$

Centripetal force an electron is due to electrostatic attraction.

So, $$m(r - x)\omega^2 = K \cdot \frac{Ze^2}{r^2}$$

$$\Rightarrow \quad m\left(r - \frac{mr}{M + m}\right)\omega^2 = \frac{KZe^2}{r^2}$$

$$\Rightarrow \quad \left(\frac{m}{M + m}\right) Mr \cdot \omega^2 = \frac{KZe^2}{r^2}$$

$$\Rightarrow \quad \left(\frac{Mm}{M + m}\right) r^3\omega^2 = KZe^2 \quad \ldots\text{(ii)}$$

where, $\frac{Mm}{M + m} = \mu$ = reduced mass of system

Now, moment of inertia of system

$$= I = Mx^2 + m(r - x)^2$$

Now, we substitute for x from Eq. (i) in above expression, we get

$$I = M\left(\frac{m}{M + m}\right)^2 \cdot r^2 + m\left(r - \frac{mr}{M + m}\right)^2$$

$$= \frac{Mm^2r^2 + mM^2r^2}{(M + m)^2}$$

$$= \left(\frac{Mm}{M + m}\right) \cdot r^2 = \mu r^2$$

By Bohr's postulate,

$$I\omega = \mu r^2\omega = \text{angular momentum}$$

$$= \frac{nh}{2\pi} \quad \ldots\text{(iii)}$$

So, from Eqs. (ii) and (iii), we get

$$\text{Angular momentum} = I\omega = \left(\frac{Mm}{M + m}\right) r^2\omega.$$

23. (b) As energy $E \propto m$

or energy = $-Km$ for some constant $-K$

$$\Rightarrow \quad E = -Km$$

So, for corrected atomic model,

$$E' = -K\left(\frac{Mm}{M + m}\right)$$

As, $$\frac{Mm}{M + m} < m$$

$$\Rightarrow \quad \frac{E'}{E} = \left(\frac{M}{M + m}\right) < 1$$

$$\Rightarrow \quad E' < E$$

As, there is a negative sign, so the new ground state energy of electron will be more than that of Bohr's atom.

DPP-3 **Electronic Transitions in Hydrogen and Hydrogen-like Atoms**

1. (i) Let r be the distance between positron (the nucleus) and electron and let r_1 and r_2 be the distances of COM from nucleus and electron.

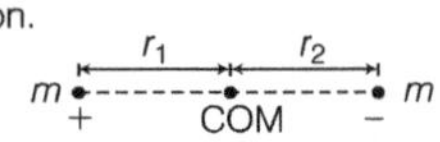

Then, $mr_1 = mr_2$ and $r_1 + r_2 = r$

$$\Rightarrow \qquad r_1 = r_2 = \frac{r}{2},$$

COM lies at centre and so, both particles orbits around COM.

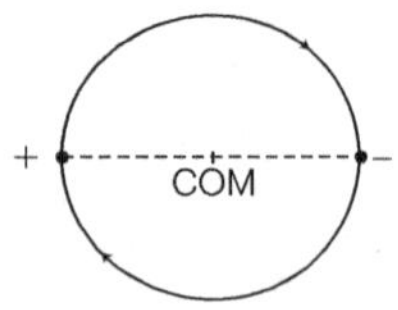

A positronium atom

As centripetal force on electron is due to electrostatic attraction.

$$\text{So,}\quad m\left(\frac{r}{2}\right)\omega^2 = \frac{1}{4\pi\varepsilon_0}\cdot\frac{e^2}{r^2}$$

$$\Rightarrow \qquad mr^3\omega^2 = \frac{2e^2}{4\pi\varepsilon_0} \qquad \ldots\text{(i)}$$

and moment of inertia of system about COM

$$= m\left(\frac{r}{2}\right)^2 + m\left(\frac{r}{2}\right)^2 = 2\cdot\frac{mr^2}{4} = \frac{mr^2}{2}$$

Applying quantisation of angular momentum,

$$L = \left(\frac{mr^2}{2}\right)\omega = I\omega = \frac{nh}{2\pi}$$

$$\Rightarrow \qquad mr^2\omega^2 = \frac{nh}{\pi} \qquad \ldots\text{(ii)}$$

Dividing Eqs. (ii) and (i), we get

$$\frac{m^2r^4\omega^2}{mr^3\omega^2} = \frac{n^2h^2}{\pi^2}\times\frac{4\pi\varepsilon_0}{2e^2}$$

$$\Rightarrow \qquad r = \frac{4n^2h^2\varepsilon_0}{2\pi e^2 m} = 2\cdot\frac{n^2h^2\varepsilon_0}{\pi e^2 m}$$

Now, $\dfrac{h^2\varepsilon_0}{\pi me^2}$ = Bohr radius (a_0)

= radius of electron of H-atom in ground state

= 5.292×10^{-11} m $\approx$ 0.53 Å

So, $\quad r = 2r_H$

[So, positronium atom has twice as large orbit as of H-atom]

where, r_H = radius of electron in nth orbit of H-atom

So, $\quad r_n = 2r_{nH} = 2n^2(0.53)$ Å

(ii) As, both particles have same mass so, we consider nuclear motion also.

Reduced mass of system is $m' = \dfrac{mM}{m+M} = \dfrac{m^2}{2m} = \dfrac{m}{2}$.

where, m = mass of electron = mass of positron.

To take account of motion of nucleus, all we have to do is to replace the electron with a particle of mass m'. The energy levels of the atom, then become

$$E'_n = \frac{-m'e^4}{8\varepsilon_0^2h^2}\cdot\left(\frac{1}{n^2}\right) = -\frac{m'}{m}\cdot\left(\frac{me^4}{8\varepsilon_0^2h^2}\right)\frac{1}{n^2}$$

where, $\quad \dfrac{-me^4}{8\varepsilon_0^2h^2} = -2.18\times10^{-18}$ J $= -13.6$ eV

So, $\quad E'_n = \left(\dfrac{m'}{m}\right)\cdot E_{nH}$

where $E_{n_H} = \dfrac{-13.6}{n^2}$ eV $= \dfrac{-2.18\times10^{-18}}{n^2}$ J

$$= \frac{-me^4}{8\varepsilon_0^2h^2n^2}$$

energy of an electron in nth stationary state of a hydrogen atom.

Now, $\quad E'_n = \dfrac{m'}{m}\cdot E_{n_H} \quad\Rightarrow\quad E'_n = \dfrac{m/2}{m}\cdot E_{n_H}$

$$\Rightarrow \qquad E'_n = \frac{E_{n_H}}{2}$$

So, energy of a positronium atom are twice that of a hydrogen atom, (because E'_n being smaller in absolute values are $\therefore$ less negative).

(iii) Using, $\quad \dfrac{1}{\lambda} = \dfrac{-E_1}{ch}\left(\dfrac{1}{n_f^2} - \dfrac{1}{n_i^2}\right)$

To compare we take, $n_i = \infty$ and $n_f = 1$, we get a smallest wavelength.

$$\frac{1}{\lambda} = \frac{-E_1}{ch}\left(\frac{1}{\infty^2} - \frac{1}{1^2}\right)$$

$$\Rightarrow \qquad \frac{1}{\lambda} = \frac{E_1}{ch} \Rightarrow \lambda = \frac{ch}{E_1}$$

From result of solution 1(ii) we get,

$$\lambda_{\text{Positronium}} = \frac{ch}{E'_1} = \frac{ch}{\left(\frac{E_{1H}}{2}\right)}$$

$$\Rightarrow \qquad \lambda_{\text{Positronium}} = 2\left(\frac{ch}{E_{1H}}\right) = 2\lambda_H$$

$$\Rightarrow \qquad \lambda_{\text{Positronium}} > \lambda_{\text{Hydrogen}}$$

(Do not get confused with this result as energy difference is higher for hydrogen, we get a lower wavelength).

(iv) Ionisation energy is the minimum energy required to move an electron from ground state to $n = \infty$ state. For a hydrogen atom, ionisation energy = 13.6 eV.

For a positronium atom, (using result of answer 1 (ii), Ionisation energy is

$$\frac{13.6}{2} = 6.8 \text{ eV}$$

So, less energy is required to ionise positronium atom than hydrogen atom.

(v) A proton-antiproton system has similar configuration of a positronium ion, two particles of same mass . But in this case particles are much more massive.

Changes observed are (we can use results of 1 (i) and 1 (ii))

$$\text{radius } r = \frac{2n^2h^2\varepsilon_0}{\pi e^2 m}$$

$\Rightarrow \qquad r$ = radius of orbit in its ground state

$$= \frac{2}{m_p}\cdot\left(\frac{h^2\varepsilon_0}{\pi e^2}\right) = \frac{2m_e}{m_p}\cdot\left(\frac{h^2\varepsilon_0}{\pi e^2 m_e}\right)$$

where, $\dfrac{h^2\varepsilon_0}{\pi e^2 m_e} = a_0$ = Bohr's radius = 0.53 Å

$$\therefore \qquad r = \frac{2\times9.1\times10^{-31}}{1.67\times10^{-27}}\times0.53\text{ Å}$$

$$\approx 11\times10^{-4}\times0.53\text{ Å}$$

$$= 5.83\times10^{-4}\times10^{-10} = 5.83\times10^{-14}\text{ m}$$

System is stable you can check this by finding ground state energy.

2. (d) For a H-like atom,

$$E_n = \frac{Z^2}{n^2} \cdot E_1, \qquad \text{(where, } E_1 = -13.6 \text{ eV)}$$

$$\Rightarrow \quad E_{4n} - E_{2n} = \frac{Z^2E_1}{16n^2} - \frac{Z^2E_1}{4n^2}$$

$$= \frac{Z^2E_1}{n^2}\left(\frac{1}{16} - \frac{1}{4}\right)$$

$$= \frac{Z^2E_1}{n^2}\left(\frac{-3}{16}\right)$$

and

$$E_{2n} - E_n = \frac{Z^2E_1}{n^2}\left(\frac{1}{4} - \frac{1}{1}\right)$$

$$= \frac{Z^2E_1}{n^2}\left(\frac{-3}{4}\right)$$

$$\text{Ratio } \frac{E_{4n} - E_{2n}}{E_{2n} - E_n} = \frac{4}{16}$$

$$= \frac{1}{4} = \text{a constant.}$$

3. (c)

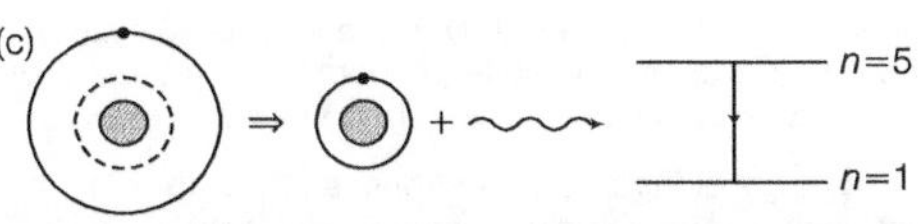

When an excited atom, is at rest in some external frame of reference, emits a photon of momentum p, conservation of momentum requires that the atom recoil with same momentum p.

For photon, $E = hf = \frac{hc}{\lambda}$

and $\lambda = \frac{h}{p}$ so, $E = pc$

or p = momentum of photon $= \frac{E}{c}$

Now, $E = E_5 - E_1 = \frac{-13.6}{25} - \frac{(-13.6)}{1}$

$$= 13.6 \times \frac{24}{25} \text{ eV}$$

$$= 13.6 \times \frac{24}{25} \times 1.6 \times 10^{-19} \text{ J}$$

$$\text{Momentum } p = \frac{13.6 \times 24 \times 1.6 \times 10^{-19}}{25 \times 3 \times 10^8}$$

$$= 6.96 \times 10^{-27} \text{ kg-ms}^{-1}$$

So, recoil speed of H-atom is

$$\frac{p}{m} \approx \frac{7 \times 10^{-27}}{1.67 \times 10^{-27}} \approx 4.2 \text{ ms}^{-1}$$

Note *We had not taken mass of electron in calculation as it is very less and does not make much of a difference.*

4. (d) For an electron in nth state of H-atom,

$$v_n = 2.2 \times 10^6 \frac{Z}{n} \text{ ms}^{-1}$$

$$r_n = 0.53 \times \frac{n^2}{Z} \text{Å}$$

So, centripetal acceleration of electron $= v_n^2/r_n$

$$= \frac{(2.2 \times 10^6 \, Z/n)^2}{0.53 \times 10^{-10}\left(\frac{n^2}{Z}\right)}$$

$$= \frac{4.84 \times 10^{12}}{0.53 \times 10^{-10}} \cdot \left(\frac{Z^2}{n^4}\right)$$

$$= 9.1 \times 10^{22} \cdot \left(\frac{Z^2}{n^4}\right)$$

So, ratio of centripetal acceleration of 3rd and 2nd orbits is

$$\frac{a_3}{a_2} = \frac{9.1 \times 10^{22} \times \frac{Z^2}{3^4}}{9.1 \times 10^{22} \times \frac{Z^2}{2^4}}$$

$$\Rightarrow \quad \frac{a_3}{a_2} = \frac{2^4}{3^4} = \frac{16}{81}.$$

5. (a) As sample is at room temperature, initially all of atoms are in ground state. In de-excitation (after being excited by wavelength λ), 10 wavelengths are emitted this indicates 10 different transitions.

Now, using $\frac{n}{2}(n-1) = 10,$

We have $n = 5$

So, incident wavelength is able to excite atom to $n = 5$ state.

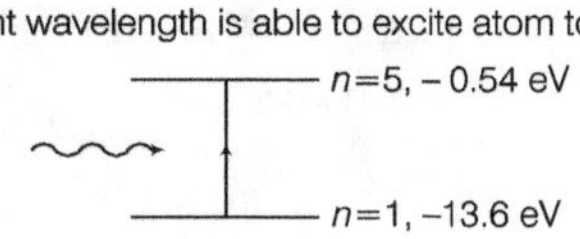

So, energy of incident photon $= [-0.54 - (-13.6)]$ eV

$$= 13.06 \text{ eV}.$$

As energy of photon $E = \frac{hc}{\lambda}$

$$= \frac{1240 \text{ eV-nm}}{\lambda_{nm}}$$

We have, $\lambda = \frac{1240}{13.06}$ nm $= 94.9 \approx 95$ nm.

6. (c) Largest wavelengths of Balmer series corresponds to $n = 3$ to $n = 2$ transition.

Thus, the incoming electron must have an energy

= energy required to reach $n = 3$ state.

$$= \left(\frac{-13.6}{3^2} - (-13.6)\right)$$

$$= 12.09 \approx 12.1 \text{ eV}.$$

7. (d) When white light is passed through hydrogen gas, photons of those wavelengths that corresponds to transitions between energy levels are absorbed. The resulting excited H-atoms re-radiate their excitation energy almost at once.

But the radiated photons come off in all random directions with only few in same direction as the original white beam.

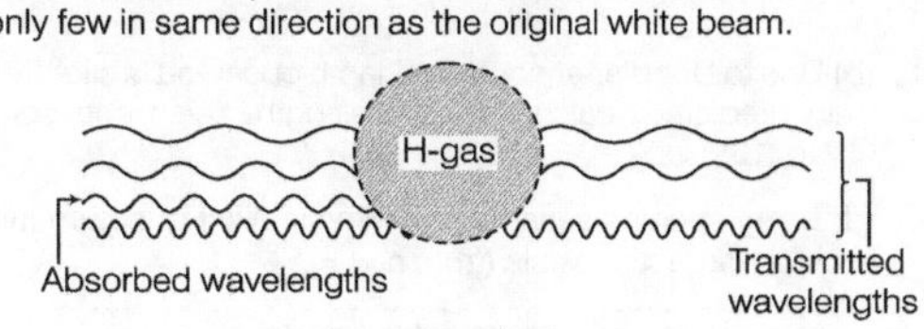

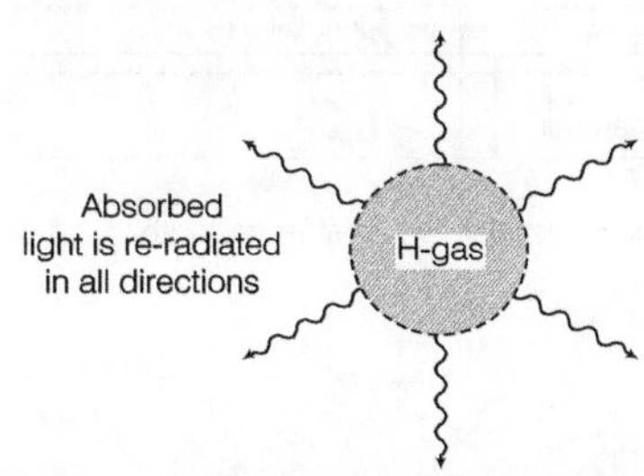

The dark lines in emission spectrum are not completely black but appears, so by contrast with the bright background. Lines in the absorption spectrum of any element, coincide with those in emission spectrum that represent transition to the ground state.

8. (c) Frequency of revolution of an electron in its orbit

$$f_1 = \frac{\text{Electron speed}}{\text{Orbit circumference}}$$

$$= \frac{v_n}{2\pi r_n} = \frac{e}{2\pi\sqrt{4\pi\varepsilon_0 m r_n^3}}$$

$$= \frac{me^4}{8\varepsilon_0^2 h^3}\cdot\left(\frac{2}{n^3}\right) \quad \text{(By substituting } r_n\text{)}$$

$$= \left(\frac{-2E_1}{h}\right)\cdot\frac{1}{n^3}$$

$$E_1 = -2.18 \times 10^{-18} \text{ J or } -13.6 \text{ eV}$$

So, f_{n_1} = frequency of revolution in state $n = 1$.

$$= \frac{2.18 \times 10^{-18}}{6.63 \times 10^{-34}} \times \frac{2}{1^3}$$

$$= 6.58 \times 10^{15} \text{ revs}^{-1}$$

f_{n_2} = frequency in state $n = 2$

$$= 0.823 \times 10^{15} \text{ revs}^{-1}$$

Also, frequency of emitted photon

$$\nu = \frac{-E_1}{h}\left(\frac{1}{n_f^2} - \frac{1}{n_i^2}\right)$$

$$= \frac{2.18 \times 10^{-18}}{6.63 \times 10^{-34}}\left(\frac{1}{1^2} - \frac{1}{2^2}\right)$$

$$= 2.88 \times 10^{15} \text{Hz}$$

Clearly, this frequency is intermediate between f_1 and f_2.

9. (c) Energy is ground state is E_1. So, wavelength of emitted photon is

$$\frac{1}{\lambda} = \frac{-E_1}{hc} = R = 1.097 \times 10^7 \text{m}$$

$$\therefore \quad \lambda = \frac{1}{R} = 0.912 \times 10^{-7} \text{m}$$

$$= 91.2 \text{ nm}$$

10. (a) Considering the Doppler's effect, we have

$$\omega = \frac{\omega_0}{1 - v/c}$$

$$\Rightarrow \quad \frac{\Delta\omega}{\omega} \approx v/c$$

$$\therefore \quad \omega' = \omega + \Delta\omega = \omega(1 + v/c)$$

11. (b) Due to Doppler-effect, wavelength observed is slightly smaller than predicted (calculated) wavelength. (Also see solution of Q. No. 10)

12. (d) Every spectral series starts with smallest quantum jump and ends on largest possible (from $n = \infty$).

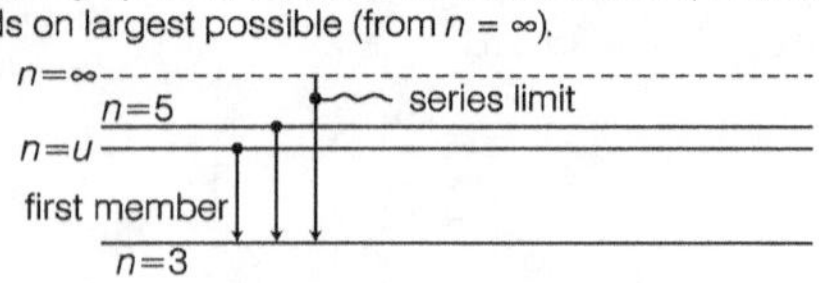

Longest jump results in shortest wavelength (λ_{min}) called series limit.

Now, for Brackett series,

$$\lambda_{min} = \frac{16}{R}$$

For Pfund series,

$$\lambda_{min} = \frac{25}{R}$$

$$\Rightarrow \text{Ratio is } \frac{16/R}{25/R} = \frac{16}{25}.$$

13. (a) At room temperature, H-gas absorbs any photon that has energy equals to that of a transition from ground ($n = 1$) to some other excited state ($n = 2, 3, \ldots$ etc.). So, few of wavelengths are absorbed. Also, emission lines lies in Lyman series which is in ultraviolet region.

14. (a,c,d) Binding energy for H-like atom is given by

$$E_B = 13.6\, Z^2$$

$$\Rightarrow \quad Z^2 = \frac{122.4}{13.6} = 9$$

$$\Rightarrow \quad Z = 3$$

So, energy levels of atom are

$$E = -13.6 \times \frac{Z^2}{n^2} = \frac{-122.4}{n^2}$$

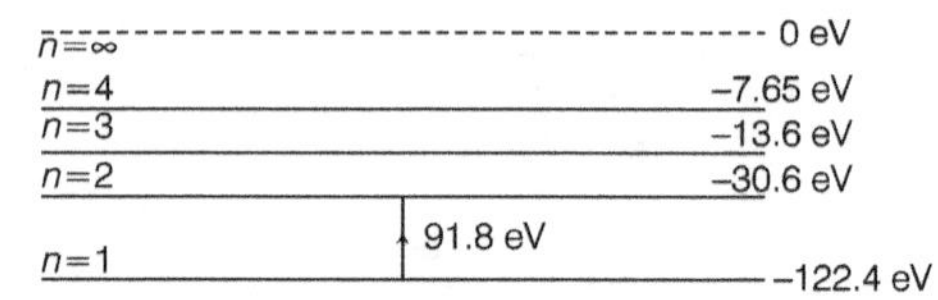

For an electron accelerated by 90V, energy is 90 eV. But to excite an atom from its ground state (at room temperature) at least 91.8 eV energy is required so, option (b) is incorrect.

15. (a,b) Energy of incoming electrons is 12.75 so, they can excite H-atom to $n = 3$ state only. (Take, ΔE = 12.09 eV)

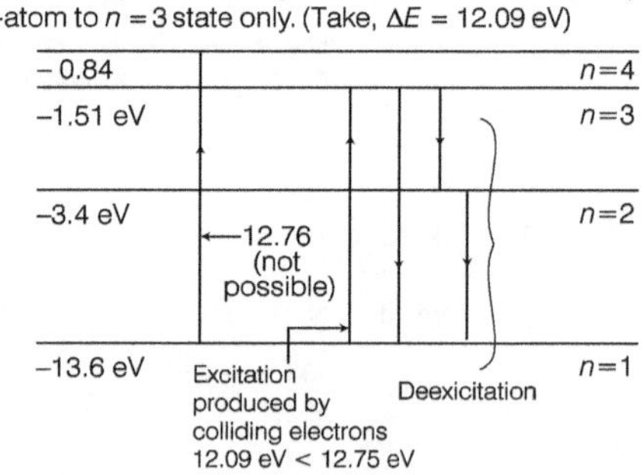

So, we can observe emissions

$3 \to 1$, $3 \to 2$, $2 \to 1$, clearly Lyman and Balmer series components.

16. (b,c) After emitting a photon of Balmer series, electron is in $n = 2$ state. So, excess of energy remains with atom.

= 10.2 eV

In emission $n = 2$ to $n = 1$, a wavelength of Lyman series is emitted.

$$\lambda = \frac{hc}{\Delta E} = \frac{1240}{10.2} \text{ (nm)}$$

$$= 121.56$$

$$\approx 122 \text{ nm.}$$

17. (b,c) For transitions 1, 2 and 3.

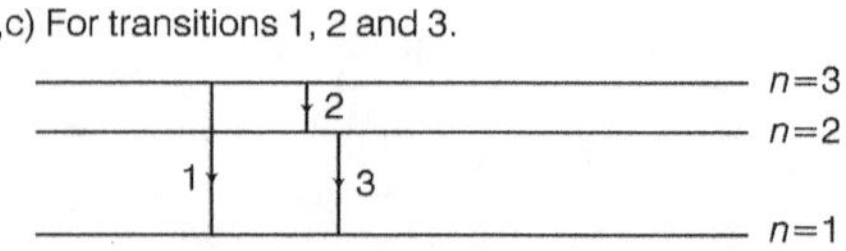

$$E_{31} = E_{32} + E_{21}$$

$$\Rightarrow \quad h \times f_{31} = h \times f_{32} + h \times f_{21}$$

$$\Rightarrow \quad f_3 = f_2 + f_1 \text{ (option c)}$$

$$\Rightarrow \quad \frac{c}{\lambda_3} = \frac{c}{\lambda_2} + \frac{c}{\lambda_1}$$

$$\Rightarrow \quad \frac{1}{\lambda_3} = \frac{\lambda_1 + \lambda_2}{\lambda_2 \cdot \lambda_1} \text{ (option b)}$$

18. (b) Force for rotation is provided by magnetic field.

$$\Rightarrow \quad \frac{mv^2}{r} = Bqv \text{ (Lorentz force)}$$

$$mvr = \frac{nh}{2\pi} \text{ (Angular momentum quantisation)}$$

Combining, we get

$$\frac{mvr}{Bq} = r^2$$

$$\Rightarrow \quad r = \sqrt{\frac{nh}{2\pi Bq}}$$

19. (a)

$$\text{KE} = \frac{1}{2}mv^2$$

$$= \frac{B^2q^2r^2}{2m} = \frac{B^2q^2}{2m} \cdot \frac{nh}{2\pi Bq}$$

$$= \frac{Bqnh}{4\pi m}$$

20. (d) An revolving electron is like a current loop. In magnetic field, its potential energy is

$$U = -\mathbf{M} \cdot \mathbf{B} = -MB\cos\theta$$

$$= -(i \cdot A)\, B\cos\theta$$

$$\therefore \quad \mathbf{M} = NIA$$

$$= -(qf\pi r^2)\, B\cos\theta,\ f = \text{frequency} = 1/T$$

$$= -\left(q\pi r^2 \cdot \frac{Bq}{2\pi m}\right) B\cos\theta = \frac{2\pi m}{Bq}$$

$$= -\left(\frac{B^2q^2r^2}{2m}\right) \cdot \cos\theta,\ \theta = 180°,$$

$$= \frac{-B^2q^2r^2}{2m} \cdot (-1)$$

or

$$\text{PE} = U = \text{KE} = \frac{Bqnh}{4\pi m}$$

So, total energy of electron in its nth state, $E_n = \frac{Bqh}{4\pi m} \cdot n$

In transaction $2 \to 1$,

Energy lost = photon energy

$$= E_2 - E_1 = \frac{Bqh}{4\pi m}(2-1)$$

$$= \frac{Bqh}{2\pi m} = hf$$

or frequency of emitted photon = Classical revolution frequency

$$= \frac{Bq}{2\pi m}.$$

21. (d) $L = \frac{nh}{2\pi} = I\omega, \quad \omega = \frac{nh}{2\pi I}$

$$\text{KE} = \frac{1}{2}I\omega^2 = \frac{1}{2}I\left(\frac{nh}{2\pi I}\right)^2 = \frac{n^2h^2}{8\pi^2 I}$$

$$\therefore \quad \text{KE} \propto n^2$$

As energy is absorbed in form of KE only

So, $\frac{K(n=2)}{K(n=1)} = \frac{2^2}{1^2} = \frac{4}{1}.$

22. (d) $hf = K_2 - K_1 = \frac{3h^2}{8\pi^2 I}$

$$\Rightarrow \quad f = \frac{3h}{8\pi^2 I}$$

$$\Rightarrow \quad f = \frac{3 \times 2\pi \times 10^{-34}}{8\pi^2 \times \frac{3}{16} \times 10^{-46}} = \frac{4}{\pi} \times 10^{11}\text{Hz}$$

23. (b) Both Statement I and Statement II are individually correct but Statement II is not a reason for Statement I. Broadening of a spectral line is due to Doppler's effect and recoil of atom.

24. (d) Statement I is incorrect but Statement II is correct.

For the given atom, energy levels are

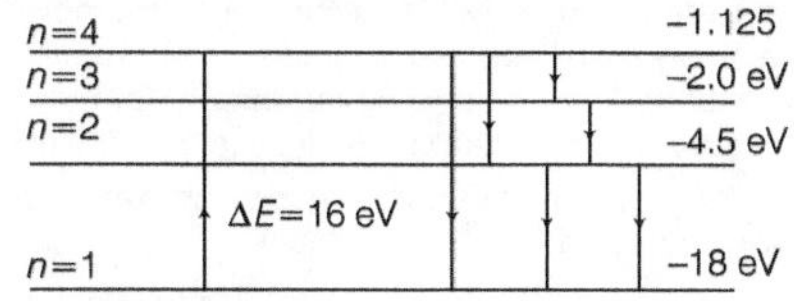

So, atom is excited to $n = 3$ state (atmost)

Clearly emission lines contains $= \frac{n}{2}(n-1)$

$$= \frac{3(3-1)}{2}$$

$$= 6 \text{ spectral lines.}$$

25. (a) Both Statement I and Statement II are correct. Absorption is not a stepwise process like emission.

26. (a) By using $\Delta E = \frac{hc}{\lambda}$

We have $\lambda_{max} = \frac{n^2(n+1)^2}{(2n+1)R}$ $\quad [n+1 \to n \text{ transition}]$

$\lambda_{min} = \frac{n^2}{R}$ $\quad [\infty \to n \text{ transition}]$

For Lyman series, $n = 1$
Balmer series, $n = 2$
Paschen series, $n = 3$
Brackett series, $n = 4$
Pfund series, $n = 5$

Hence, (i) → (s), (ii) → (t), (iii) → (p), (iv) → (q), (v) → (r).

27. (a) Radius $= 0.53 \cdot \frac{n^2}{Z}$ (Å)

$$\text{Speed} = 2.2 \times 10^6 \frac{Z}{n} \text{ ms}^{-1}$$

$$\text{Current} = ef = \frac{me^5}{4\pi\varepsilon_0^2 h^3} \cdot \frac{Z^2}{n^3}$$

$$\text{Magnetic field, } B = \frac{\mu_0 i_n}{2r_n}$$

$$= \frac{\pi m^2 e^7 \mu_0}{8\varepsilon_0^3 h^5} \cdot \frac{Z^3}{n^5}$$

Hence, (i) → (q), (ii) → (r), (iii) → (s), (iv) → (p).

28. (3) As all of atoms are in same nth state, they release equal energy. (but number of transitions may be different for each)

Released energy $= \frac{4800}{49} \times Rch$

Energy released per atom

$$= \frac{48}{49} Rch = \frac{48}{49} \times 13.6 \text{ eV}$$

$$= \left(\frac{49-1}{49}\right) \times 13.6$$

$$= \left(\frac{1}{1^2} - \frac{1}{7^2}\right) \times 13.6 \text{ eV}$$

So, atoms are in 7th excited state initially.

So, each atom can undergoes a maximum of

$$\frac{n(n-1)}{2} = \frac{7 \times 6}{2} = 21 \text{ transitions (per atom).}$$

So, maximum possible number of photons emitted, by the sample of 100 atoms.

$$= 21 \times 100 = 2100$$

$$= 7 \times 3 \times 100$$

$$\therefore \quad N = 3$$

DPP-4 X-Rays

1. (a) X-rays are produced when an energetic (fast moving) electron trikes a heavy metal target and looses part of (or completely) its kinetic energy. This lost energy appears in form of radiation of λ-range : 0.001 nm to 1 nm (1×10^{-12} m to 1×10^{-9} m).

Recoil energy of atom is very less (due to large mass of atom) and is usually neglected.

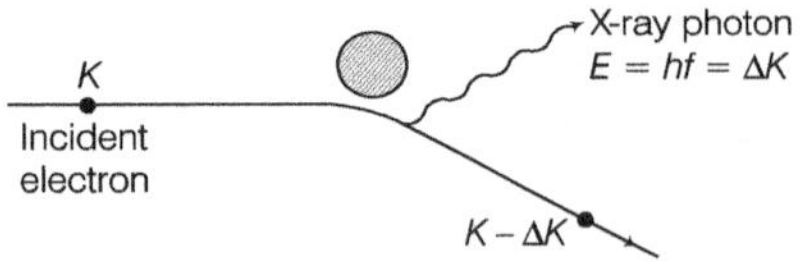

Scattered electron (whose energy is now less K_0) may have a second and then a third collision generating other photons of less energy (more wavelength).

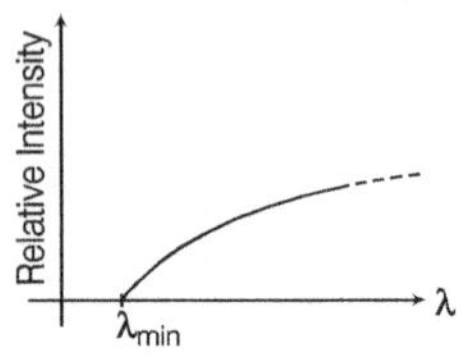

(Due to sucessive collisions different wavelengths of radiations, X-rays are produced.)

2. (d) X-rays are electromagnetic rays and all formulae for X-ray photons are valid.

So, energy of an X-ray photon is

$$E = hf = \frac{hc}{\lambda}$$

$\Rightarrow$ $E \cdot \lambda = hc = \text{constant}$

$\Rightarrow$ $E_1\lambda_1 = E_2\lambda_2$

$\Rightarrow$ $\frac{E_1}{E_2} = \frac{\lambda_2}{\lambda_1} = \frac{0.5}{0.01} = \frac{50}{1}.$

3. (d) When a X-ray beam passes through a material intensity of emergent X-rays is

$$I = I_0 e^{-\mu x}$$

Where, μ = absorption coefficient, x = thickness of medium

I_0 = incident Intensity.

And, the absorption coefficient depends on

(i) $\mu \propto \lambda^3$ (wavelengths of X-rays)

(ii) $\mu \propto \frac{1}{f^3}$ (frequency of X-rays)

(iii) $\mu \propto Z^4$ (atomic number of medium)

(iv) $\mu \propto \rho$ (density of medium)

4. (a) Characteristics X-rays are prorduced when bombarding electron knocks out an electron from an inner shell (K or L) and this vacancy produced is filled by an electron from outer shell. K_α wavelength is produced due to electronic transition of an electron ($n = 2$ to $n = 1$) of target metal atom.

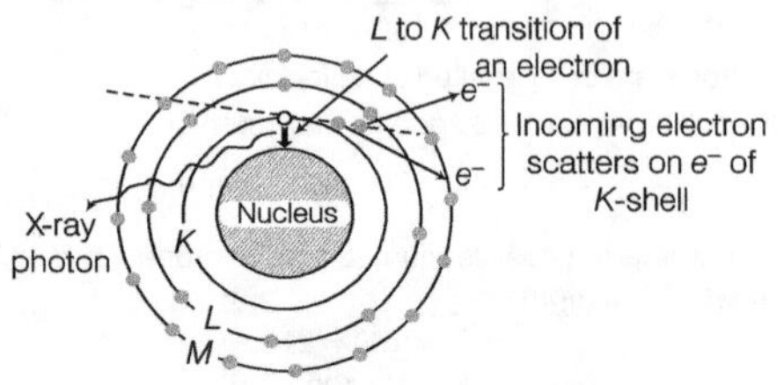

So, $\lambda_{K\alpha} = X_K = \frac{hc}{\Delta E_{2\to 1}} = \frac{hc}{10.2\,(Z-1)^2}$

Note *$E_{K\alpha}$ = Energy of K_α photon $= 10.2(Z-1)^2$ and also refer to passage for Q. 16, 17 and 18.*

Also, minimum wavelength is produced when the bombarding electron lost complete KE in only one collision.

Then, $\lambda_{min} = \frac{hc}{E_{electron}} = \frac{hc}{eV_0}$

where, V_0 = accelerating voltage for bombarding electron.

So, when V_0 is increased, λ_{min} is reduced but λ_K is constant. So, $\lambda_K - \lambda_C$ increases.

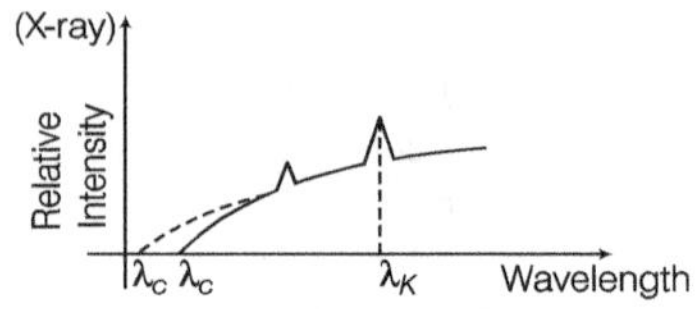

5. (c) Energy of electron $= eV = 1.6 \times 10^{-19} \times 1.5 \times 10^5$

$= 2.4 \times 10^{-14}$

1% of this energy becomes photon energy

$\therefore$ $hf = \frac{hc}{\lambda} = \Delta E = eV \times \frac{1}{100}$

$\Rightarrow$ $\lambda = \frac{hc \times 100}{eV}$

$= \frac{6.6 \times 10^{-34} \times 3 \times 10^8 \times 100}{2.4 \times 10^{-14}}$

$= 8.25 \times 10^{-10} = 8.25$ Å

6. (b) For K_α wavelength, electron transits from $L(n=2)$ to $K(n=1)$ state

$\therefore$ $\lambda_{K\alpha} = \frac{hc}{E_{2\to 1}}$

For K_β wavelength, electron transits from $M(n=3)$ to $K(n=1)$ state

$\therefore$ $\lambda_{K\beta} = \frac{hc}{E_{3\to 1}}$

So, $\frac{\lambda_{K\alpha}}{\lambda_{K\beta}} = \frac{E_{3\to 1}}{E_{2\to 1}}$ = a constant for a given material.

Different materials have different value of this ratio.

7. (a) For K_α, $2 \to 1$ transition, there is only 1 screening electron so, energy of K_α photon is – $K_{K\alpha} = 10.2\,(Z-1)^2$

and $\lambda_{K\alpha} = \frac{hc}{E_{K\alpha}} = \frac{hc}{10.2\,(Z-1)^2}$

So, $\frac{\lambda_{K\alpha}(\text{Cu})}{\lambda_{K\alpha}(\text{Mo})} = \frac{(Z-1)^2, \text{Mo}}{(Z-1)^2, \text{Cu}} = \frac{(42-1)^2}{(29-1)^2}$

$\Rightarrow$ $\lambda_{K\alpha}(\text{Cu}) = \lambda_{K\alpha(\text{Mo})} \times \frac{41^2}{29^2} = 0.71 \times \frac{41^2}{29^2} = 1.52$ Å

Also, energy of K_β photon $= E_{K\beta} = (Z-9)^2 \times 10.2$

As there are 8 e^- of L-shell and 1 electron of K-shell are producing screening. So, effective nuclear charge is "$Z - 9$".

Hence, $\lambda_{K\beta} = \frac{hc}{10.2(Z-9)^2}$

Now, $\left.\frac{\lambda_{K\beta}}{\lambda_{K\alpha}}\right|_{\text{For copper}} = \frac{(Z-1)^2}{(Z-9)^2} = \frac{28^2}{20^2}$

$\Rightarrow$ $\lambda_{K\beta} = \lambda_{K\alpha} \cdot \frac{(Z-1)^2}{(Z-9)^2}$

$= 1.52 \times \frac{28^2}{20^2} = 1.52 \times \frac{784}{400}$

$= 2.9792 \approx 3$Å

8. (d) Minimum wavelength occurs when energy corresponding to the transition is maximum. $E_{max}: n = \infty$ to $n = 1$.

9. (d)
$$\frac{\lambda_1}{\lambda_2} = \frac{(Z-1)^2}{(Z_1-1)^2}$$
$$\Rightarrow \quad (Z_2-1)^2 = \frac{0.21}{0.1537} \times 73^2$$
$$\Rightarrow \quad Z_2 = 28, \text{ metal is nickel.}$$

10. (d) Moseley's law is
$$\sqrt{f} = a(Z-b)$$
a and b are constants.

Clearly, for $Z = 0$, $\sqrt{f} \neq 0$

So, line does not pass through origin.

11. (b,c,d) Transition $2 \longrightarrow 1$ gives K_α

Transition $3 \longrightarrow 1$ gives K_β

Transition $4 \longrightarrow 1$ gives K_γ

So, $E_{K_\alpha} < E_{K_\beta} < E_{K_\gamma}$

Also, $\lambda \propto \frac{1}{E} \Rightarrow \lambda_{K_\alpha} > \lambda_{K_\beta} > \lambda_{K_\gamma}$

Also, transition $4 \longrightarrow 3$ gives M_α and transtiion $3 \longrightarrow 2$ gives L_α
$$\Rightarrow \quad E(M_\alpha) < E(L_\alpha) < E(K_\alpha)$$
$$\Rightarrow \quad \lambda(M_\alpha) > \lambda(L_\alpha) > \lambda(K_\alpha)$$

12. (a,c,d) $\lambda_{min} = \frac{hc}{K}$,

K = KE of bombarding electron

= eV, V = accelerating voltage

$= \frac{hc}{eV} \Rightarrow \lambda_{min} \propto V^{-1}$

And, $\lambda_{K\alpha} = \frac{hc}{10.2(Z-1)^2}$

$\Rightarrow$ λ(peak) decreases with Z.

Penetrating power $\propto f$. And $f \propto \nu$ also an increase of filament current results in increase of bombarding electrons per unit time and hence, the number of X-rays photons also increases.

13. (a,d)
$$\lambda_{K\alpha} = \frac{hc}{10.2(Z-1)^2}$$
or $Z_1 = Z_2 = Z_3$

$\Rightarrow \quad \lambda_1 = \lambda_2 = \lambda_3$

$\Rightarrow \quad \lambda_2 = \sqrt{\lambda_1\lambda_3}$

14. (b,c) Supplied power = $P = VI$
$$\Rightarrow \quad P = 50 \times 10^3 \times 20 \times 10^{-3} \text{ W}$$
$$= 1 \times 10^{-3} \text{ W}$$
This power is divided into

1% to X-rays

99% as heat

Hence, rate of heat generation
$$= \frac{d}{dt} ms\theta = \frac{99}{100} \times 1 \times 10^{-3} \text{ W}$$
$$\Rightarrow \quad \frac{d\theta}{dt} = \text{rate of rise of temperature}$$
$$= \frac{99 \times 10^{-3}}{100 \times m \times s}$$
$$= \frac{99 \times 10^{-3}}{100 \times 1 \times 495}$$
$$= \frac{10^{-3}}{500} = 2 \times 10^{-6} \text{ °Cs}^{-1}$$
Also, $\lambda_{min} = \frac{hc}{eV} = \frac{12400}{50 \times 10^3}$ (Å)
$$= 0.25 \times 10^{-10} \text{ m}$$

15. (c,d) For a given operating voltage maximum frequency (f_{max}) (or λ_{min}) occurs, when complete energy of bombarding electron is liberated as X-rays.
$$hf_{max} = K$$
$$\Rightarrow \quad f_{max} = \frac{K}{h} = \frac{eV}{h}$$
$$\Rightarrow \quad f_{max} = \frac{1.6 \times 10^{-19} \times 6.6 \times 10^3}{6.6 \times 10^{-34}}$$
$$= 1.6 \times 10^{18} \text{ Hz}$$
Frequencies above than this is absent.

16. (c)
$$\lambda \propto \frac{1}{(Z-1)^2}$$
$$\Rightarrow \quad \frac{\lambda_2}{\lambda_1} = \frac{(Z_1-1)^2}{(Z_2-1)^2}$$
$$\Rightarrow \quad \frac{\lambda_2}{\lambda} = \left(\frac{43-1}{29-1}\right)^2$$
$$\Rightarrow \quad \lambda_2 = \frac{9}{4}\lambda$$

17. (b) As, $E(K_\alpha) = 10.2\,(Z-1)^2$ and $E = hf$.

So, $\sqrt{f} \propto (Z-1)$ correct graph is option (b).

18. (d) Energy of an electron in K-shell of tungsten is around
$$E_K = -(74-1)^2 \times 13.6 \text{ eV} = -72474 \text{ eV}$$
An electron in the M-shell is subjected to an effective nuclear charge $(Z-9)$ as there are 8 screening electrons in L-shell and 1 remaining in K-shell. So, energy of electron in M-shell is
$$E_m = -\frac{13.6 \cdot Z_{eff}^2}{n^2} = \frac{-13.6 \times (74-9)^2}{3^2} = -6384 \text{ eV}$$
So, emitted X-ray photon has an energy of
$$\Delta E = E_M - E_K$$
$$= (-6384 - (-72474))$$
$$= 66090 \text{ eV}$$

19. (c) $\lambda_{min} = \frac{hc}{eV} = \frac{1240 \text{ eV} \cdot \text{nm}}{12.4 \times 10^3 \text{ eV}} = 10^{-1} \text{ nm} = 1\text{Å}$

20. (d)
$$\lambda_{K\alpha} = \frac{hc}{10.2\,(Z-1)^2}$$
$$\lambda_{K\beta} = \frac{hc}{10.2(Z-9)^2}$$
$$\Rightarrow \quad \text{Ratio, } \frac{\lambda_{K\alpha}}{\lambda_{K\beta}} = \frac{(Z-9)^2}{(Z-1)^2}$$
which cannot be determined without knowing Z.

21. (d) $\lambda_{K\alpha}$ depends on atomic number of target atom. Also, no X-rays are emitted from a H-atom.
$$\lambda_{min} \propto \frac{1}{V} \Rightarrow f_{max} \propto V$$
Penetrating power $\propto f$

Absorption in a medium $\propto f^{-3}$

Also, $\lambda_{K\alpha} \propto \frac{1}{Z}$ and absorption $\propto Z^4$

So, correct match is

Hence, (i) → (q,s,v), (ii) → (p,r,u,v), (iii) → (t).

22. (c) Statement I is correct and Statement II is incorrect. X-rays are not used in Radar systems as X-rays are not reflected from the surface.

23. (c) X-rays does not form an interference pattern as any slit size is very large compared to wavelength of X-rays.

Statement I is correct but Statement II is incorrect. Coherence length for X-rays is much large compared to visible light.

Revisal Problems (JEE Main)

1. (d) $\lambda = \frac{hc}{\Delta E_{2\to1}} = \frac{hc}{\left(\frac{-13.6\,Z^2}{2^2}\right) - \left(\frac{-13.6Z^2}{1^2}\right)}$

$$= \frac{hc}{13.6\,Z^2\left(1-\frac{1}{4}\right)} = \frac{1240 \times 4}{13.6 \times 3Z^2}$$

$\Rightarrow \quad \lambda \propto \frac{1}{Z^2}$

So, shortest wavelength is produced, when Z is maximum which is in case of Lithium ion ($Z = 3$).

2. (d) In Coolidge tube a bombarding electron can loose its kinetic energy either in a single collision or in multiple collisions.

$\therefore \lambda = \frac{hc}{\Delta K}$, clearly for a given value of electron energy, ΔK is always less than or equals to kinetic energy of electron.

So, λ has least value,

$$\lambda_{min} = \frac{hc}{\text{KE of electron}}.$$

3. (a) X-rays have wavelengths from 0.1 Å to 100Å.

$\therefore$ Energy of X-ray photons is from

$$E_1 = \frac{12400}{100}\text{ eV} = 1240\text{ eV} = 1.2\text{ keV}$$

and is upto $\quad E_2 = \frac{12400}{0.1}\text{ eV} = 12\text{ keV}$

Energy of a gas molecule is a function of its temperature.

Energy of a gas molecule at temperature T is

$$E = \frac{f}{2}k_B T \approx 0.02\text{ eV},$$

$$k_B = 1.38 \times 10^{-23}\text{ JK}^{-1}$$

For a solid molecule,

$$f = 6$$

$\therefore \quad E = 3k_BT \approx 0.06\text{ eV}$

Work functions of photosensitive metals lies from 1 to 10 eV.

Hence, (i) → (p), (ii) → (s), (iii) → (r), (iv) → (q).

4. (c) To produce a K_α photon an e^- must transit.

as $\quad \underset{n=2}{L} \longrightarrow \underset{n=1}{K}$

$\Rightarrow \quad E_L - E_K = \frac{hc}{\lambda_{K\alpha}} = \frac{1240}{0.021}$

$$= 59047\text{ eV} \approx 59\text{ keV}$$

5. (d) Ground state energy of a doubly ionised lithium atom (ion) is

$$E_1 = -13.6\left(\frac{Z^2}{n^2}\right), Z = 3\text{ (Li)}$$

$$= -13.6 \times \frac{9}{1}$$

$$= -122.4\text{ eV}$$

So, to ionise it, 122.4 eV energy must be supplied.

6. (c) $\quad \lambda_{min} = \frac{hc}{K}$

$\Rightarrow \quad \frac{12400}{K} = 155 \times 10^{-3}$

$\Rightarrow \quad K = \frac{12400}{155 \times 10^{-3}} = 80\text{ keV}$

7. (c) Electronic configuration for U_{92} is

$\Rightarrow \quad$ 2, 8, 18, 32, 32

and atomic sub levels in last orbit is

$\Rightarrow \quad 1s^2, 2s^2p^6, 3s^2p^6, 4s^2 3d^{10} 4p^2.$

So, for last electron of U_{92}, $n = 4$.

and $\quad r_n = 0.53\frac{n^2}{Z}$, Bohr's orbit

$$= 0.53 \times \frac{16}{92}$$

$$= 0.092\text{ Å}$$

8. (b) Energies of incoming photons are

$$-E = \frac{hc}{\lambda} \approx \frac{1240\text{ eV}}{\lambda}$$

$\Rightarrow \quad E_1 = \frac{1240}{100} = 12.4\text{ eV}$

$$E_2 = \frac{1240}{80} = 15.5\text{ eV}$$

Now photons of energy E_1 excites electrons of few atoms upto 3rd ($n = 3$) stationary state.

So, number of different wavelengths emitted

$$= \frac{n}{2}(n-1)$$

$$= \frac{3}{2}(3-1)$$

$$= 3\text{ different wavelengths}$$

And photons of energy E_2 produces ionisation in few atoms. Released atoms have remaining energy as KE

$$= 15.5 - 13.6 = 1.9\text{ eV}$$

9. (d) Energy of an electron in a stationary state of Li^{++} is

$$E_n = -13.6\frac{Z^2}{n^2}, Z = 3\text{ for Li}$$

$\Rightarrow \quad E_1 = -13.6 \times \frac{9}{1^2} = -122.4\text{ eV}$

$$E_2 = -13.6 \times \frac{9}{4} = -30.6\text{ eV}$$

$$\Delta E_{2\to1} = (-30.6) - (-122.4)$$

$$= 91.8\text{ eV}$$

As, photon's energy is insufficient to produce an excitation, no emission line is observed.

10. (b) $r_n = a_0\frac{n^2}{Z}$

For Beryllium, $\quad Z = 4$

$\therefore \quad r_{(n=2)} = a_0 \times \frac{2^2}{4} = a_0$

11. (d) For photoemission from sodium,

$$E = \text{KE}_{max} + \phi_0$$

$\Rightarrow \quad E = 1.82 + 0.73$

$$= 2.55\text{ eV}$$

So, H-atom produces a photon of 2.55 eV in transition process.

$\Rightarrow \quad E_n - E_{n+1} = 2.55$

So, transition made by electron is from 4th to 2nd level.

So, change in angular momentum

$$\Delta L = L_2 - L_4$$

$$= \frac{2h}{2\pi} - \frac{4h}{2\pi}$$

$$= \frac{-h}{\pi}.$$

12. (a) Helium atom formed is in 4th excited state (n = 5)

Now, for helium atom,

$$E_5 = -13.6\frac{Z^2}{n^2} = -2.2 \text{ eV}$$
$$E_4 = -3.4 \text{ eV}$$
$$E_3 = -6.04 \text{ eV}$$
$$E_2 = -13.6 \text{ eV}$$
$$E_1 = -54.4 \text{ eV}$$

Transitions possible occurs are

$$E_{(5\to4)} = E_5 - E_4 = 1.2 \text{eV}$$
$$E_{(5\to3)} = E_5 - E_3 = 3.84 \text{ eV}$$
$$E_{(5\to2)} = E_5 - E_2 = 11.4 \text{ eV}$$
$$E_{(5\to1)} = E_5 - E_1 = 52.2 \text{eV}$$
$$E_{(4\to3)} = E_4 - E_3 = 2.64 \text{ eV}$$
$$E_{(4\to2)} = E_4 - E_2 = 10.2 \text{ eV}$$
$$E_{(4\to1)} = E_4 - E_1 = 51.0 \text{ eV}$$
$$E_{(3\to2)} = E_3 - E_2 = 7.56 \text{ eV}$$
$$E_{(3\to1)} = E_3 - E_1 = 48.36 \text{ eV}$$
$$E_{(2\to1)} = E_2 - E_1 = 40.80 \text{ eV}$$

These are 10 emission lines when a helium atom in n = 5 state makes its way down to n = 1 state.

From given values of photon energies, 3.4 eV photon is not emitted in the transition. So, it must be from following process :

$${}^4_2\text{He} + {}^{0}_{-1}e \longrightarrow \underset{\substack{\text{(Helium ion in}\\ n = 5 \text{ state)}}}{{}^4_1\text{He}} + \underset{\substack{\text{(photon}\\ \text{of 3.4 eV)}}}{hf}$$

Now, energy released in this process = 3.4 eV.

So, Eenergy of electron + Energy difference of He^{++} and He^{+}

= Released energy

$\Rightarrow$ $E_{\text{electron}} + (2.2) = 3.4$

$\Rightarrow$ $E(\text{electron}) = 3.4 - 2.2$

$= 1.2$ eV

13. (a) $\lambda = \frac{1500\,n^2}{n^2 - 1}$, $n = 2, 3, \ldots$

Least energetic photon is emitted in $2 \to 1$ transition

$\Rightarrow$ λ_{max} = Maximum wavelength

$$= \frac{1500 \times 4}{(4-1)} = \frac{1500 \times 4}{3}$$

= 2000 Å

Most energetic photon is emitted in transition $\infty \to 1$

$\Rightarrow$ λ_{min} = minimum wavelength

$$= \underset{n\to\infty}{lt}\left\{\left(\frac{1500\,n^2}{n^2-1}\right)\right\} = \underset{n\to\infty}{lt}\left(1500 + \frac{1}{n^2-1}\right)$$

= 1500 Å

$\therefore$ Ratio of momentum

$$= \frac{p_{max}}{p_{min}} = \frac{h/\lambda_{max}}{h/\lambda_{min}} = \frac{\lambda_{min}}{\lambda_{max}}$$

$$= \frac{1500}{2000} = \frac{3}{4}.$$

14. (b) 1 Rydberg = 2.2×10^{-18} J = Rhc

Ionisation energy of a H-like atom = 4 Rydberg

$$= 4 \times 2.2 \times 10^{-18} \text{ J}$$
$$= \frac{4 \times 2.2 \times 10^{-18}}{1.6 \times 10^{-19}} \text{ eV}$$
$$= 55 \text{ eV}$$

So, energy in first orbit = $-$ 55 eV = $-13.6 \times Z^2$

$\Rightarrow$ $Z^2 = \frac{-55\text{eV}}{-13.6 \text{ eV}} = 4$

$\Rightarrow$ $Z = 2$

Now, radius of ground state = $\frac{0.53}{Z}$ Å = 0.265 Å.

15. (a) Let's consider a perfectly inelastic collision. For this, neutron and hydrogen atom will be moving together after collision (speed let v').

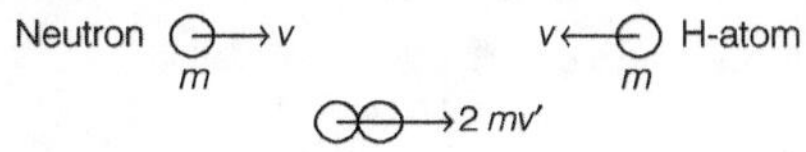

conservation of momentum gives $v' = 0$.

So, complete energy is converted into excitation energy. This must be at least 10.2 eV.

$\Rightarrow$ $2 \times \frac{1}{2}mv^2 = 10.2 \times 1.6 \times 10^{-19}$

$\Rightarrow$ $v^2 = \frac{10.2 \times 1.6 \times 10^{-19}}{1.6 \times 10^{-27}}$

$= 10.2 \times 10^8$

$= 4.0 \times 10^4 \text{ ms}^{-1}$ (approximately)

16. (c) Statment I is correct, the hot interior of Sun emits light which passes through some what 'cooled' atmosphere. Absorption lines of Sun are called Fraunhofer lines. Statement II is incorrect.

17. (d) $F_n \propto^{1/n^3}$, frequency of rotation in $n = 2$ is $f_1/8$.

$\therefore$ Number of rounds = $8 \times \frac{f_1}{8} = f_1$.

18. (c) In n = 4 state, an electron behaves like a wave of such wavelength, which covers complete circumference in 4 wavelengths.

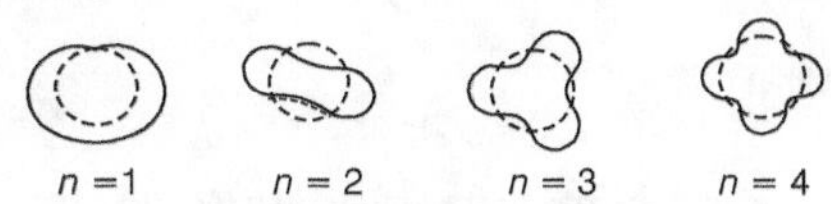

$\therefore$ $2\pi r = 4\lambda$

$\Rightarrow$ $\lambda = \frac{1}{4} \times 2\pi r$

19. (d) $\frac{\lambda_{max}(\text{Brackett})}{\lambda_{max}(\text{Pfund})} = \frac{\frac{400}{9R}}{\frac{900}{11R}} = \frac{400}{9R} \times \frac{11R}{900} = \frac{44}{81}.$

20. (d) All options are correct.

Revisal Problems (JEE Advanced)

1. (c) As per given condition,

$$2\left(\lambda_{K_\alpha} - \frac{12.4}{12.4}\right) = \left(\lambda_{K_\alpha} - \frac{12.4}{24.8}\right)$$

$$\Rightarrow \quad \lambda_{K_\alpha} = 2 - \frac{1}{2} = \frac{3}{2} = 1.5$$

2. (c) Energy of X-ray photon

$$= \Delta(IP)$$
$$= (1400 - 240) \times 10^{-17} \text{J}$$
$$= 1.16 \times 10^{-14} \text{ J}$$

3. (d) Photon energy = KE of products + Mass energy of products

$$\Rightarrow \quad K_{\text{products}} = E_{\text{photon}} - 2mc^2 \text{ products}$$
$$= 2.67 - 2 \times 0.511$$
$$= 1.65 \text{ MeV}$$

4. (c) We use Wien's law of black body radiation

$$T = \frac{2.90 \times 10^{-3}}{\lambda_{\text{peak}}}$$
$$= \frac{2.90 \times 10^{-3}}{460 \times 10^{-9}}$$
$$= 6300 \text{ K}$$

5. (a) Transition 1 ⟶ 3 is largest energy, it is a absorption and so cannot occur in steps.

Transition 4 ⟶ 5 is least energy transition, it cannot occurs in steps.

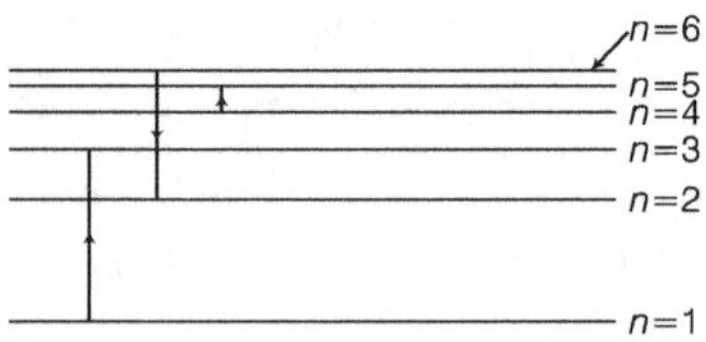

Transition 5 ⟶ 2 is emission, it can occur in steps.

6. (b,c,d)

As, $$E_n = -13.6\left(\frac{Z^2}{n^2}\right)$$

For Lithium, $Z = 3$

$$E_1 = -13.6 \times 9 = -122.4 \text{ eV}$$

and $$E_2 = -13.6 \times \frac{9}{4} = -30.6 \text{ eV}$$

Also, in stable state,

$$\text{KE} = |E_n|$$

and $$\text{PE} = 2 \times E_n$$

7. (b,d) Any electron can absorbs a photon of energy more than of ionisation energy of electron.

When, absorbed energy is greater than $|E_n - E_\infty|$, then electron carriers maximum part of remaining energy as its KE.

8. (a,b,c,d) Use : Number of spectral lines

$n_i = n_2, n_f = n_1$ is

$$N = \frac{(n_2 - n_1 + 1)(n_2 - n_1)}{2}$$

Recoil momentum of photon = recoil moment of atom

$$= \frac{h}{\lambda} = hRZ^2\left(\frac{1}{n_1^2} - \frac{1}{n_2^2}\right)$$

9. (a,b,c,d) Frequency,

$$f_n = \frac{\omega_n}{2\pi} = \frac{mZ^2e^4}{4\pi\varepsilon_0^2 h^3} \cdot \frac{1}{n^3}$$

$$f_n \propto \frac{1}{n^3}, \text{ graph } (B)$$

Magnetic field, $$B_n = \frac{\mu_0 i_n}{2\pi r_n}$$

$$= \frac{\pi m^2 Z^3 e^7 \mu_0}{8\varepsilon_0^3 h^5} \cdot \frac{1}{n^5}$$

$$\Rightarrow \quad B_n \propto \frac{1}{n^5}, \text{ graph } (C)$$

Energy $$E_n \propto \frac{1}{n^2}, \text{ graph } (A)$$

Magnetic moment $M_n \propto n$, graph (D).

10. (d) With $n = 2$, $l = 0$ or 1

With $l = 0$, $m_l = 0$

With $l = 1$, $m_l = -1$, 0 or +1

So, all of the given states are allowed.

11. (c) For transitions $3p \rightarrow 1s$ and $2p \rightarrow 1s$, $\Delta l = -1$

But , $3p \rightarrow 2p$ transition is forbidden, $\Delta l = 0$

But $3p \rightarrow 2p$ also occurs although very rarely with very low probability compared to other transitions.

12. (d) For SHO,

$$U_{\text{SHO}} = \frac{1}{2}kx^2$$

and $U_{\text{SHO}} = 0$ at $x = 0$ $(r = r_0)$,

Taking $x^2 = 1.6 \times 10^{22}$ m^2

$$k = \frac{2U_{\text{SHO}}}{x^2} = \frac{2 \times (0.5 \text{ eV}) \times (1.6 \times 10^{-19} \text{ J/eV})}{1.6 \times 10^{22} \text{m}^2}$$

$$= 1000 \text{ Nm}^{-2}$$

13. (c) Reduced mass $$\mu = \frac{m_1 m_2}{m_1 + m_2}$$

$$= \frac{m}{2} = \frac{1}{2} \times 1.0\, u$$

$$= \frac{1}{2} \times 1.0 \times 1.66 \times 10^{-27} \text{kg}$$

$$= 0.83 \times 10^{-27} \text{kg}$$

and $$\lambda = \frac{c}{f} = 2\pi c\sqrt{\frac{\mu}{k}}$$

$$\Rightarrow \quad \lambda = 2\pi(3 \times 10^8 \text{ms}^{-1})\left(\sqrt{\frac{0.83 \times 10^{-27}}{1000}}\right)$$

$$= 2\pi \times 3 \times 10^8 \times \sqrt{0.83 \times 10^{-30}}$$

$$= 2\pi \times 3 \times 10^8 \times \sqrt{0.83} \times 10^{-15}$$

$$\approx 2\pi \times 3 \times 0.9 \times 10^{-7}$$

$$\approx 5.4\pi \times 10^{-7} \text{m} \approx 540\,\pi \times 10^{-9} \text{m}$$

$$\approx 540\pi \text{ nm}$$

14. (b) As selection rule allows vibrational transitions for $\Delta v = \pm 1$,

So, allowed transitions have same energies $\Delta E_{\text{vib}} = \frac{1}{2}hf$

or $\Delta E_{\text{vib}} \approx 0.27$ eV (Practically with an accurate calculation).

So, every vibrational transition results in a photon of same wavelength.

15. (d) As nucleus is heavy recoil momentum of atom is neglected.

Now, E_n = energy of nth state,

$$= \frac{-Z^2R}{n}$$

R = Rydberg constant and $Z = 24$

So, $\Delta E = (Z \to 1)$

$$= Z^2R \cdot \left(1 - \frac{1}{4}\right)$$

$$= \frac{3}{4} Z^2R$$

16. (c) Energy required to remove $n = 4$ electron

$$= |E_u| = \frac{Z^2R}{16}$$

So, KE of emitted electron $= Z^2R\left(\frac{3}{4} - \frac{1}{16}\right)$

$$= \frac{11}{16} Z^2R$$

17. (a) Atomic density for a sample consisting of only one type of atoms

= Number of atoms per unit volume

= No. of atoms/kg × mass per m^3

$$= \frac{N_A}{M} \cdot \rho$$

$$= \frac{6.023 \times 10^{23} \times 3 \times 10^3}{26.98 \times 10^{-3}}$$

$$= 6.6 \times 10^{28}\ m^{-3}$$

18. (d) Electron density $= \frac{N_A}{M} \cdot Z \cdot \rho$

$$= 7.8 \times 10^{29}\ m^{-3}$$

19. (d) With some gas atoms in tube, electrons strike atoms and ionise them.

Ions are moved towards cathode and produces heat of collision, causing it to melt.

20. (c) $1/\lambda_{K_\alpha} = \left(\frac{1}{1^2} - \frac{1}{2^2}\right) \cdot R \cdot (Z-1)^2$

$$\therefore \quad \frac{\lambda_{K_\alpha}(Z = 43)}{\lambda_{K_\alpha}(Z = 29)} = \frac{1/(43-1)^2}{1/(29-1)^2}$$

$$\Rightarrow \quad \frac{\lambda}{\lambda'} = \frac{28^2}{42^2}$$

$$\Rightarrow \quad \lambda' = \frac{42^2}{28^2} \cdot \lambda = \frac{9}{4}\lambda$$

21. (a) In presence of an external magnetic field, the electron the magnetic dipole, (having dipole moment, $\mu_e = \frac{e}{2m_e} \cdot \left(\frac{nh}{2\pi}\right)$

with $\frac{e}{2m_e} \cdot \frac{h}{2\pi} = 9.274 \times 10^{-24}$ J/T

The Bohr magneton, interacts with field and have additional magnetic energy

$$U_B = -\mu \cdot \mathbf{B}$$

$$= \frac{e}{2m_e}(\mathbf{L} \cdot \mathbf{B})$$

$$= \left(\frac{eBm_e}{2m_e}\right)\frac{nh}{2\pi} = \frac{nh}{2\pi} \cdot \omega_L \cdot m_l$$

So, spectral lines have additional energy with

$$m_l = 0, +1 \text{ or } -1$$

So, in $n = 2 \to n = 1$ transition, there are three spectral lines which are equally spaced. (This is called Zeeman effect). Both Statement I and Statement II are correct with II being correct reason for Statement I.

22. (a) Due to partial screening,

$Z_{eff} = (Z-1)$ and so for

$n = 2 \to n = 1$ transition,

$$E_{K_\alpha} = \frac{-ke^2}{2a_0}\frac{(Z-1)^2}{Z^2} - \left(\frac{-ke^2}{2a_0}\frac{(Z-1)^2}{1^2}\right)$$

$$= \frac{ke^2}{2a_0} \times \frac{3}{4}(Z-1)^2 = 13.6 \times \frac{3}{4}(Z-1)^2$$

JEE Main & AIEEE Archive

1. (c) For hydrogen atom, we get

$$\frac{1}{\lambda} = R\,Z^2\left(\frac{1}{1^2} - \frac{1}{2^2}\right)$$

$$\frac{1}{\lambda_1} = R(1)^2\left(\frac{3}{4}\right)$$

$$\Rightarrow \quad \frac{1}{\lambda_2} = R(1)^2\left(\frac{3}{4}\right)$$

$$\frac{1}{\lambda_3} = R(2)^2\left(\frac{3}{4}\right)$$

$$\frac{1}{\lambda_4} = R(3)^2\left(\frac{3}{4}\right)$$

$$\therefore \quad \frac{1}{\lambda_1} = \frac{1}{4\lambda_3} = \frac{1}{9\lambda_4} = \frac{1}{\lambda_2}$$

2. (d) $\Delta E = h\nu$

where, ΔE is energy of radiation, h is Planck's constant and ν is frequency.

$$\Rightarrow \quad \nu = \frac{\Delta E}{h} = k\left[\frac{1}{(n-1)^2} - \frac{1}{n^2}\right]$$

$$= \frac{k2n}{n^2(n-1)^2} \approx \frac{2k}{n^3} \propto \frac{1}{n^3}$$

3. (d) In emission spectrum, number of bright lines is given by

$$\frac{n(n-1)}{2} = \frac{4(4-1)}{2} = 6$$

4. (d) Rotational kinetic energy of the two body system rotating about their centre of mass is

$$\text{R KE} = \frac{1}{2}\mu\omega^2r^2$$

where, $\mu = \frac{m_1m_2}{m_1 + m_2}$ = reduced mass

and angular momentum,

$$L = \mu\omega r^2 = \frac{nh}{2\pi}$$

$$\therefore \quad \text{R KE} = \frac{1}{2}\mu\omega^2r^2$$

$$= \frac{1}{2}\mu \cdot \left(\frac{nh}{2\pi\mu r^2}\right)^2 r^2$$

$$= \frac{n^2h^2}{8\pi^2\mu r^2} = \frac{n^2h^2}{2\mu r^2}$$

$$= \frac{(m_1 + m_2)n^2h^2}{2m_1m_2r^2}$$

5. (b) $\Delta E = 13.6\, Z^2 \left(\frac{1}{n_1^2} - \frac{1}{n_2^2}\right)$ $\left[\because \Delta E = Rhc\left(\frac{1}{n_1^2} - \frac{1}{n_2^2}\right)\right]$

$$= 13.6(3)^2 \left(\frac{1}{1^2} - \frac{1}{3^2}\right)$$

$$= 108.8 \text{ eV}$$

6. (d) IR corresponds to least value of $\left(\frac{1}{n_1^2} - \frac{1}{n_2^2}\right)$

i.e., from Paschen, Brackett and Pfund series. Thus, the transition corresponds to $5 \rightarrow 4$.

7. (b) $\frac{mv^2}{r_n} = \frac{k}{r_n}$ [given]

$$mvr_n = \frac{nh}{2\pi} \quad \text{[from Bohr's theory]}$$

On solving, $r_n \propto n$ and T_n is independent of n.

8. (c) Emission spectrum would rises, when electron makes a jump from higher energy level to lower energy level.

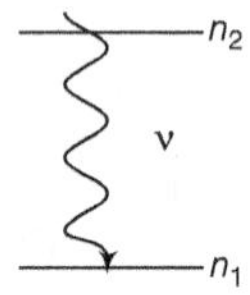

Frequency of emitted photon is proportional to change in energy of two energy levels, i.e.

$$\nu = RcZ^2 \left(\frac{1}{n_1^2} - \frac{1}{n_2^2}\right)$$

9. (a) $E = Rhc\left[\frac{1}{n_1^2} - \frac{1}{n_2^2}\right]$

$$E_{(4 \rightarrow 3)} = Rhc\left[\frac{1}{3^2} - \frac{1}{4^2}\right] = Rhc\left[\frac{7}{9 \times 16}\right] = 0.05\, Rhc$$

$$E_{(4 \rightarrow 2)} = Rhc\left[\frac{1}{2^2} - \frac{1}{4^2}\right]$$

$$= Rhc\left[\frac{3}{16}\right] = 0.2\, Rhc$$

$$E_{(2 \rightarrow 1)} = Rhc\left[\frac{1}{(1)^2} - \frac{1}{(2)^2}\right]$$

$$= Rhc\left[\frac{3}{4}\right] = 0.75\, Rhc$$

$$E_{(1 \rightarrow 3)} = Rhc\left[\frac{1}{(3)^2} - \frac{1}{(1)^2}\right]$$

$$= -\frac{8}{9} Rhc = -0.9\, Rhc$$

Thus, III transition gives most energy. I transition represents the absorption of energy.

10. (a) $E = -Z^2 \frac{13.6}{n^2}$ eV

For first excited state,

$$E_2 = -3^2 \times \frac{13.6}{4} \quad [\because n = 2, \text{ for Ist excited state}]$$

$$= -30.6 \text{ eV}$$

Ionisation energy for first excited state of Li^{2+} is 30.6 eV.

JEE Advanced & IIT JEE Archive

1. (b) K_α transition takes place from $n_1 = 2$ to $n_2 = 1$

$\therefore$ $$\frac{1}{\lambda} = R(Z - b)^2 \left[\frac{1}{(1)^2} - \frac{1}{(2)^2}\right]$$

For K-series, $b = 1$

$\therefore$ $$\frac{1}{\lambda} \propto (Z - 1)^2$$

$\Rightarrow$ $$\frac{\lambda_{Cu}}{\lambda_{Mo}} = \frac{(Z_{Mo} - 1)^2}{(Z_{Cu} - 1)^2} = \frac{(42 - 1)^2}{(29 - 1)^2}$$

$$= \frac{41 \times 41}{28 \times 28} = \frac{1681}{784} = 2.144$$

2. (a)

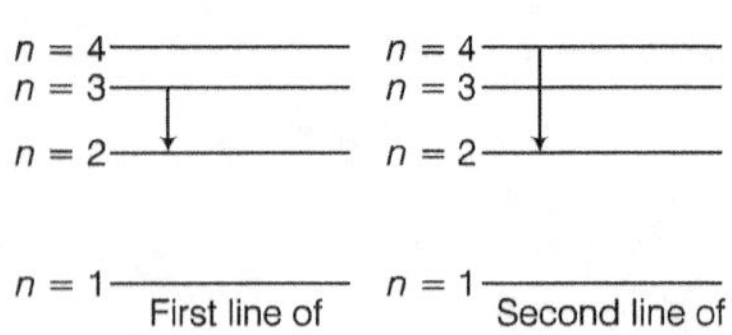

First line of Balmer series — Second line of Balmer series

For hydrogen or hydrogen type atoms

$$\frac{1}{\lambda} = RZ^2 \left(\frac{1}{n_f^2} - \frac{1}{n_i^2}\right)$$

In the transition from $n_i \longrightarrow n_f$

$\therefore$ $$\lambda \propto \frac{1}{Z^2\left(\frac{1}{n_f^2} - \frac{1}{n_i^2}\right)}$$

$\therefore$ $$\frac{\lambda_2}{\lambda_1} = \frac{Z_1^2\left(\frac{1}{n_f^2} - \frac{1}{n_i^2}\right)_1}{Z_2^2\left(\frac{1}{n_f^2} - \frac{1}{n_i^2}\right)_2},$$

$$\lambda_2 = \frac{\lambda_1 Z_1^2\left(\frac{1}{n_f^2} - \frac{1}{n_i^2}\right)_1}{Z_2^2\left(\frac{1}{n_f^2} - \frac{1}{n_i^2}\right)_2}$$

Substituting the values, we have

$$= \frac{(6561 \text{ Å})(1)^2\left(\frac{1}{2^2} - \frac{1}{3^2}\right)}{(2)^2\left(\frac{1}{2^2} - \frac{1}{4^2}\right)}$$

$$= 1215 \text{ Å}$$

3. (d) $L = I\omega = \frac{nh}{2\pi}$

$\therefore$ $$\omega = \frac{nh}{2\pi I}$$

$$K = \frac{1}{2} I\omega^2$$

$$= \frac{1}{2} I\left(\frac{nh}{2\pi I}\right)^2$$

$$= \frac{n^2h^2}{8\pi^2 I}$$

4. (b) $h\nu = K_2 - K_1 = \dfrac{3h^2}{8\pi^2 I}$

$\therefore \quad I = \dfrac{3h}{8\pi^2 f} = \dfrac{3 \times 2\pi \times 10^{-34} \times \pi}{8 \times \pi^2 \times 4 \times 10^{11}}$

$= 1.87 \times 10^{-46}$ kg-m^2

5. (c) $I = \mu r^2$ (where μ = reduced mass)

$\mu = \dfrac{m_1 m_2}{m_1 + m_2} = \dfrac{48}{7}$ amu

$= 11.43 \times 10^{-27}$ kg

Substituting in $I = \mu r^2$ we get,

$r = \sqrt{\dfrac{I}{\mu}} = \sqrt{\dfrac{1.87 \times 10^{-46}}{11.43 \times 10^{-27}}}$

$= 1.28 \times 10^{-10}$ m

6. (a,c) $L = 3\left(\dfrac{h}{2\pi}\right)$

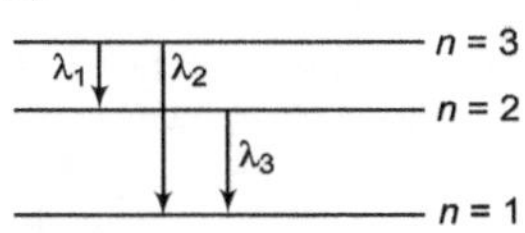

$\therefore \quad n = 3$, as $L = n\left(\dfrac{h}{2\pi}\right)$

$r_n \propto \dfrac{n^2}{z}$

$r_3 = 4.5\, a_0$

$\therefore \quad z = 2$

$\dfrac{1}{\lambda_1} = Rz^2\left(\dfrac{1}{2^2} - \dfrac{1}{3^2}\right) = 4R\left(\dfrac{1}{4} - \dfrac{1}{9}\right)$

$\therefore \quad \lambda_1 = \dfrac{9}{5R}$

$\dfrac{1}{\lambda_2} = Rz^2\left(\dfrac{1}{1^2} - \dfrac{1}{3^2}\right) = 4R\left(1 - \dfrac{1}{9}\right)$

$\Rightarrow \quad \lambda_2 = \dfrac{9}{32R}$

$\dfrac{1}{\lambda_3} = Rz^2\left(\dfrac{1}{1^2} - \dfrac{1}{2^2}\right) = 4R\left(1 - \dfrac{1}{4}\right)$

$\Rightarrow \quad \lambda_3 = \dfrac{1}{3R}$

7. (b) Cut-off wavelength depends on the applied voltage not on the atomic number of the target. Characteristic wavelengths depend on the atomic number of target.

8. (c)

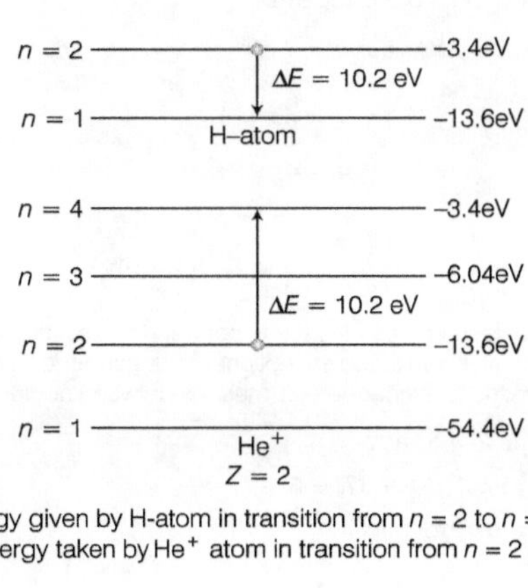

Energy given by H-atom in transition from $n = 2$ to $n = 1$ is equal to energy taken by He^+ atom in transition from $n = 2$ to $n = 4$.

9. (c) Visible light lies in the range, $\lambda_1 = 4000$Å to $\lambda_2 = 7000$Å Energy of photons corresponding to these wavelengths (in eV) would be :

$E_1 = \dfrac{12375}{4000} = 3.09$ eV

$E_2 = \dfrac{12375}{7000} = 1.77$ eV

From energy level diagram of He^+ atom we can see that in transition from $n = 4$ to $n = 3$, energy of photon released will lie between E_1 and E_2.

$\Delta E_{43} = -3.4 - (-6.04)$

$= 2.64$ eV

Wavelength of photon corresponding to this energy,

$\lambda = \dfrac{12375}{2.64}$ Å

$= 4687.5$ Å

$= 4.68 \times 10^{-7}$m

10. (a) Kinetic energy $K \propto Z^2$

$\therefore \quad \dfrac{K_H}{K_{He^+}} = \left(\dfrac{1}{2}\right)^2 = \dfrac{1}{4}$

11. (b) Hence, (i) → (p,r), (ii) → (p,q,s), (iii) → (p), (iv) → (q).

12. (b) The series in UV region is Lymen series. Longest wavelength corresponds to minimum energy which occurs in transition from $n = 2$ to $n = 1$.

$\therefore \quad 122 = \dfrac{\frac{1}{R}}{\left(\frac{1}{1^2} - \frac{1}{2^2}\right)}$ ...(i)

The smallest wavelength in the infrared region corresponds to maximum energy of Paschen series.

$\therefore \quad \lambda = \dfrac{\frac{1}{R}}{\left(\frac{1}{3^2} - \frac{1}{\infty}\right)}$ ...(ii)

Solving Eqs. (i) and (ii), we get

$\lambda = 823.5$ nm

13. (a) $\dfrac{1}{\lambda} \propto (Z-1)^2$

$\therefore \quad \dfrac{\lambda_1}{\lambda_2} = \left(\dfrac{Z_2 - 1}{Z_1 - 1}\right)^2$

or $\quad \dfrac{1}{4} = \left(\dfrac{Z_2 - 1}{11 - 1}\right)^2$

Solving this, we get $Z_2 = 6$

14. (i) From the relation $r \propto A^{1/3}$,

we have, $\dfrac{r_2}{r_1} = \left(\dfrac{A_2}{A_1}\right)^{1/3}$

or $\quad \left(\dfrac{A_2}{4}\right)^{1/3} = (14)^{1/3}$

$\therefore \quad A_2 = 56$

(ii) $Z_2 = A_2$ − number of neutrons

$= 56 - 30 = 26$

$\therefore \; f_{K\alpha} = Rc\,(Z-1)^2\left(\dfrac{1}{1^2} - \dfrac{1}{2^2}\right)$

$= \dfrac{3Rc}{4}(Z-1)^2$

Substituting the given values of R, c and Z, we get

$f_{K\alpha} = 1.55 \times 10^{18}$ Hz.

3. Nuclei

DPP-1 **Atomic Nucleus**
(Composition, Size, Atomic Masses, Isotopes, Isobars, Isotones and General Stability Trends of Nucleus)

1. (i) To resolve two objects, wavelength of the probing signal must be less than distance between them.

$$\lambda << d$$

Now for an electron to have an associated wavelength 10^{-15}m,

$$\lambda = \frac{h}{p} \text{ and KE} = pc$$

$$\Rightarrow \quad \text{KE} = pc = \frac{hc}{\lambda}$$

$$= \frac{6.6 \times 10^{-34} \times 3 \times 10^{8}}{1.6 \times 10^{-19} \times 10^{-15}} \text{ (eV)}$$

$$= 10^{9}\text{eV} = 1\,\text{GeV}$$

(ii) Atomic mass of $^{12}_{6}$C is 12*u* (amu). Neglecting the masses and binding energies of the six electrons we have,

$$\text{Nuclear density } \rho = \frac{\text{Mass}(m)}{\text{Volume}\left(\frac{4}{3}\pi R^3\right)}$$

$$= \frac{12 \times 1.66 \times 10^{-27}}{\frac{4}{3} \times \pi \times 2.7 \times 10^{-15}}$$

$$= 2.4 \times 10^{17}\,\text{kgm}^{-3}$$

(Nearly 2-billion tons per cubic centimeter)

Nuclear density is same for all nuclei.

(iii) We use coulomb's formula

$$F = \frac{1}{4\pi\varepsilon_0} \cdot \frac{e^2}{r^2}$$

$$= \frac{9 \times 10^{9} \times (1.6 \times 10^{-19})^2}{(2.4 \times 10^{-15})^2}$$

$$= 40\text{ N}$$

The attractive forces that bind protons into nucleus are much more stronger.

(iv) For a stable nucleus, its nucleons must be tightly bound and the binding energy must be maximum for given mass number.

For this, we solve $\frac{dE_b}{dZ} = 0$.

$$\Rightarrow \quad \frac{dE_b}{dZ} = \frac{d}{dZ}\left\{a_1 A - a_2 A^{2/3} - \frac{a_3 Z(Z-1)}{A^{1/3}} - \frac{a_4 \cdot (A-2Z)^2}{A} \pm \frac{a_5}{A^{3/4}}\right\}$$

(A = constant, Z = variable)

$$\Rightarrow \quad \frac{dE_b}{dZ} = \frac{-a_3}{A^{1/3}}(2Z-1) + \frac{4a_4}{A}(A-2Z) = 0$$

$$\Rightarrow \quad Z = \frac{a_3 A^{-1/3} + 4a_4}{2a_3 A^{-1/3} + 8a_4 A^{-1}}$$

$$= \frac{0.595\,A^{-1/3} + 76}{1.19\,A^{-1/3} + 152\,A^{-1}}$$

For $A = 25$, this formula gives $Z = 11.7$, which conclude that $Z = 12$. This stable nuclide is $^{25}_{12}$Mg which is the only stable isobar with $A = 25$. The other isobars $^{25}_{11}$Na and $^{25}_{13}$Al are radioactive.

2. (c) In silicon,

$$A = N + Z = 28$$
$$Z = 14$$
$$\Rightarrow \quad N_{si} = A - Z = 28 - 14 = 14$$

In Rubidium,

$$N_R = A - Z = 85 - 37 = 48$$

$$\text{Ratio} = \frac{N_{Si}}{N_R} = \frac{14}{48} = \frac{7}{24}$$

3. (d) As, $A = N + Z$ so, given nuclei may have different neutron number (N) and different atomic number (Z), but they may have same A.

Also, nucleus radius

$$R \propto A^{1/3}$$

$\Rightarrow$ Both A and B nuclei may have same radii.

4. (c) In a chemical reaction, energies involved are in eV's so, this is not possible to involve any nucleon for this less energy.
Isotopes have same atomic number (Z) and similar electronic configuration, they differ in terms of their nuclei. As energies involved are less so, chemical properties of isotopes are similar.

5. (a) Two isobars with same mass number (A) have more neutron number (N) or more atomic number (Z).

An isobar with more atomic number is slight unstable with respect to the isobar which have more N.

As, neutrons are slightly heavier than protons, a more stable isobar is heavier (But both isobars have same mass number).

6. (d) Option (d) is incorrect as nuclear force becomes weaker than electrostatic force at large separations. Nuclear force is essentially short range.

7. (c) Most stable, $N = Z(B)$. Nucleus with more neutron number (N) is slightly more stable than nucleus which have more atomic number (Z).

So, order of stability is $B > A > C$.

8. (d) For $Z > 83$, all naturally occurring elements are radioactive.
As size of nucleus grows, nuclear forces, being short range does not able hold protons together.

9. (d) Volume of nucleus $V = \frac{4}{3}\pi R^3$

$$V = \frac{4}{3}\pi R_0^3 A,$$

$$R = R_0 A^{1/3}$$

$$R_0 = \text{Probe constant} = 1.2\text{ fm}$$

$$\Rightarrow \quad \frac{V}{A} = \frac{4}{3}\pi R_0^3 = \text{a constant}$$

10. (d) Nuclear forces are not central forces, these are non conservative and these are not charge dependent. Nuclear forces occurs due to exchange of π-mesons between nucleons.

11. (d) In $^{7}_{3}$Li, $Z = 3$ and $N = 4$

In $^{7}_{4}$Be, $Z = 4$ and $N = 3$

These are mirror nuclei.

12. (d) Velocity vector of positively charged nucleus can be resolved in to parallel and perpendicular velocity components.

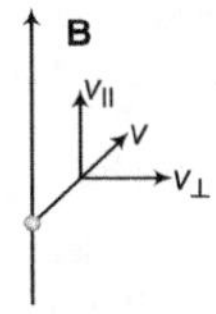

Due to perpendicular velocity component $v_{\perp}$, nucleus moves in circular path of radius $r = \frac{mv_{\perp}}{Bq}$.

Pitch of helical path $p = T \cdot v_{||}$,

T = Time period of rotation.

$$= \frac{2\pi m}{Bq} \cdot v_{||}$$

For two nuclei of two different isotopes, Z = same but m may be different. So, a heavier nuclei go in a large circular helix of larger pitch.

13. (a,b,c) For a stable nucleus, $\frac{N}{Z} = 1$, for a high $\frac{N}{Z}$ value, nucleus is unstable.

Low binding energy means loosely bound nucleus and hence, less stable nucleus.

A high packing fraction means unstable nucleus.

14. (a,b,c,d) All options are correct.

Same A = Isobars, Same Z = Isotopes

Same N = Isotones

15. (b,c,d) Radius of a nucleus depends on mass number A as

$$R = R_0 A^{1/3}$$

Also increase in atomic number (Z) does not necessarily means an increase in atomic size.

16. (b,c,d) Only option (a) is correct. $^1H_1$ and $^2H_1$ have same atomic number (Z).

$\therefore$ These are isotopes.

17. (a,b,d) For each proton, if there is a neutron, then nucleus is more stable.
Also, each nuclear level must have two protons (of opposite spins) and 2-neutrons of opposite spin, in an stable nucleus.

18. (a) Size of a nucleus is estimated by measuring distance of closest approach by a probe. Both Statement I and Statement II are correct and Statement II is a correct statement for Statement I.

19. (a) Average atomic mass

$$= \frac{m_1 x_1 + m_2 x_2}{x_1 + x_2}$$

m_1, m_2 are the masses of isotopes in sample with x_1 and x_2 as their relative abundance.

20. (b) For a charged particle, radius of path

$$r = \frac{mv}{qB}$$

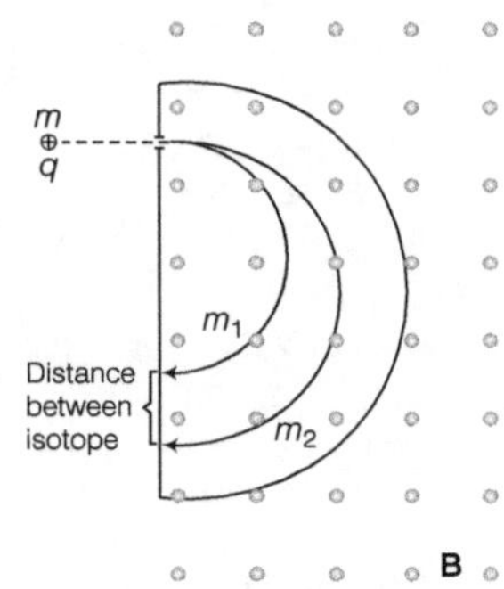

$$\Rightarrow \quad r = \frac{mv}{qB} = \frac{\sqrt{2mK}}{Bq}$$

When particles (nuclei of 2-different isotopes) have same energies, then

$$r_1 = \frac{\sqrt{2m_1K}}{BZ}$$

and $$r_2 = \frac{\sqrt{2m_1K}}{BZ}$$

Separation between two traces,

$$d = 2r_2 - 2r_1 \quad \text{(see figure)}$$

$$= \frac{2\sqrt{2m_2K}}{BZ} - \frac{2\sqrt{m_1K}}{BZ}$$

$$= \frac{2\sqrt{2K}}{BZ}(\sqrt{m_2} - \sqrt{m_1})$$

21. (c) For same speed separation

$$= d = \frac{2v(m_2 - m_1)}{qB}$$

$$= \frac{2v(m_2 - m_1)}{BZ}$$

DPP-2 **Mass Defect, Binding Energy, Nuclear Stability Criteria for Fission and Fusion Processes**

1. (i) $3(^2H) - \frac{1}{2}(^{12}C) = 4.2300 \times 10^{-2}$

$$\Rightarrow \quad 3(^2H) - \frac{1}{2} \times 12 = 4.2300 \times 10^{-2}$$

$$\Rightarrow \quad 3(^2H) = 6 + 0.0423$$

$$\Rightarrow \quad 3(^2H) = 6.0423$$

$$\Rightarrow \quad ^2H = \frac{1}{3} \times 6.0423 = 2.0141 \text{ u}$$

Similarly solving given equations, we get

$$m(^2H) = 2.0141 \text{ u}$$

$$m(^1H) = 1.0078 \text{ u}$$

$$m(^{16}O) = 15.9949 \text{ u}$$

(ii) $$m(n) + m(^{235}_{92}U) = 236.0529 \text{ u}$$

$$m(^{95}_{42}M_O) + m(^{139}_{57}La) + 2(n) = 235.8299 \text{ u}$$

So, excess mass

= mass initial – mass final

= 0.223 u

$\therefore$ Energy released in fission

= 0.223×931.5 MeV

= 207.16 MeV

This energy appears a kinetic energy of fission products.

2. (d) Relation is $\Delta M \times 931.5 = \Delta E$

ΔM in u, ΔE in MeV.

3. (d) We obtain a reverse equation,

$$^{41}_{20}\text{Ca} \longrightarrow\ ^{40}_{20}\text{Ca} +\ ^{1}_{0}n$$

Now, $\Sigma[E(\text{products})] = 37225.154 + 939.57145 = 38164.72545$

And ΔE = energy required to remove one neutron

$= \Sigma E(\text{reactants}) - \Sigma E(\text{products})$

$= (38164.72545 - 38156.362) = 8.36345$ MeV

4. (d) As, $3P \longrightarrow Q + E_1$

$\Rightarrow \quad Q - 3P = E_1$

$\Rightarrow \quad E_1 = E_Q - 3E_P \quad \ldots\text{(i)}$

Aslo, $2Q \longrightarrow R + E_2$

$\Rightarrow \quad E_2 = E_R - 2E_Q \quad \ldots\text{(ii)}$

And $R - 6P = E_3$

$\Rightarrow \quad E_R - 6E_P = E_3 \quad \ldots\text{(iii)}$

Now, $2E_1 + E_2 = 2(E_Q - 3E_P) + (E_R - 2E_Q)$

$= E_R - 6E_P = E_3$

5. (a) Fission rate $= \dfrac{\text{Total energy required per second}}{\text{Energy per fission}}$

$$= \frac{5}{200 \times 1.6 \times 10^{-19} \times 10^{6}}$$

$= 1.56 \times 10^{11}\ \text{s}^{-1}$

6. (c) Energy released $= E_{\text{product}} - E_{\text{reactants}}$

$= z - x + y = z - x - y$

7. (a) α-particle is $^4_2$He (2 protons and 2 neutrons).

$\Rightarrow \{m(2p) + m(2n) - m(^4_2\text{He})\} \times 931 = BE$ (in MeV)

$\Rightarrow \quad \text{BE} = (2 \times 1.008 + 2 \times 1.009 - 4.003) \times 931.5$

$= (2.016 + 2.018 - 4.003) \times 931.5$

$= 0.03 \times 931.5 = 0.031 \times 931.5$

$= 28.8765 \approx 28.8$ MeV

8. (a) ΔE = energy released

= energy initial – energy final

$= 2 \times 118 \times 8.54 - 235 \times 7.6$

$= 2015.44 - 1786 \approx 229$ MeV

9. (c) The complete process is a shown in the figure.

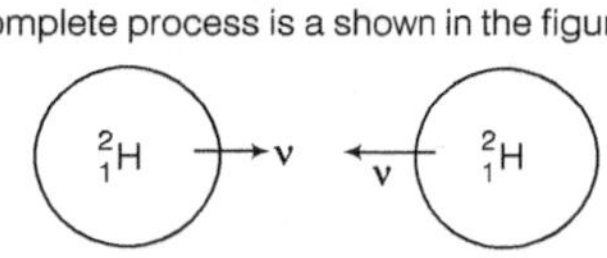

Potential energy barrier for 2 fm separation $= U = \dfrac{kq_1q_2}{r}$

$$= \frac{9 \times 10^{9} \times (1.6 \times 10^{-19})^2}{2 \times 10^{-15}}$$

$= 1.15 \times 10^{-13}$ J

On equating this with available KE, we get

$$2 \times \frac{1}{2} mv^2 = 1.15 \times 10^{-13}$$

$$\Rightarrow \quad v^2 = \frac{1.15 \times 10^{-13}}{2 \times 1.67 \times 10^{-27}}$$

$$\Rightarrow \quad v \approx 5.8 \times 10^{6}\ \text{ms}^{-1}$$

10. (d) Majority of atoms in a human body is of low atomic number (Z), such as H,C,O etc. Also 70% of our body is of water (H_2O) with 8 neutrons and 10 photons.

Let half of 50 kg mass composed of protons and neutrons.

So, $\quad N = 25 \times \dfrac{1}{1.6 \times 10^{-27}} \approx 1.5 \times 10^{28}$

11. (d) Number of protons in a class XII students is of order 10^{28} (Refer to solution of Q. No. 10).

12. (b,c) Option (a) is incorrect, $^4_2$He is a more stable species among its neighbours due to its high BE per nucleon.

$$^2_1\text{H} +\ ^2_1\text{H} \longrightarrow\ ^3_1\text{H} +\ ^1_1p$$

is not possible as nucleon number is not conserved.

13. (a,b) option (c) and option (d) are possible but in option (a) and option (b), nutrions must appear.

14. (a,c,d) Due to electrostatic repulsion proton removal is considerably easier. BE is different for different nucleons.

15. (a,b) For fusion, two nuclei must overcome coulomb barrier, which is very higher. Due to short range of nuclear forces, attraction of nuclei is not possible for distances more than 2 fm.

16. (a,b,c,d) All option are correct. In lighter species, we can consider shell model and magic numbers. So few lighter species with 2, 8 ... etc. nucleons are more stable and have large values of BE per nucleon than others.

17. (d) Statement I is incorrect as BE $\propto A$ and Statement II is correct as a nucleon intracts with its immediate neighbours.

18. (a) Statement I is correct, Statement II is correct and is also a correct reason for Statement I. Pauli's exclusion principle allows 4-nucleons, 2-protons of opposite spin and 2 neutron of opposite spin to occupy the same energy level. If a nucleon is added, it is not allowed in same level. So, it go to the next level and would not be so tightly bound. Thus, most stable nuclides are those which have their neutron number equal to atomic number ($N = Z$).

19. (a) Σ(mass of nucleons) $> M$(nucleus)

A higher mass means a higher energy state or configuration, a lower mass means a lower energy state which is stable.

20. (d)

(i) Large BE = Stable

(ii) Large mass defect ≡ large BE ≡ Stable

(iii) Large packing fraction ≡ less densely packed ≡ Unstable

(iv) Large number of e^- ≡ large number of protons ≡ Unstable

(v) Large number of ^1_0n ≡ Stable

(vi) Large radius ≡ less stable.

21. (5)

$$\Delta M = \{m(\text{U}^{235}) + m(n')\} - \{m(\text{Zr}^{99}) + m(\text{Te})^{134} + 3m(n')\}$$

$= 0.1985$ u

So, Q value of reaction is

$Q = \Delta M \times 931.5$

$= 185$ MeV

$= 5 \times 37$ MeV

Hence, (i) → (p), (ii) → (p), (iii) → (q), (iv) → (q), (v) → (p), (vi) → (q).

DPP-3 **Radioactivity and Law of Radioactive Disintegration**

1. (i) $N = N_0 e^{-\lambda t}$

and Activity, R = decay

$$\text{Rate} = \frac{dN}{dt} = -\lambda \cdot N_0 e^{-\lambda t}$$

$$\Rightarrow \quad R = \frac{dN}{dt} = -\lambda N$$

So, activity is proportional to number of active nucleus

$$\text{So,} \quad \frac{R(t_2)}{R(t_1)} = \frac{-\lambda N(t_2)}{-\lambda N(t_1)} = \frac{N_0 e^{-\lambda t_2}}{N_0 e^{-\lambda t}}$$

$$\Rightarrow \quad \frac{R(t_2 = t)}{R(t_1 = 0)} = \frac{\text{Final activity}}{\text{Initial activity}} = \frac{e^{-\lambda t}}{e^{-\lambda \times 0}} = e^{-\lambda t}$$

$$\Rightarrow \quad \frac{9}{15} = e^{-\lambda t}$$

$$\Rightarrow \quad t = \frac{1}{\lambda} \log\left(\frac{5}{3}\right)$$

As, in the specimen replenishment of C^{14} is stopped, it shows a smaller activity due to decay. That's why we are taking its value as final activity.

$$\Rightarrow \quad t = \frac{T_{1/2}}{0.693} \times \log\frac{5}{3}$$

$$\therefore \quad \lambda = \frac{0.693}{T_{1/2}}$$

$\Rightarrow \quad t = 4224$ yr (Approximately)

(ii) No, age of an old rock cannot be found by carbon dating technique, as a rock may not contains C^{14} isotope whose replenishment is stopped at the time of formation of rock.

Carbon dating can be used only for biological specimen as they keep on replenishing C^{14} isotope till they are alive and replenishment is stopped when the specimen dies.

2. (c) ${}^{27}_{13}\text{Al} + {}^{1}_{0}n \longrightarrow {}^{28}_{13}\text{Si}^*$

$${}^{28}_{13}\text{Si}^* \longrightarrow {}^{28}_{14}\text{Si} + \beta^-$$

$${}^{1}_{0}n \longrightarrow \underset{\text{remains in nucleus}}{{}^{1}_{1}p} + \underset{\beta^-\text{ radiation}}{{}^{0}_{-1}e} + \underset{\text{Antinutrino}}{\bar{\nu}}$$

3. (d) As, $m({}^{2}_{1}\text{H}) = 2.014082$ u

and $m({}^{3}_{2}\text{He}) = 3.016029$

and $m({}^{1}_{0}n) = 1.008665$

Δm = mass defect

$= 2m({}^{2}_{1}\text{H}) - m^3({}^{3}_{2}\text{He}) - m({}^{1}_{0}n)$

$= 2 \times 2.014082 - 3.016029 - 1.008665$

$\Rightarrow$ Q-value = Energy librated in reaction

= [mass (reactants) – mass (products)] × 931.5

$= \Delta m \times 931.5$ MeV = 3.232 MeV

As Q-value is positive

So, no threshold energy is required.

4. (b) As, $m({}^{1}_{0}n) = 1.008665$ u

$m({}^{24}_{12}\text{Mg}) = 23.985042$ u

$m({}^{23}_{11}\text{Na}) = 23.985042$ u

$m({}^{2}_{1}\text{H}) = 2.014082$

We find mass effect

$$\Delta M = m({}^{1}_{0}n) + m({}^{24}_{12}\text{Mg}) - m({}^{23}_{11}\text{Na}) - m({}^{2}_{1}\text{H})$$

Q = {mass (reactants) – mass (product)}× 931.5

= – 9.449 MeV

So, an energy of more than Q-value is required (Q-value is negative) for reaction to occur. But KE of neutrons is more than Q-value

∴ Reaction can proceed and,

16 – 9.449 = 6.55 MeV of energy is released.

5. (a) From figure, we see that collision occurs, when

$$d \le R_1 + R_2$$

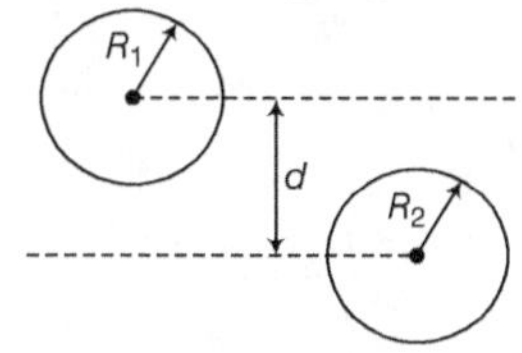

⇒ Area of effective cross-section is

$\pi(d)^2 = \pi(R_1 + R_2)^2$

$= \pi(R_0 A_1^{1/3} + R_0 A_2^{1/3})^2$

$= \pi(1.2\text{ fm}(A_1^{1/3} + A_2^{1/3})^2$

$= \pi[(1.2(27^{1/3} + 125^{1/3})]^2$

$= \pi \times [1.2(3+5)]^2 = \pi \times (1.2 \times 8)^2$

$= 92.16\,\pi\,(\text{fm})^2$

6. (b) From fission of ${}^{238}_{92}\text{U}$, a total of 200 MeV per fission is released.

$$\Rightarrow \quad \frac{950 \times 10^6\text{ J}}{0.38\text{s}} \times \frac{1\text{ atom}}{200 \times 10^6\text{ eV}} \times \frac{1\text{ eV}}{1.6 \times 10^{-19}\text{ j}} \times \frac{0.235\text{ kg}}{6.02 \times 10^{23}}$$

= Mass of ${}^{235}\text{U}$ required per second

So, for 1 year, amount of ${}^{235}\text{U}$ required = 962 kg,

(Take, 1 Ton ≡ 1000 kg)

7. (b) We use statistical relation

$$E_n = E_0\left(\frac{1}{2}\right)^n$$

which is analogous to activity rule.

n = number of collisions.

$$\Rightarrow \quad E_n = E_0\left(\frac{1}{2}\right)^n$$

$$\Rightarrow \quad n = \frac{\ln\frac{E_n}{E_0}}{\ln\frac{1}{2}} = 24.58 \approx 25$$

Hence, 25 collisions are to be make.

8. (c) $C > B > A$

α-particles have large ionisation power, they have same energy as γ-rays. So, α-rays are more dangerous than γ-rays.

9. (c) Conservation of momentum gives $p_{pr} = p_b$

$$\sqrt{2m\,pr \cdot K_{pr}} = \sqrt{2m_b \cdot K_{br}}$$

$$\Rightarrow \quad K_{pr} = \left(\frac{m_b}{m_{pr}}\right) K_b = Q + K_b$$

$$\Rightarrow \quad K_b = \frac{-Qm_{pr}}{m_{pr} - m_b}$$

10. (a) Decay constant, $\lambda = \dfrac{\log 2}{T_{1/2}}$

As, $\quad T_{1/2}(\text{Te}) > T_{1/2}(\text{Sr})$

$\Rightarrow \quad \lambda(\text{Te}) < \lambda(\text{Sr})$

(All other options are incorrect)

11. (d) 200 g of carbon is nearly all of ${}^{6}_{12}\text{C}$.

12 g of ${}^{12}C_6 = 6.02 \times 10^{23}$ atoms

⇒ 200 g contains a total of

$$= \left(\frac{200}{12}\right) \times 6.02 \times 10^{23}$$

$= 1 \times 10^{25}$ atoms

When animal is alive, ratio of ${}_6^{14}C$ to ${}_6^{12}C$ is 1.3×10^{-12} in the bone.
Number of ${}_6^{14}C$ nuclei at the time.

$$= 1 \times 10^{25} \times 1.3 \times 10^{-12} = 1.3 \times 10^{13} \text{ atoms}$$

As, $$\lambda = \frac{(dN/dt)\,t = 0}{N_0}$$

So, $$\left(\frac{dN}{dt}\right)_{t=0} = 3.83 \times 10^{-12} \times 1.3 \times 10^{13} = 50 \text{ s}^{-1}$$

So, $$e^{\lambda t} = \frac{\left(\frac{dN}{dt}\right)_{t=0}}{\left(\frac{dN}{dt}\right)_{t=t}} = \frac{50 \text{ s}^{-1}}{16 \text{ s}^{-1}}$$

$$\Rightarrow \quad t = \frac{1}{\lambda} \ln\left(\frac{50}{16}\right) = \frac{1}{3.83 \times 10^{-12}} \times \ln\left(\frac{50}{16}\right)$$

$\approx$ 9400 yr (About 10000 yr)

12. (a,b,c,d) α-decay occurs because strong nuclear force is unable to hold very large nuclei together. The large atomic number (Z) implies a large electric force but nuclear force being short range does not exceeds, resulting in α-decay.

Q-value of α-decay is 931.5 $\{M_P - (M_D + M_\alpha)\}$

Here, M_P = mass of parent nucleus
M_D = mass of daughter nucleus
M_α = mass of α-particle

When $M_P < (M_D + M_\alpha)$, Q = negative and conservation of energy is not followed so, the decay cannot occur spontaneously.

13. (a,c) In case of a β^--decay.

$${}_0^1n \longrightarrow {}_1^1p + {}_{-1}^{\ 0}e + \bar{\nu}$$

In case of a β^+-decay

$${}_1^1p \longrightarrow {}_0^1n + {}_{+1}^{\ 0}e + \bar{\nu}$$

Energy is carried mostly by β and ν (or $\bar{\nu}$) $\therefore$ β has a continuous energy distribution.

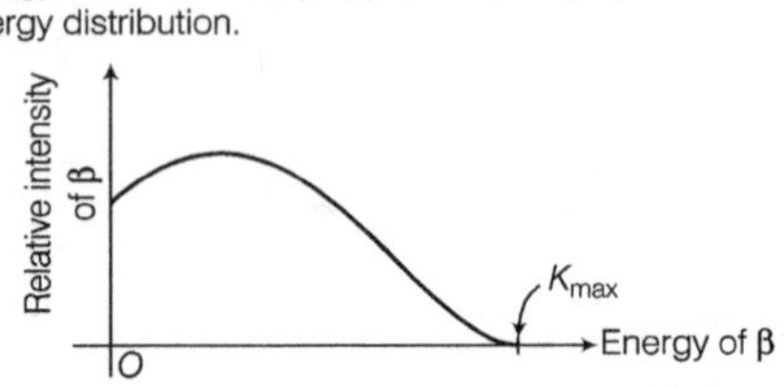

Option (b) is incorrect as part of energy and momentum is carried by nutrino so following momentum conservation recoil direction of nucleus is not same as β.

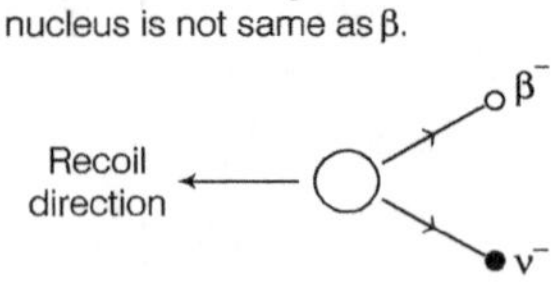

Option (d) is incorrect as neutrions are not massless, also they intracts with matter in certain conditions.

14. (a,d) Option (b) and option (c) are incorrect. An excited nucleus radiates energy and jumps to a lower or ground state. As nuclear energy levels are much further apart than those of an atom, frequency of emitted photon is greater than 10^{19} Hz and wavelength smaller than 3×10^{-12} m.

15. (a,c) ${}_{92}^{232}U \longrightarrow {}_{90}^{228}Th + {}_2^4He$

$$Q = \Delta M \times 931.5 = 5.40 \text{ MeV}$$

This energy is distributed between the α and daughter nucleus in the inverse ratio of their masses (4/228) in this case.

So, $$\frac{K_\alpha}{K_{Th}} = \frac{228}{4}$$

$\Rightarrow$ $K_\alpha = 5.30$ MeV and $K_{Th} = 0.10$ MeV

But if ${}_{90}^{238}Th$ is in excited state, then energy K_α is less.

$$K_\alpha \text{ (max)} = 5.30 \text{ MeV}$$

16. (a,c,d) ${}_{84}^{210}Po \longrightarrow {}_2^4He + {}_{82}^{206}X$

So, daughter nucleus is ${}_{82}^{206}Pb$

Disintegration energy

$$Q = \frac{A}{A-4} \times (KE_\alpha)$$

$$= \left(\frac{210}{210-4}\right) \times 5.30$$

$$= 5.40 \text{ MeV}$$

Mass equivalence of disintegration energy,

$$M_Q = \frac{5.40}{931} = 0.0058 \text{ u}$$

$$\Rightarrow \quad m({}_{82}^{206}Pb) = m_{Po} - m_\alpha - m_Q$$

$$= 209.9829 - 4.0026 - 0.0058$$

$$= 205.9745 \text{ u}$$

17. (d) Probabiltiy of survival = $\dfrac{\text{Number of undecayed nuclei}}{\text{Total number}}$

$$= \frac{N(t)}{N_0} = \frac{N_0 e^{-\lambda t}}{N_0} = e^{-\lambda t}$$

18. (a) Probability of disintegration = 1 – probability of survival

$$= 1 - e^{-\lambda t}$$

Explanations for Q. Nos. (19 and 20)

Frequency of an emitted gamma photon = $\dfrac{\Delta E}{h}$

$$\Rightarrow \quad f_{\gamma_1} = 2.608 \times 10^{20} \text{ Hz}$$

$$f_{\gamma_2} = 1.633 \times 10^{20} \text{ Hz}$$

$$f_{\gamma_3} = 9.950 \times 10^{19} \text{ Hz}$$

Also, maximum KE of emitted β-particle.

$$K_{max}(\beta_1) = (m\,({}_{79}^{198}Au) - \text{mass of second excited state of } {}_{80}^{198}Hg)$$

$$= 931.5 \left\{m\,({}_{79}^{198}Au) - m\,({}_{80}^{198}Hg) - \frac{1.088}{931.5}\right\}$$

$$= 0.281 \text{ MeV}$$

Similarly, $K(\beta_2) = 0.957$ MeV

19. (a) **20.** (c)

21. (b) Both Statement I and Statement II are correct but reason for Statement I is law of conservation of energy.

22. (c) Statement I is correct but Statement II is incorrect. Sea water is suitable because D_2O (heavy water) occurs naturally in sea water.

23. (d) Statement I is incorrect and Statement II is correct. In the process of slowing down, neutrons passes through a critical energy interval of 1 to 100 eV in which they are particularly succeptible to non-fission capture by ${}_{92}^{238}U$ nuclei.

Such resonance capture removes neutron from fission chain. To minimise such non-fission capture, the uranium fuel is not intimately mixed. But are clumped together in different regions of reactor volume.

24. (d) Statement I is incorrect but Statement II is correct. When $K < 1$, reactor is sub critical so, neutrons of next generation have less number than the previous number. So reactor is slowing down and if K is not made one, then reaction will eventually stop.

25. (a) For α-decay, electrostatic repulsion must be greater than nuclear force, this happens. When nucleus is large. In β^+ decay position is emitted and a proton is changed to neutron.

In β^--decay a neutron is changed to a proton. In K capture, a proton is changed to a neutron.

Hence, (i) $\rightarrow$ (p,t), (ii) $\rightarrow$ (r,t), (iii) $\rightarrow$ (q,t), (iv) $\rightarrow$ (s), (v) $\rightarrow$ (r,t).

Revisal Problems (JEE Main)

1. (a) Atomic mass of an element is its average atomic mass. Due to occurance of isotopes in different proportions atomic masses of an element may have fractional value.

2. (a) Relative abundence of different elements in our universe is

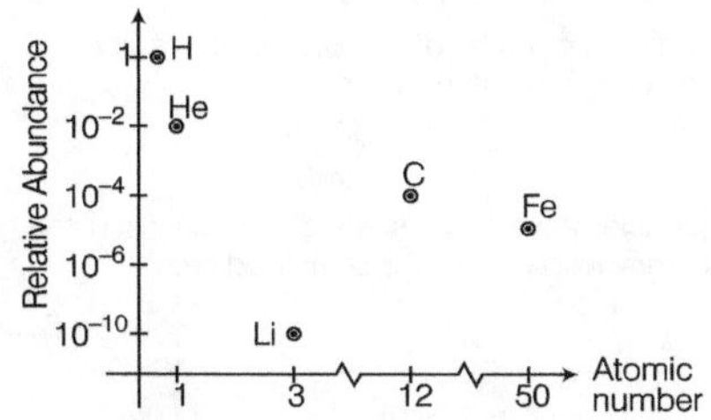

Only few elements are shown, but H_2 has maximum abundence.

3. (c) $A_1 = (466.56)\pi \times 10^{-30}\ m^2$

$= 324 \times \pi(1.2 \times 10^{-15})^2 = 4\pi(1.2 \times 10^{-15} \times 9)^2$

As, radius of a nucleus, $R_1 = R_0 A_1^{1/3}$

and nuclear-surface area $= 4\pi R_1^2 = 4\pi R_0^2 \times A_1^{2/3}$

So we have for first nucleus,

$$A^{2/3} = 9 \Rightarrow A = 27$$

For X, area $= 4 \times 4\pi R_0^2 \times A_1^{2/3}$

$\Rightarrow \quad 4\pi R_2^2 = 4 \times 4\pi R_0^2 \times A_1^{2/3}$

$\Rightarrow \quad 4\pi R_0^2 \times A_2^{2/3} = 4 \times 4\pi R_0^2 \times A_1^{2/3}$

$\Rightarrow \quad A_2^{2/3} = 4 \times A_1^{2/3}$

$\Rightarrow \quad A_2^{1/3} = 2 \times A_1^{1/3}$

$\Rightarrow \quad A_2 = 8 \times A_1 = 8 \times 27 = 216$

4. (c) $T_{1/2} = (\log_e 2)\tau \Rightarrow T =$ mean life time $= 1.44\ T_{1/2}$

Hence, mean life time is about 44% more than half-life time.

5. (c) $\Delta m = \frac{\Delta E}{c^2} = \frac{mS\Delta Q}{c^2} = \frac{1.0 \times 4.2 \times 10^3 \times 1}{(3 \times 10^8)^2}$

$= 4.7 \times 10^{-14}$ kg

6. (b) Reaction is $^4_2\text{He} + ^{14}_7\text{N} \longrightarrow ^{17}_8\text{O} + ^1_1\text{H}$

Q-value of reaction is

$Q = 931.5 \times [m(^{14}_7\text{N}) + m(^4_2\text{He})] - \{m(^{17}_8\text{O}) + m(^1_1\text{H})\}$

$= 931.5\,(14.00307 + 4.00260 - 1.00783 - 16.99913)$

$= -1.20$ MeV

As *Q*-value is negative hence, reaction is endothermic and cannot proceed without external energy.

Minimum KE of α-particle to initiate this reaction would be

$$K_{min} = |Q| \times \left(\frac{m_\alpha}{m_N} + 1\right) = 1.20 \times \left(\frac{4.00260}{14.00307} + 1\right) = 1.54 \text{ MeV}$$

7. (d) By momentum conservation, $\Sigma p_x =$ constant

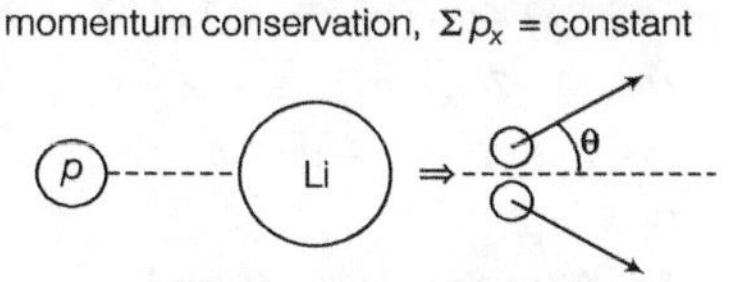

$p_p + p_{Li} = 2p_\alpha \cos\theta$

$\sqrt{2m_p K_p} + 0 = 2(\sqrt{2m_\alpha K_\alpha})\cos\theta$

$\Rightarrow \quad 2m_p K_p = 2^2 \times 2m_\alpha K_\alpha \cos^2\theta$

$\Rightarrow \quad m_p K_p = 4 \times 4\, m_p \times K_\alpha \times \frac{1}{16}$

$\Rightarrow \quad K_p = K_\alpha$

8. (c) Let at time *t*, number of active nuclei are *N*, then

$$-\frac{dN}{dt} = \left(-\frac{dN}{dt}\right)_\alpha + \left(-\frac{dN}{dt}\right)_\beta$$

$\Rightarrow \quad \lambda N = \lambda_\alpha N + \lambda_\beta N$

$\Rightarrow \quad \lambda = \lambda_\alpha + \lambda_\beta$

$\Rightarrow \quad \lambda = \frac{1}{1620} + \frac{1}{405} = \frac{1}{324}$

Now, $\quad \frac{N_0}{4} = N_0 e^{-\lambda t} \quad \left\{\because \frac{3}{4}\text{th sample is decayed}\right\}$

$\Rightarrow \quad -\lambda t = \log\left(\frac{1}{4}\right) = -\log 4$

$\Rightarrow \quad \lambda t = 2\log 2 \quad \Rightarrow \quad \frac{1}{324}t = 1.386$

$\Rightarrow \quad t = 449$ yr

9. (d) $(T_{1/2})_A = (\tau)_B$

$$\frac{0.693}{\lambda_A} = \frac{1}{\lambda_B}$$

$\Rightarrow \quad \lambda_A = \lambda_B \times 0.693$

$\Rightarrow \quad \lambda_B > \lambda_A$

As, rate of decay $= \lambda N$

So, *B* decays faster than *A*.

10. (d) After *t* seconds,

$$\frac{N_A}{N_B} = \frac{N_0 e^{-10\lambda t}}{N_0 e^{-\lambda t}} = \frac{e^{-10\lambda \times \lambda/9}}{e^{-\lambda \times \lambda/9}} = \frac{e^{-\frac{10}{9}}}{e^{-\frac{1}{9}}} = e^{\left(-\frac{10}{9}+\frac{1}{9}\right)}$$

$= e^{-9/9} = e^{-1}$

11. (c) Binding energy = Mass defect $\times c^2$

$= (Zm_p + Nm_n -$ mass of nucleus$)\cdot c^2$

$= \{Zm_p + Nm_n - (M_{atom} -$ mass of electrons$)\}\cdot c^2$

$= \{Zm_p + Nm_n - M + Ze\}\cdot c^2$

A more accurate expression will be

BE $= (Zm_p + Nm_n - M + Ze)\cdot c^2 -$ (Electronic binding energy of electrons).

12. (c) *Q*-value is positive for $A \geq 150$ and $Q \approx 6$ MeV for $A = 200$. Experimentally α-decay arises only for $A \geq 200$.

13. (d) Fusion involves 2 lighter nuclei, all other are option correct except option (d).

14. (a) Velocity of beam

$$= \sqrt{\frac{2E}{m}} = \sqrt{\frac{2 \times 0.08 \times 1.6 \times 10^{-19}}{1.6 \times 10^{-27}}}$$

$= \sqrt{2 \times 0.08 \times 10^8} = \sqrt{16 \times 10^6} = 4 \times 10^3\ ms^{-1}$

Time required to travel a distance of 4 km

$$= \frac{\text{Distance}}{\text{Speed}} = \frac{4 \times 1000}{4 \times 10^3} = 1\text{ s}$$

Now, decay rate $\frac{dN}{dt} = \lambda N \Rightarrow \frac{dN}{N} = \lambda(\Delta t)$

Fraction of neutrons that decay in 1s

$$= \frac{dN}{N} = \frac{\Delta N}{N} = \lambda \Delta t = \lambda = \frac{\log 2}{T_{1/2}} = \frac{0.693}{693} = 10^{-3}$$

$\Rightarrow \quad \frac{N_2 - N_1}{N} = 10^{-3}$

$\Rightarrow \quad N_2 - N_1 = 10^{-3} \times N = 10^{-3} \times 10^4 = 10$

So, 10 neutrons will decay.

15. (b) Number of active nuclei after n-half-lives

$$N = N_0\left(\frac{1}{2}\right)^n = N_0\left(\frac{1}{2}\right)^{(t/T_{1/2})}$$

$$\Rightarrow \quad \left(\frac{N}{N_0}\right) = \left(\frac{1}{2}\right)^{\left(\frac{t}{T_{1/2}}\right)} \Rightarrow t = \frac{T_{1/2}\log\left(\frac{N_0}{N}\right)}{\log_e 2}$$

$$\Rightarrow \quad \frac{t_1}{t_2} = \frac{\log_e\left(\frac{N_0}{N_1}\right)}{\log_e\left(\frac{N_0}{N_2}\right)}$$

$$\Rightarrow \quad \frac{6}{10} = \frac{\log_e (8/1)}{\log_e (N_0/N)}$$

$$\Rightarrow \quad \log_e = \frac{N_0}{N} = \frac{10}{6}\log_e 8 = \log_e 32$$

$$\Rightarrow \quad \frac{N_0}{N} = 32$$

So, fraction decayed $= 1 - \frac{1}{32} = \frac{31}{32}$.

16. (c) Closest separation for fusion will be around 2 fm as nuclear force is not effective for distances more than this.

For deuteron nuclei at this separation, the coulomb barrier is

$$U = \frac{Kq_1q_2}{r} = \frac{9\times 10^9 \times (1.6\times 10^{-19})^2}{2\times 10^{-15}} = 5.7\times 10^{-14}\ \text{J}$$

If gas is at temperature T, the average KE of random motion of each nuclei will be $3/2\ k_BT$. When this energy is equal to energy to overcome potential barrier, then only there is a fusion reaction.

So, $\frac{3}{2}K_BT = 5.7\times 10^{-19}$

$$\Rightarrow \quad T = \frac{5.7\times 10^{-14}}{1.5\times 1.38\times 10^{-23}} = 2.8\times 10^9\ \text{K}$$

17. (a) Both Statement I and Statement II are correct and Statement II is a correct reason for Statement I. In case of β-decay, reactions are also produced so, there may be various possible energy values carried by β-particles.

In case of α-decay, only two particles are produced, the daughter nuclei and α-particle. Also as daughter nucleus is more massive, maximum KE is carried by α-particles.

18. (a) Radius of path of α-particle in region of a uniform perpendicular magnetic field is

$$r = \frac{\sqrt{2\,Km}}{2eB}$$

As α-particles have only discreate and quantised energy values (K), so, they follow circular path of fixed radii.

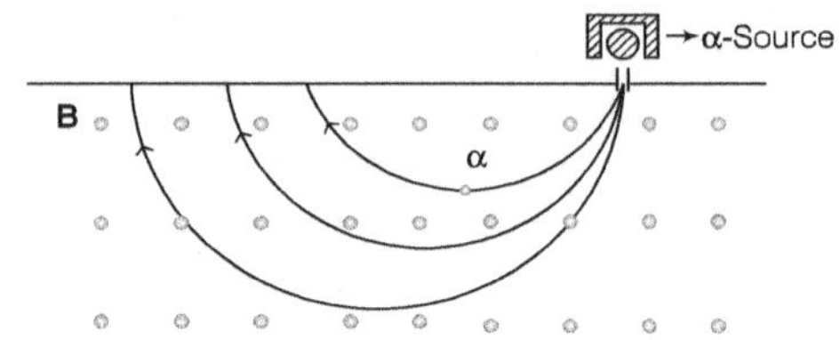

19. (a) ${}^6_2\text{H} \longrightarrow {}^6\text{H}_3 + {}^{0}_{-1}e + \bar{\nu}$

${}^6\text{H}_3$ is more stable as $\frac{N}{Z}$ ratio is exactly 1. This is possible due to β^- decay. So Both Statement I and Statement II are correct and Statement II is a correct reason for Statement I.

20. (b) Both Statement I and Statement II are correct but reason for Statement I is that each water molecule has two protons which have mass nearing to that of neutron so, energy exchange is efficiently possible.

Revisal Problems (JEE Advanced)

1. (d) Let, we have initially 100 g of substance.

$$\text{Then,}\quad 100\text{ gm} \xrightarrow{5\text{ days}} \frac{100\times 10}{100} = 10\text{ g decayed}$$

$$90\text{ gm} \xrightarrow{5\text{ days}} \frac{90\times 10}{100} = 9\text{ g decayed}$$

$$81\text{ gm} \xrightarrow{5\text{ days}} \frac{81\times 10}{100} = 8.1\text{ g decayed}$$

$$73\text{ gm} \xrightarrow{5\text{ days}} \frac{73\times 10}{100} = 7.3\text{ g decayed}$$

So, amount left after 20 days ≈ 65 g

So, $\quad \text{Probability} = \frac{65}{100} = 65\%$

2. (b) Reaction is

$${}^2_1\text{H} + {}^2_1\text{H} \rightarrow {}^4_2\text{He}$$

BE of deutron = 1.1 × 2 = 2.2 MeV

BE of Helium = 7 × 4 = 28 MeV

Energy balance for reaction

$$E = 28\times(-2\times 2.2)$$

$$\Rightarrow \quad E = 23.6\ \text{MeV}$$

3. (a) For stability, BE per nucleon must be as large as possible. So, A is most stable.

4. (a,c,d) Spontaneous decay will occur only with $Q > 0$.

Where Q = (mass of reactants – mass of products) × c^2

5. (a,b,c,d) For decay of a free neutron at rest, $n \rightarrow p^+ + e^-$

So, momentum of products must be $|p_{p^+}| = |p_{e^-}|$.

So, KE of electron is $T_e = \frac{p_e^2}{2m_e}$.

which must be a constant.

This is opposite of observed result. So, electron can have a continuous energy distribution, when a third particle is present.

Correct equation for β^--decay is

$$n \rightarrow p + e^- + \bar{\nu}$$

$\bar{\nu}$: The antineutrinos are particles with very little mass and spins of $+\frac{1}{2}$.

Neutrinos which are emitted with β^+ have $-\frac{1}{2}$ spin.

6. (b,c) $Q_2 = \{m_N({}^A_ZX) - m_N({}^A_{Z-1}Y) - m_e\}c^2$

$$= \{m({}^A_ZX) - Zm_e - m({}^A_{Z-1}Y) - (Z-1)m_e - m_e\}\cdot c^2$$

$$= \{m({}^A_ZX) - m({}^A_{Z-1}Y) - 2m_e\}\cdot c^2$$

$$= Q_1 - 2m_ec^2$$

and $Q_1 = \{m_N({}^A_ZX) + m_e - m_N({}^A_{Z-1}Y)\}\cdot c^2$

$$= c^2\cdot\{m({}^A_ZX) - Zm_e + m_e - m({}^A_{Z-1}Y) + (Z-1)m_e\}$$

$$= \{m({}^A_ZX) - m({}^A_{Z-1}Y)\}\cdot c^2$$

So, when $Q_2 > 0 \Rightarrow Q_1 > 0$

But $Q_1 > 0$ does not necessarily means that $Q_2 > 0$.

7. (a,d) Frequency of γ's is $f = \frac{\Delta E}{h}$.

This gives, $f(\gamma_1) = 2.608 \times 10^{20}$ Hz

$f(\gamma_2) = 9.950 \times 10^{19}$ Hz

$f(\gamma_3) = 1.633 \times 10^{20}$ Hz

Explanations for Q. Nos. (8-10)

Decay series is as follows.

$$A \xrightarrow{\lambda_1} B \xrightarrow{\lambda_2} C \text{ (stable)}$$

	A	B	C
at $t = 0$,	N_0	0	0
at $t = t$,	N_1	N_2	N_3

Let number of nuclei of A, B and C at time t be N_1, N_2 and N_3. As, N_0 nuclei of A decays to form N_2 nuclei of B so, number of active nuclei of A remain undecayed after time t is

$$N_1 = N_0 e^{-\lambda_1 t}$$

For element B,

$$\frac{d}{dt} N_2 = \lambda_1 N_1 - \lambda_2 N_2$$

or $$\frac{d}{dt} N_2 + \lambda_2 N_2 = \lambda_1 N_1$$

This is a linear differential equation in N_2 and integrating factor.

$$\text{IF} = e^{\lambda_2 t}$$

$$\Rightarrow e^{\lambda_2 t} \cdot \frac{dN_2}{dt} + e^{\lambda_2 t} \cdot \lambda_2 N_2 = \lambda_1 N_1 \cdot e^{\lambda_2 t}$$

$$\Rightarrow \int d(N_2 e^{\lambda_2 t}) = \int \lambda_1 N_1 e^{\lambda_2 t} \cdot dt$$

$$\Rightarrow N_2 e^{\lambda_2 t} = \lambda_1 N_0 \left(\frac{e^{(\lambda_2 - \lambda_1)t}}{\lambda_2 - \lambda_1} \right) + C$$

at $t = 0$, $N_2 = 0$

$$\Rightarrow C = \frac{\lambda_1 N_0}{\lambda_1 - \lambda_2}$$

So, $$N_2 = \frac{\lambda_1 N_0}{\lambda_2 - \lambda_1} (e^{-\lambda_1 t} - e^{-\lambda_2 t})$$

and $$N_3 = N_0 - (N_1 + N_2)$$

$$= N_0 - N_0 e^{-\lambda_1 t} - \frac{\lambda_1 N_0}{\lambda_2 - \lambda_1} (e^{-\lambda_1 t} - e^{-\lambda_2 t})$$

8. (a)

9. (a)

10. (c)

11. (c) Efficiency $\eta = \frac{\text{Electric power output}}{\text{Reactor power input}}$

$\Rightarrow$ Reactor power $P_r = \frac{P_e}{\eta}$

$$\Rightarrow P_r = \frac{1000}{0.3} = 3333 \text{ MW}$$

Rate of fission events $R = \frac{\text{Rate of energy produced}}{\text{Energy per fission}}$

$$= \frac{3.333 \times 10^9 \text{ Js}^{-1}}{200 \text{ MeV} \times 1.6 \times 10^{-13} \text{ J/MeV}}$$

or $$R = 1.04 \times 10^{20} \text{ events/s}$$

12. (a) Mass of ^{235}U consumed per year

$$= \frac{R \times 235 \times 10^{-5} \text{ kg/mole}}{6.02 \times 10^{23} \text{ nuclei/mole}} \times \text{time}$$

$$= \frac{1.04 \times 10^{20} \times 235 \times 10^{-3}}{6.02 \times 10^{23}} \times 60 \times 60 \times 24 \times 365$$

$$= 1.28 \times 10^3 \text{ kg/yr}$$

13. (d) Although the smaller nuclei have higher BE per nucleon, but total nuclear BE of products is less or they have fewer nucleons.

This difference of energy is evolved in form of KE of products which is converted into heat by collisions.

14. (d) Mass number conservation yields

$$236 = 140 + 94 + x$$

$$\Rightarrow x = 2$$

And proton number conservation gives,

$$92 = 54 + 38$$

So, resulting particles must be neutrons.

15. (c) Possibility of neutron absorption is high as a proton experiences an extra coulomb barrier, which is absent in case of neutrons.

16. (c) Q = energy released = $(m_X - m_Y - m_\alpha)c^2$

With X at rest,

$$Q = T_Y + T_\alpha$$

Also, conservation of momentum gives,

$$p_Y = p_\alpha$$

$$\Rightarrow T_\alpha = \frac{Q}{\left(1 + \frac{m_\alpha}{m_Y}\right)}$$

As Y is a heavy nucleus,

$$\frac{m_\alpha}{m_Y} \approx \frac{4}{A - 4}$$

$$T_\alpha = Q\left(1 - \frac{4}{A}\right)$$

17. (b) X is at rest and, $p_\alpha = p_Y$

$\therefore$ α and Y moves along same straight line but in opposite directions.

18. (b) As total energy = PE + KE

and KE = TE − PE

Also, KE $\geq$ 0

$\Rightarrow$ TE $\geq$ PE

A particle is not allowed where its PE exceeds its TE

So, α-particle cannot be found in $r = a$ to $r = b$, regions.

19. (a) There are 3-natural series as,

Fourth is an artificial series.

$$^{238}\text{U} \xrightarrow{8\alpha,\ 6\beta} {}^{206}\text{Pb}$$

$$^{235}\text{U} \xrightarrow{7\alpha,\ 4\beta} {}^{207}\text{Pb}$$

$$^{232}\text{Th} \xrightarrow{6\alpha,\ 4\beta} {}^{208}\text{Pb}$$

Fourth is an artificial series.

$$^{237}\text{Np} \xrightarrow{8\alpha,\ 6\beta} {}^{209}\text{Bi}$$

Hence, (i) $\rightarrow$ (p), (ii) $\rightarrow$ (q), (iii) $\rightarrow$ (r), (iv) $\rightarrow$ (s).

20. (d) In every type of decay (α, β and γ), mass of products is always less than mass of reactants remaining mass energy is released in form of Q value of reaction.

Also, BE per nucleon increases in every type of decay.

Hence, (i) $\rightarrow$ (p,q,r,s), (ii) $\rightarrow$ (p,q,r,s), (iii) $\rightarrow$ (q,r), (iv) $\rightarrow$ (p).

21. (d) α-decay occurs in a heavy nucleus.

γ-decay occurs when an excited nucleus radiates excess energy.

In both K capture and β^+-decays, then neutron number increases.

Hence, (i) $\rightarrow$ (s), (ii) $\rightarrow$ (r), (iii) $\rightarrow$ (q), (iv) $\rightarrow$ (p), (v) $\rightarrow$ (p).

JEE Main & AIEEE Archive

1. (a) According to given data, mass of neutron and proton are equal which do not permit the breaking up of neutron and proton. But if we take standard mass of neutron as 1.6750×10^{-27} kg, then

Energy released

$= \text{Mass defect} \times c^2$

$= (m_n - m_p - m_e)\times c^2$

[$\because$ energy released mass of reactant − mass of product × c^2]

$$= \frac{(1.6750\times 10^{-27} - 1.6725\times 10^{-27} - 9\times 10^{-31})}{1.66\times 10^{-27}}\times 931.5 \text{ MeV}$$

[$\because$ 1amu = 931.5 MeV]

$$= \frac{16\times 10^{-31}}{1.66\times 10^{-27}}\times 931.5 \text{ MeV}$$

$$= \frac{1.6\times 0.9315}{1.66} \text{ MeV} \approx 0.9 \text{ MeV}$$

2. (c) In particle situation, atleast three particles take place in transformation, so

Energy of β-particle + Energy of third particle = $E_1 - E_2$

Hence, energy of β-particle ≤ $E_1 - E_2$

3. (b) $N_1 = N_0 - \frac{1}{3}N_0 = \frac{2}{3}N_0$

$$N_2 = N_0 - \frac{2}{3}N_0 = \frac{1}{3}N_0$$

$$\frac{N_1}{N_2} = \left(\frac{1}{2}\right)^n$$

$\Rightarrow$ $n = 1$

$\therefore$ $t_2 - t_1$ = One half-life = 20 min

4. (c) After decay, the daughter nuclei will be more stable, hence binding energy per nucleon will be more than that of their parent nucleus.

5. (b) By conservation of momentum,

$$0 = \frac{M}{2}v_1 - \frac{M}{2}v_2$$

$$v_1 = v_2 \quad ...(i)$$

$\Rightarrow$ $\Delta mc^2 = \frac{1}{2}\cdot\frac{M}{2}v_1^2 + \frac{1}{2}\cdot\frac{M}{2}\cdot v_2^2$ ...(ii)

$\Rightarrow$ $\Delta mc^2 = \frac{M}{2}v_1^2$

$\Rightarrow$ $\frac{2\Delta mc^2}{M} = v_1^2$

$\Rightarrow$ $v_1 = c\sqrt{\frac{2\Delta m}{M}}$

6. (b) In positive β - decay, a proton is transformed into a neutron and a positron is emitted.

$$p^+ \longrightarrow n^0 + e^+$$

Number of neutrons initially was $A - Z$.

Number of neutrons after decay $(A - Z) - 3\times 2$ (due to α - particles) + 2 × 1 (due to positive β - decay)

The number of protons will reduce by 8.

[as 3 × 2 (due to α- particles) + 2 (due to positive β - decay)]

Hence, atomic number reduces by 8.

So, the ratio of number of neutrons to that of protons

$$= \frac{A - Z - 4}{Z - 8}$$

7. (a) 1st reaction is fusion and 4th reaction is fission.

8. (b) Statement I is correct and Statement II is incorrect which can be directly concluded from binding energy/nucleon curve.

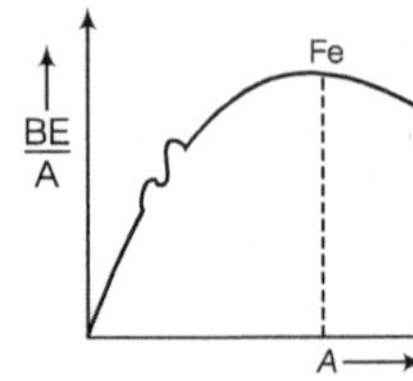

9. (b) $T_{1/2}(X) = \tau(Y)$

$\Rightarrow$ $\frac{0.693}{\lambda_X} = \frac{1}{\lambda_Y}$

or $\lambda_Y = \frac{\lambda_X}{0.693}$

$\Rightarrow$ $\lambda_Y > \lambda_X$

So, Y will decay faster than X.

10. (b) Binding energy, BE = $(M_{\text{nucleus}} - M_{\text{nucleons}})c^2$

$= (M_o - 8M_p - 9M_n)c^2$

11. (b) Since, nuclear target is heavy, it can be assumed safely that it will remains stationary and will not move due to the Coulombic interaction force.

At distance of closest approach, relative velocity of two particles is v.

Here, target is considered as stationary, so α-particle comes to rest instantaneously at distance be closest approach. Let required distance be r, then from work-energy theorem,

$$0 - \frac{mv^2}{2} = -\frac{1}{4\pi\varepsilon_0}\frac{Ze\times 2e}{r}$$

$\Rightarrow$ $r \propto \frac{1}{m}$

or $r \propto \frac{1}{v^2}$

or $r \propto Ze^2$

12. (b) From the equation,

Energy of proton + (7 × 5.60) = 2 × [4 × 7.06]

$\therefore$ Energy of proton = 17.28 MeV

13. (c) Energy spectrum of emitted β-particles from radioactive source is drawn as shown in the figure.

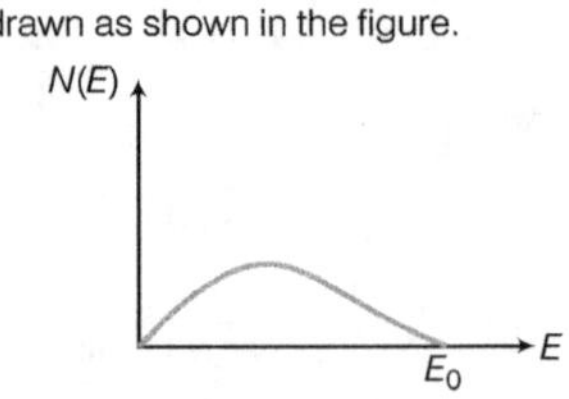

14. (c) The nuclear reaction can be represented as

$${}^7_3\text{Li} + {}^1_1\text{H} \longrightarrow {}^8_4\text{Be} + {}^A_Z X$$

Applying conservation of atomic number (charge),

$$3 + 1 = 4 + Z$$

$\Rightarrow$ $Z = 0$

Applying conservation of atomic mass,

$$7 + 1 = 8 + A$$

$\Rightarrow$ $A = 0$

Thus, the emitted particles are γ-photons (${}^0_0X$).

15. (c)
$$N = N_0(1 - e^{-\lambda t})$$
$$\Rightarrow \quad \frac{N_0 - N}{N_0} = e^{-\lambda t}$$
$$\therefore \quad 1/8 = e^{-\lambda t}$$
or $\quad 8 = e^{\lambda t}$

or $\quad 3 \ln 2 = \lambda t \qquad [\because \log_e m^n = n \log_e m]$

or $\quad \lambda = \dfrac{3 \times 0.693}{15}$

Half-life period, $t_{1/2} = \dfrac{0.693}{3 \times 0.693} \times 15 \qquad \left[\because t_{1/2} = \dfrac{0.693}{\lambda}\right]$

$= 5$ min

16. (a) $R = R_0(A)^{1/3}$
$$\frac{R_{Al}}{R_{Te}} = \frac{R_0(A_{Al})^{1/3}}{R_0(A_{Te})^{1/3}}$$
or $\quad \dfrac{R_{Al}}{R_{Te}} = \dfrac{(A_{Al})^{1/3}}{(A_{Te})^{1/3}}$

or $\quad \dfrac{R_{Al}}{R_{Te}} = \dfrac{(27)^{1/3}}{(125)^{1/3}} = \dfrac{3}{5}$

or $\quad R_{Te} = \dfrac{5}{3} \times 3.6$

or $\quad R_{Te} = 6$ fm

17. (b) ${}^{A}_{Z}X + {}^{1}_{0}n \longrightarrow {}^{7}_{3}Li + {}^{4}_{2}He$

It implies that, $A + 1 = 7 + 4$

$\Rightarrow \quad A = 10$

and $\quad Z + 0 = 3 + 2$

$\Rightarrow \quad Z = 5$

Thus, it is Boron ${}^{10}_{5}B$.

18. (d) Law of conservation of momentum gives
$$m_1 v_1 = m_2 v_2$$
$$\Rightarrow \quad \frac{m_1}{m_2} = \frac{v_2}{v_1}$$
But $\quad m = \dfrac{4}{3}\pi r^3 \rho$

or $\quad m \propto r^3$

$\therefore \quad \dfrac{m_1}{m_2} = \dfrac{r_1^3}{r_2^3} = \dfrac{v_2}{v_1}$

$\Rightarrow \quad \dfrac{r_1}{r_2} = \left(\dfrac{1}{2}\right)^{1/3}$

or $\quad r_1 : r_2 = 1 : 2^{1/3}$

19. (c) As given ${}^{2}_{1}H + {}^{2}_{1}H \longrightarrow {}^{4}_{2}He +$ energy

The binding energy per nucleon of a deuteron (${}^{2}_{1}H$)

$= 1.1$ MeV

$\therefore$ Total binding energy of one deuteron nucleus

$= 2 \times 1.1$
$= 2.2$ MeV

The binding energy per nucleon of helium (${}^{4}_{2}He$)

$= 7$ MeV

$\therefore$ Total binding energy $= 4 \times 7 = 28$ MeV

Hence, energy released in the above process

$= 28 - 2 \times 2.2$
$= 28 - 4.4$
$= 23.6$ MeV

20. (c) According to law of conservation of energy,

Kinetic energy of α-particle = Potential energy of α-particle at distance of closest approach

i.e., $\dfrac{1}{2} mv^2 = \dfrac{1}{4\pi\varepsilon_0} \dfrac{q_1 q_2}{r}$

$\therefore \quad 5 \text{ MeV} = \dfrac{9 \times 10^9 \times (2e) \times (92e)}{r} \quad \left[\because \dfrac{1}{2} mv^2 = 5 \text{ MeV, given}\right]$

$\Rightarrow \quad r = \dfrac{9 \times 10^9 \times 2 \times 92 \times (1.6 \times 10^{-19})^2}{5 \times 10^6 \times 1.6 \times 10^{-19}}$

$\Rightarrow \quad r = 5.3 \times 10^{-14} \text{ m} \approx 10^{-12}$ cm

21. (c)

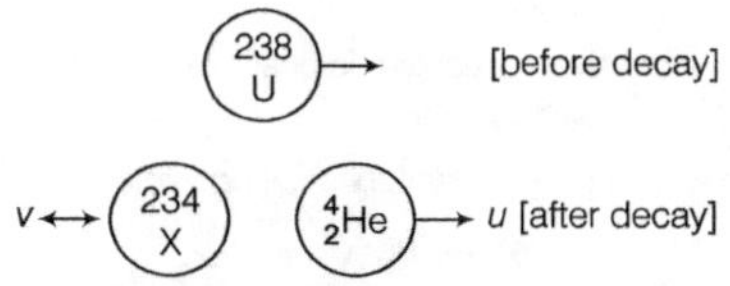

Apply conservation of linear momentum.

$0 = 4u - 234v \qquad [\because Mu = m_1 v_1 + m_2 v_2]$

$\Rightarrow \quad v = \dfrac{4u}{234}$

The residual nucleus will recoil with a velocity of $\dfrac{4u}{234}$ unit.

Note *If they will ask the recoil velocity, then answer remains same i.e.,* $\dfrac{4u}{234}$ *and not* $-\dfrac{4u}{234}$ *as the word 'recoil' itself is signifying the direction of motion of residual nucleus.*

22. (a) Given, $N_0\lambda = 5000, \; N\lambda = 1250$
$$N = N_0 e^{-\lambda t} = N_0 e^{-5\lambda}$$
where, λ is decay constant per minute.
$$N\lambda = N_0 \lambda e^{-5\lambda}$$
$$\Rightarrow \quad 1250 = N_0 \lambda e^{-5\lambda}$$
$\therefore \quad \dfrac{N_0 \lambda}{N_0 \lambda e^{-5\lambda}} = \dfrac{5000}{1250} = 4 \quad \Rightarrow e^{5\lambda} = 4$

$\Rightarrow \quad 5\lambda = 2 \log_e 2$

$\Rightarrow \quad \lambda = 0.4 \ln 2$

23. (b) Since, 8 α-particles 4β^--particles and 2β^+-particles are emitted, so new atomic number [$\alpha \rightarrow {}^{4}_{2}He$ and $\beta \rightarrow -{}^{0}_{1}n$].
$$Z' = Z - 8 \times 2 + 4 \times 1 - 2 \times 1$$
$$= 92 - 16 + 4 - 2 = 92 - 14 = 78$$

24. (a) Protons cannot be emitted by radioactive substances during their decay.

25. (d) $\quad \dfrac{3}{2} T = 7.7 \times 10^{-14}$ J

or $\quad T = \dfrac{2 \times 7.7 \times 10^{-14}}{3 \times 1.38 \times 10^{-23}} = 3.7 \times 10^9$ K

26. (a) N_0 is the initial amount of substance and N is the amount left after decay.

Thus, $\quad N = N_0 \left(\dfrac{1}{2}\right)^n$

n = number of half-lives $= \dfrac{t}{t_{1/2}} = \dfrac{15}{5} = 3$

Therefore, $\quad N = N_0 \left(\dfrac{1}{2}\right)^3 = \dfrac{N_0}{8}$

JEE Advanced & IIT JEE Archive

1. (c)

(a) $_3Li^7 \rightarrow {_2He^4} + {_1H^3}$

$$\Delta m = [M_{Li} - M_{He} - M_{H^3}]$$
$$= [6.01513 - 4.002603 - 3.016050]$$
$$= -1.003523\,u$$

Δm is negative so, reaction is not possible.

(b) $_{84}Po^{210} \rightarrow {_{83}Bi^{209}} + {_1P^1}$

Δm is negative so, reaction is not possible.

(c) $_1H^2 \rightarrow {_2He^4} + {_3Li^6}$

Δm is positive so reaction is possible.

(d) $_{30}Zn^{70} + {_{34}Se^{82}} \rightarrow {_{64}Gd^{152}}$

Δm is positive so, reaction is not possible.

2. (a) $_{84}Po^{210} \longrightarrow {_2He^4} + {_{82}Pb^{206}}$

Mass defect $\Delta m = (m_{Po} - M_{He} - m_{Pb}) = 0.005818$ u

$\therefore \quad Q = (\Delta m)(931.48)$ MeV $= 5.4193$ MeV $= 5419$ keV

From conservation of linear momentum, $p_{Pb} = p_\alpha$

$\therefore \sqrt{2m_{Pb}k_{Pb}} = \sqrt{2m_\alpha k_\alpha}$ or $\dfrac{k_\alpha}{k_{Pb}} = \dfrac{m_{Pb}}{m_\alpha} = \dfrac{206}{4}$

$$\therefore \quad k_\alpha = \left(\frac{206}{206+4}\right)(k_{total}) = \left(\frac{206}{210}\right)(5419)$$
$$= 5316 \text{ keV} \approx 5319 \text{ keV}$$

3. (4) Number of nuclei decayed in time t,

$$N_d = N_0(1 - e^{-\lambda t})$$

$$\therefore \quad \%\text{ decayed} = \left(\frac{N_d}{N_0}\right) \times 100 = (1 - e^{-\lambda t}d) \times 100 \quad \ldots\text{(i)}$$

Here, $\lambda = \dfrac{0.693}{1386} = 5 \times 10^{-4}\,s^{-1}$

$\therefore$ % decayed $\approx (\lambda t) \times 100 = (5 \times 10^{-4})(80)(100) = 4$

Explanations for Q. Nos. (4 and 5)

Maximum kinetic energy of anti-neutrino is nearly (0.8×10^6) eV.

4. (d)

5. (c)

6. (1) Activity $\left(-\dfrac{dN}{dt}\right) = \lambda N = \left(\dfrac{1}{t_{mean}}\right) \times N$

$\therefore N = \left(-\dfrac{dN}{dt}\right) \times t_{mean}$ = Total number of atoms

Mass of one atom is 10^{-25} kg $= m$ (say)

$\therefore$ Total mass of radioactive substance

$$= \text{(number of atoms)} \times \text{(mass of one atom)}$$
$$= \left(-\frac{dN}{dt}\right)(t_{mean})(m)$$

Substituting the values, we get

Total mass of radioactive substance = 1 *mg*

7. (8) $\left|\dfrac{dN}{dt}\right| = |\text{Activity of radioactive substance}| = \lambda N = \lambda N_0 e^{-\lambda t}$

Taking log both sides

$$\ln\left|\frac{dN}{dt}\right| = \ln(\lambda N_0) - \lambda t$$

Hence, $\ln\left|\dfrac{dN}{dt}\right|$ *versus* t graph is a straight line with slope $-\lambda$.

From the graph, we can see that,

$$\lambda = \frac{1}{2} = 0.5\ \text{yr}^{-1}$$

Now applying the equation,

$$N = N_0e^{-\lambda t} = N_0\,e^{-0.5 \times 4.16}$$
$$= N_0\,e^{-2.08} = 0.125\,N_0 = \frac{N_0}{8}$$

i.e. nuclei decreases by a factor of 8.

8. (d)

9. (a) From conservation of mechanical energy, we have

$$U_i + K_i = U_f + U_f$$
$$0 + 2(1.5kT) = \frac{1}{4\pi\varepsilon_0} \cdot \frac{(e)(e)}{d} + 0$$

Substituting the values, we get

$$T = 1.4 \times 10^9 \text{ K}$$

10. (b) As, given in the paragraph, a reactor is termed successful, if $nt_0 > 5 \times 10^{14}$ s cm^{-3}.

11. (a) Activity of $S_1 = \dfrac{1}{2}$(Activity of S_2)

or $\quad \lambda_1 N_1 = \dfrac{1}{2}(\lambda_2 N_2)$ or $\dfrac{\lambda_1}{\lambda_2} = \dfrac{N_2}{2N_1}$

or $\quad \dfrac{T_1}{T_2} = \dfrac{2N_1}{N_2} \qquad \left(T = \text{half-life} = \dfrac{\ln 2}{\lambda}\right)$

Given $\quad N_1 = 2N_2$

$\therefore \quad \dfrac{T_1}{T_2} = 4$

12. (c)

(i) In α- decay mass number decreases by 4 and atonic number decreases by 2.

(ii) In β^+ - decay mass number remains unchanged while atomic number decreases by 1.

(iii) In fission, parent nucleus breaks into all most two equal fragments.

(iv) In proton emission both mass number and atomic number decreases by 1.

Hence, (i) $\rightarrow$ (q), (ii) $\rightarrow$ (p), (iii) $\rightarrow$ (s), (iv) $\rightarrow$ (r).

13. (b,d) In fusion, two or more lighter nuclei combine to make a comparatively heavier nucleus.

In fission, a heavy nucleus breaks into two or more comparatively lighter nuclei.

Further, energy will be released in a nuclear process if total binding energy increases.

14. (a) Rest mass of parent nucleus should be greater than the rest mass of daughter nuclei. Therefore, option (a) will be correct.

15. (b) After two half-lives $\dfrac{1}{4}$th fraction of nuclei will remain undecayed. Or, $\dfrac{3}{4}$th fraction will decay. Hence, the probability that a nucleus decays in two half lives is $\dfrac{3}{4}$.

16. (a)

(i) Net force on Y is zero.

(ii) In (s) and (t) gravitational potential energy of Y is decreasing and that of X is increasing.

(iii) In (t), Y is moving with constant speed.

(iv) No explanation is required.

Hence, (i) $\rightarrow$ (p,t), (ii) $\rightarrow$ (q,s,t), (iii) $\rightarrow$ (p,r,t), (iv) $\rightarrow$ (q).

17. (c) $4\,({_2He^4}) = {_8O^{16}}$

Mass defect, $\Delta m = \{4\,(4.0026) - 15.9994\} = 0.011$ amu

$\therefore$ Energy released per oxygen nuclei

$$= (0.011)(931.48) \text{ MeV} = 10.24 \text{ MeV}$$

4. Electronic Devices

DPP-1 Semiconductors

1. (a) For two opposite charges at distance r

$$U = \frac{kq_1(-q_2)}{r} = kq_1 q_2\left(-\frac{1}{r}\right)$$

So, potential energy U is $-\infty$ as $r = 0$.

2. (b) Binding energy (which is the energy required to separate two-atoms to infinity) does not correspond to the bottom of potential energy curve, but to the lowest quantum energy state, slightly above the bottom.

3. (d) The classical separation between atoms is the lowest energy state.

So, $$F_{net} = -\frac{dU}{dr} = 0$$

$$\Rightarrow \quad \frac{1}{4\pi\varepsilon_0}\cdot\frac{-e^2}{r^2} - B\cdot\left(-\frac{6}{r^7}\right) = 0$$

$$\Rightarrow \quad r^5 = \frac{6B(4\pi\varepsilon_0)}{e^2} = 2.6\times 10^{-50}\text{m}^5$$

$$\Rightarrow \quad r = 1.2\times 10^{-10}\text{ m} = 1.2\text{ fm}$$

4. (d) For a classicial ideal gas, all the particles are in the lowest state with zero KE $\left(\text{KE} = \frac{3}{2}kT = 0 \text{ at } 0\text{ K}\right)$.

But electrons do not obey classical statistics but quantum statistics (The Fermi-Dirac statistics). At $T = 0$K, the possible energy levels will be filled, two electrons each one with spin up and other with spin down $\left(m_s = \pm\frac{1}{2}\right)$, up to a maximum level called Fermi level.

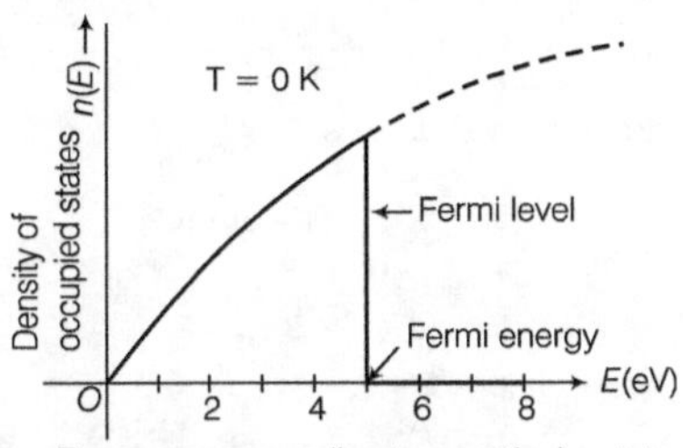

Energy corresponding to a particular state

5. (d) Fermi energy is maximum possible allowable energy value for an electron at 0 K.

Fermi energy $E_F = \frac{h^2}{8m}\left(\frac{3}{\pi}\frac{N}{V}\right)^{2/3}$

6. (d) A crystalline solid may have ionic, covalent or metallic bonding.

7. (a) When a system of N atoms combined, we get N levels closely packed to form an energy band, or a continuous distribution of energies.

8. (d) In few solids, allowed energy band may not overlap and the intervals between them represents the energies their electron cannot have. Such intervals are called forbidden energy bands or energy gaps.

9. (c) Each sodium atom has a single 3S valence electron. Each $S(l = 0)$ atomic level can have $2(2l + 1) = 2$, electrons. So, each S-band formed by N-atoms can hold 2N electrons. Thus 3S band in solid sodium is only half filled by electrons and hence, fermi energy level lies in middle of the band.

10. (c) In a semiconductor energy gap of valence and conduction band is small.

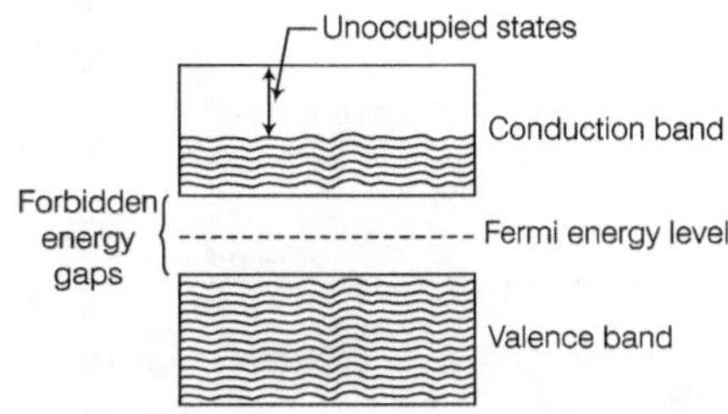

Few electrons near top of valence band can acquire enough energy to reach the conduction band. The fermi energy level is so, in middle of gap.

11. (a,b)

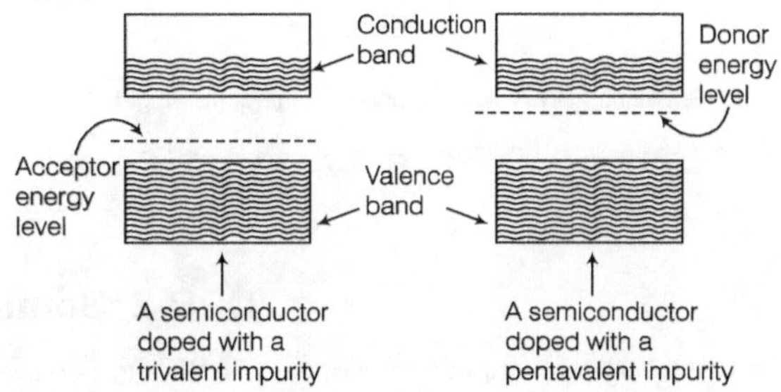

12. (a,b,c,d)

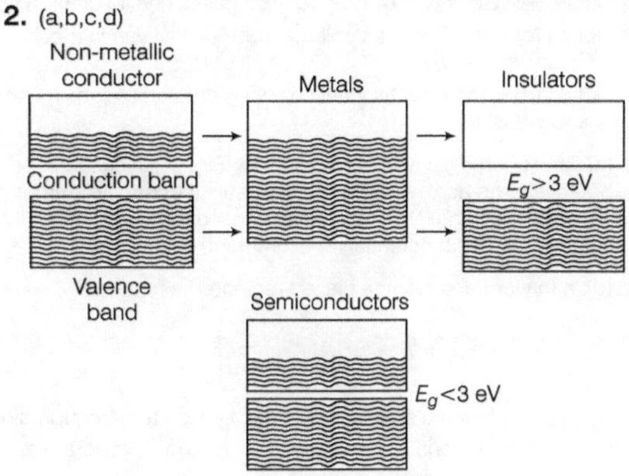

13. (b,d) Number of aluminium atoms per m^3

$$= 6.02\times 10^{23}\times\frac{1}{0.02698}\times 2700$$

$$= 6.024\times 10^{28}\text{ atoms/m}^3$$

Electron density, $n = 3\times 6.024\times 10^{28}\,\frac{\text{electrons}}{\text{m}^3}$

Electron mobility, $\mu_e = \frac{\sigma}{nq} = \frac{3.54\times 10^7}{6.024\times 10^{28}\times 1.6\times 10^{-19}}$

$$= 3.67\times 10^{-3}\text{ m}^2\text{V}^{-1}\text{s}^{-1}$$

Hence, electron mobility of metals decreases with increase of temperature.

14. (a,b,c,d) All options are correct.

Number of electrons per unit volume

$$n_e = C \cdot e^{-\frac{E_g}{2KT}}$$

where, C = material constant

and $n_e = n_h = n_i$

15. (a,b,c) Only option (d) is incorrect.

In this, conductivity of a pure semiconductor is given by

$$\sigma(\text{pure}) = n_i(\mu_e + \mu_h)q$$

$$\sigma(\text{doped}) = (n_e\mu_e + n_h\mu_h)q$$

Also, $n_e n_h = n_i^2$ for doped.

16. (d) As, $$U = U_{\text{coloumb}} + U_{\text{repulsive}} = \frac{-\alpha e^2}{4\pi\varepsilon_0 r} + \frac{B}{r^m} \qquad \ldots(i)$$

At equilibrium separation, $r = r_0$ potential energy must be minimum.

so, $$\frac{dU}{dr} = 0$$

$$\Rightarrow \quad \frac{\alpha e^2}{4\pi\varepsilon_0 r_0^2} - \frac{mB}{r_0^{m+1}} = 0$$

$$\Rightarrow \quad B = \frac{\alpha e^2}{4\pi\,\varepsilon_0 m} \cdot r_0^{m-1}$$

Substituting the value of B and $r = r_0$ in Eq. (i), we get

$$U = U_0 = -\frac{\alpha e^2}{4\pi\varepsilon_0 r_0}\left(1 - \frac{1}{m}\right)$$

17. (d) As, $\frac{dU}{dr} = -F$ and at $r = r_0$, U = minimum

$$\Rightarrow \quad \frac{dU}{dr} = 0 = F \text{ at } r = r_0.$$

18. (b) Insulators have resistivites of order of $10^8\ \Omega$-m.

Semiconductors have resistivities of order of 10^5 to $10^0\ \Omega$-m.

Insulators have energy gap more than 3 eV.

Semiconductors have energy gap less than 3 eV.

For a semiconductor to be optically sensitive, i.e. light can change its conductivity, its energy gap must be from 2.8 eV to 1.8 eV.

So, correct match is

(i) $\rho = 10^8\ \Omega\text{m}, E_g = 6$ eV

(ii) $\rho = 10^5\text{m}, \rho = 1\ \Omega\text{m}, E_g = 1$ eV or 2eV

(iii) $\rho = 1\ \Omega\text{m}, E_g = 2$ eV

(iv) $\rho = 10^5\ \Omega\text{m}, E_g = 1$ eV or 6 eV

(v) $\rho = 1\ \Omega\text{m}, E_g = 2$ eV

Hence, (i)$\rightarrow$ (p,s), (ii) $\rightarrow$ (q,r,s), (iii) $\rightarrow$ (r,t), (iv) $\rightarrow$ (q,r,s), (v) $\rightarrow$ (r,t)

19. (c) Statement I is correct as conductivity increases due to more electron-hole pairs are generated at elevated temperature. Statement II is incorrect as E_g is not affected by heating.

20. (d) Statement I is incorrect as $\frac{hc}{\lambda}$ or hf is not greater than E_g in the given case.

Hence, Statement II is correct.

DPP-2 Semiconductor Diodes

1. (b) Only option (b) as correct, a p-n junction is formed over a single Si or Ge crystal by doping each half of crystal differently.

Option (c) is incorrect as depletion zone exists even in a p-n junction with zero bias.

Option (d) is incorrect as junction field is created by ions formed due to diffusion.

2. (a) Both figures (A) and (B) are correct. Fig. (A) is suitable for low potential drop across a diode (to measure it correctly, ammeter is separated so that voltmeter measures only V_{diode}). Fig. (B) is suitable for a high potential measurement.

3. (d) In forward and reverse biased regions.

$$I_D = I_S \cdot \left(e^{\left(\frac{V_D \cdot K}{T}\right)} - 1\right)$$

where, I_S = reverse saturation current, I_D = current through diode V_D = potential drops across diode, T = absolute temperature, k = a constant $\frac{11600}{2}$ for silicon and 1600 for germenium for relativey low diode currents.

For negative values of V_D, first term quickly reduces resulting in $I_D = -I_S$

For positive values of V_D, I_D will be positive and its growth is similar to e^x. In this way we can know the result even if we are not familiar with the formula.

4. (a) With increase of temperature, more electron-hole pairs are generated and so, forward characteristics are more ideal. But in reverse bias region, reverse saturation current also increases with corresponding increase in reverse breakdown voltage.

So, $$T_1 > T_2 > T_3$$

5. (a) DC resistance of given diode is

$$r_{DC} = \frac{V}{I} = \frac{0.7}{2 \times 10^{-3}}\ \Omega$$

AC resistance of given diode is

$$r_{AC} = \frac{\Delta V}{\Delta I} = \frac{0.76 - 0.65}{(4-0) \times 10^{-3}}$$

$$= \frac{0.11}{4 \times 10^{-3}}\ \Omega$$

So, $$\text{ratio} = \frac{r_{DC}}{r_{AC}}$$

$$= \frac{0.7}{2 \times 10^{-3}} \times \frac{4 \times 10^{-3}}{0.11}$$

$$= 12.73$$

6. (c) ***Case* I** Diode is in reversed bias so,

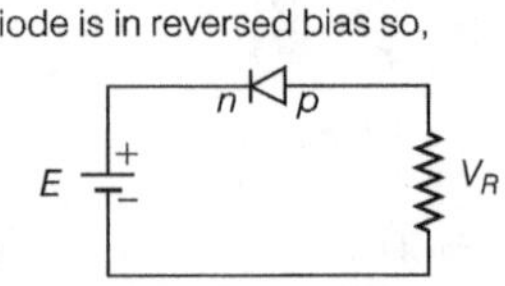

$$E - V_D - V_R = 0 \qquad \text{(by Kirchhoff's law)}$$

$$V_D = E - V_R$$

$$V_R = 0 \text{ or } I = 0$$

$$V_D = E = 8 \text{ V}$$

***Case* II** Diode is in forward bias. Since, the diode is around knee voltage = 0.7 V

So, $$\text{ratio} = \frac{8}{0.7} = \frac{80}{7}$$

7. (a) Let V_D = potential drop across diode.

Then, $V_D = 0.7$ V, as diode is in forward bias.

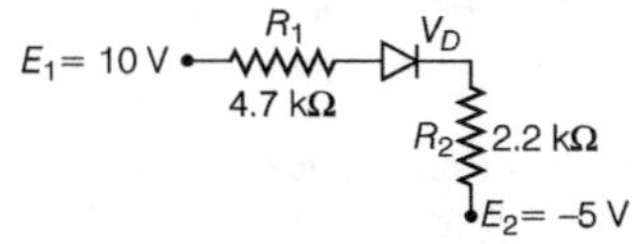

Now, current in the circuit is

$$I = \frac{E_1 + E_2 - V_D}{R_1 + R_2} = \frac{10 + 5 - 0.7}{4.7 + 2.2}$$

$$= \frac{14.3\ (\text{V})}{6.9\ (\text{k}\Omega)} \approx 2.07 \text{ mA}$$

So, potential drops across resistances are

$$V_1 = IR_1 = 2.07 \times 4.7 \approx 9.74 \text{ V}$$

$$V_2 = IR_2 = 2.07 \times 2.2 \approx 4.56 \text{ V}$$

Applying Kirchhoff's law in output region, we have

$$-E_2 + V_2 - V_0 = 0$$

$$\Rightarrow \quad V_0 = 4.56 - 5 = -0.44 \text{ V}$$

8. (a) As supply voltage grows from 0 to 12 V over a period of time in miliseconds, as potential drops of 0.3 V appears across germanium diode, it becomes ON so, the silicon diode remains in open circuit.

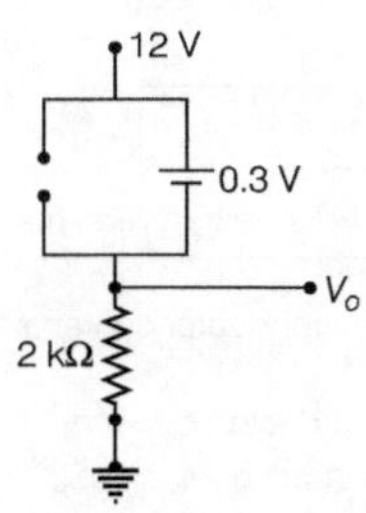

$$\therefore \quad V_o = 12 \text{ V} - 0.3 \text{ V} = 11.7 \text{ V} \approx 12 \text{ V}$$

9. (d) Input voltage, $V_i = V_m \sin \omega t$ and output voltage is

$$V_o = \begin{cases} V_i, & t = 0 \text{ to } t = \frac{T}{2} \\ 0, & t = \frac{T}{2} \text{ to } t = T \end{cases}$$

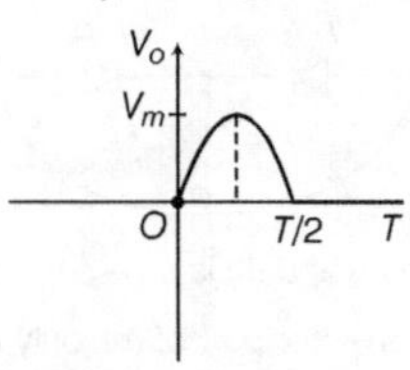

So, average value of output voltage is

$$\bar{V}_o = \frac{1}{T}\int_0^T V_o \cdot dt$$

$$= \frac{1}{T}\left[\int_0^{T/2} V_o\, dt + \int_{T/2}^{T} V_o \cdot dt\right]$$

$$= \frac{V_m}{T}\int_0^{\pi/2} \sin \omega t\, dt = \frac{V_m}{\pi}$$

10. (d) PIV or peak inverse voltage is the maximum voltage, which a diode an resist without breakdown.

In case of half-wave rectification, PIV $\geq V_m$

In case of CT, full wave rectification

$$\text{PIV} \geq 2V_m$$

(Only half of voltage appears across each diode)

So the ratio is

$$\frac{\text{PIV (Half wave)}}{\text{PIV (Full wave)}} = \frac{V_m}{2V_m} = 1:2$$

11. (a,b,c,d)

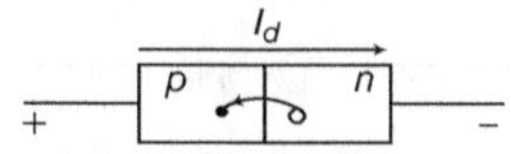

A LED is a *p-n* junction in forward bias under applied bias, e^- moves from n to p side and combines with a hole, releasing energy in form of electromagnetic wave.

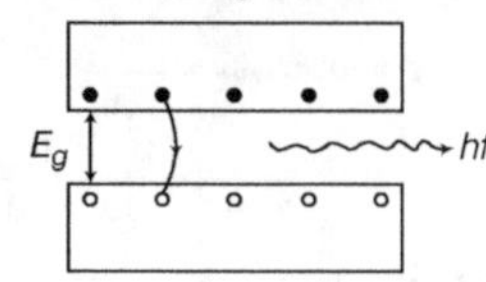

As, $$E_g = hf = \frac{hc}{\lambda}$$

$$\Rightarrow \quad \lambda = \frac{hc}{E_g}$$

For LED, $450 \leq \lambda \leq 750$ nm in visible region.

$$\therefore \quad \frac{hc}{750} \leq E_g \leq \frac{hc}{450}$$

$$\Rightarrow \quad 1.8 \leq E_g \leq 2.8 \text{ eV}$$

12. (a,b,c) For a zener breakdown(breakdown of bands in depletion zone due to excessive high field), field must be high. So, a large doping is done in zener breakdown diodes. This produces a narrow depletion zone with high junction field. In case of avalanche breakdown (breakdown of a band due to collision of a high speed thermal electron), field may not be very high. In avalanche breakdown diodes, doping concentration is not large producing a wider depletion zone.

13. (a,b,d) When a photon falls in depletion region and electron-hole pair is created if, $hf \geq E_g$.

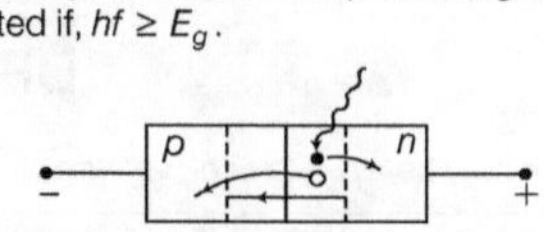

The photogenerated electron-hole are is separated by the field and moved in opposite directions. This lowers the resistivity or increases conductivity and so, increases reverse saturation current.

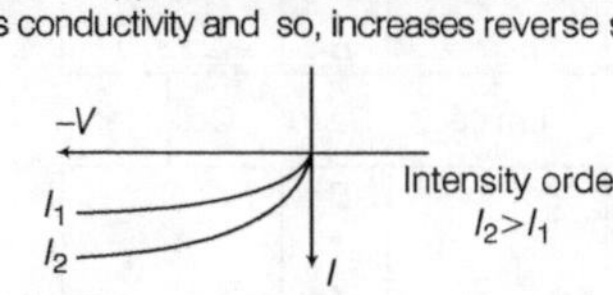

14. (b,d) Photogenerated e^- hole pairs are generated everywhere but in depletion zone they are separated and moved in different directions by the junction field.

But except depletion zone, rate of generation ≈ rate of recombination.

15. (b,d) Zone width is not directly proportional to the impurity concentration, but it is decreased if impurity concentration is increased due to high diffusion of e^- initially. Junction field is due to ionised dopant atoms.

16. (a) Resisitivity of pure Si is $\rho = \frac{1}{\sigma} = \frac{1}{en_i(\mu_e + \mu_h)}$.

⇒ Intrinsic carrier concentration

$$n_i = \frac{1}{\rho e(\mu_e + \mu_h)}$$

$$= \frac{1}{3000 \times 1.6 \times 10^{-19} \times (0.12 + 0.045)}$$

$$= 1.437 \times 10^{16} \text{ m}^{-3}$$

When 10^{19} donor atoms of phosphorus are added to per m^3 of specimen,

$$n_e n_h = n_i^2 = (1.437 \times 10^{16})^2 = 2.066 \times 10^{32}$$

and $\quad n_e - n_h = N_d - N_a = 10^{19}$

As, $\quad n_e >> n_h \Rightarrow n_e = 10^{19}$

$$\rho = \text{resistivity} = \frac{1}{en_e\mu_e}$$

$$= \frac{1}{1.6 \times 10^{-19} \times 10^{19} \times 0.12} = 5.21\ \Omega$$

17. (b) When 2×10^{19} boron (acceptor) atoms are further added,

$$n_h - e_e = N_a - N_d = 2 \times 10^{19} - 10^{19} = 10^{19}$$

$$n_h >> n_e$$

⇒ $\quad n_h = 10^{19}$

So, resistivity $= \rho = \frac{1}{en_h\mu_h} = \frac{1}{1.6 \times 10^{-19} \times 10^{19} \times 0.025} = 25\ \Omega\text{m}$

18. (d) For transformer, $N_s \times V_p = N_p \times V_s$

⇒ $\quad 50 \times 220 = 1000 \times V_s$

⇒ $\quad V_s = \frac{50 \times 220}{1000} = 11\text{V}$

Also, $\quad N_p \times I_p = N_s \times I_s$

⇒ $\quad I_s = \frac{N_p \times I_p}{N_s} = \frac{1000 \times 0.25}{50}$

$= 5\text{A} = I_{rms}$ in secondary coil.

R_L gets only half wave supply

∴ $\quad I_{av} = \frac{I_{max}}{\pi} = \frac{\sqrt{2}\, I_{rms}}{\pi} = \frac{\sqrt{2}}{\pi} \times 5\text{A}$

19. (d) $I_{rms}\ (X) = \frac{1}{T}\int_0^T I^2 dt = \frac{I_0}{2} = 2.5$ A

20. (d) Statement I is incorrect as in a *p-n* junction or in a *n*-type seminconductor phosphorus atoms are not isolated and there are enough free charge carriers in a semiconductor at room temperature.

Last or 5th electron of pentavalent phosphorus is rotating around an ion with net +1 unit charge. Energy of this electron is nearly $E = -\frac{13.6}{(\varepsilon_r)^2}\text{eV}$.

ε_r for silicon is 12 so, $E = -\ 0.1$ eV which is quite large and hence, this electron is not 'bounded' to the core. So, ionisation is possible at room temperature.

Statement II is correct.

21. (c) Both Statements are separately correct. Rate of recombination is governed by laws of thermodynamics and at same temperature both are equal.

22. (d) Statement I is incorrect. When current flows through a *p-n* junction, an e^- is diffused from *n* to *p*-side.

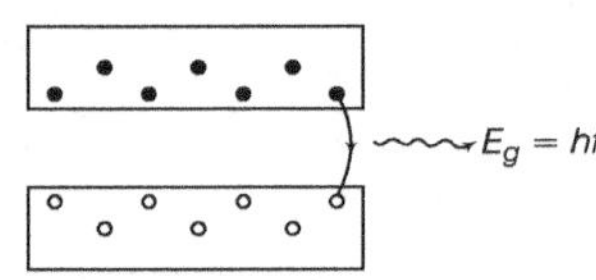

A radiative or non-radiative transition may or may not result.
Statement II is correct.

23. (c) An instrinsic semiconductor contains few thermally agitated e^- in conduction band.

For a *p*-type semiconductor, $n_h >> n_e$

For a *n*-type semiconductor, $n_e >> n_h$

Hence, (i) → (p), (ii) → (q,r,s,t), (iii) → (r), (iv) → (s,t), (v) → (t)

DPP-3 Junction Transistor

1. (b) A transistor may be a *p-n-p* or *n-p-n* type,

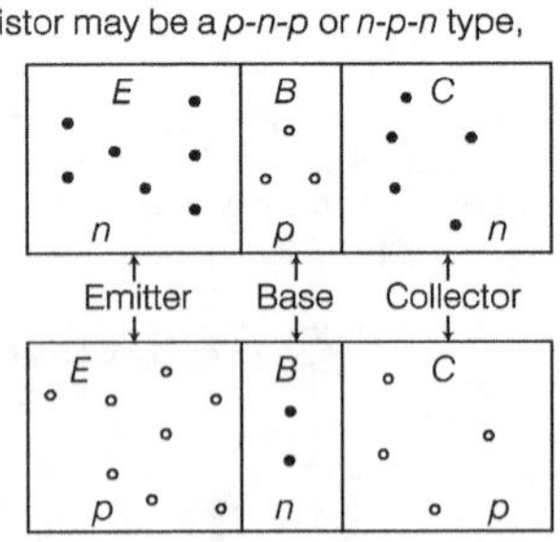

∴ It is equivalent to 2 *p-n* junctions.

2. (c) In a transistor, first layer *E-B* is forward biased

∴ Depletion layer thickness decreases due to biasing.

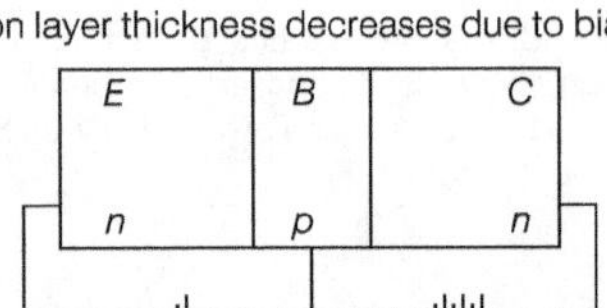

The base-collector region is in reverse bias, hence depletion layer thickness increases due to biasing.

3. (a) We put biasing batteries to clear the figure.

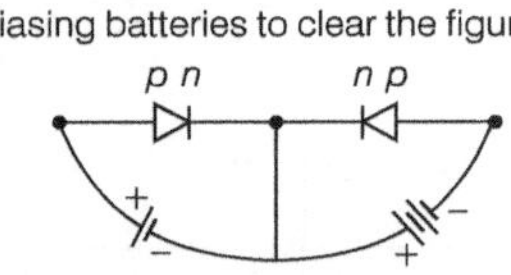

So, (I) is a *p-n-p* type and (II) is an *n-p-n* type configuration.

4. (d) In an *n-p-n* transistor part of majority carriers (electrons) are lost due to electron-hole combination in base region and rest of them (90 to 95%) passes into collector region.

5. (d) $\quad I_C = \frac{90}{100} \times I_E$

⇒ $\quad I_E = \frac{100}{90} \times I_C = \frac{100}{90} \times 10 \approx 11\text{mA}$

∴ $\quad I_B = I_E - I_C = 11 - 10 = 1$ mA

6. (c) In case (B), emitter-base junction is forward biased and collector-emitter is reversed biased. So, lamp will glow due to current flow.

7. (c) All of charge carriers entering base region will be neutralised.

8. (d) $\quad \beta = \frac{I_C}{I_B} = 50$

$$\alpha = \frac{\beta}{1 + \beta} = \frac{I_C}{I_E} = \frac{50}{51} = 0.98$$

9. (b) $$A_V = \frac{V_o}{V_i} = \beta \frac{R_C}{R_B}$$

$\Rightarrow$ $$V_o = V_i \cdot \beta \cdot \frac{R_C}{R_B}$$

$$= 0.01 \times 50 \times \frac{5 \times 10^3}{1 \times 10^3} = 2.5 \text{ V}$$

10. (a) For first transistor, $\frac{V_o}{V_i} = A_V$

$\Rightarrow$ $$V_o = 10 \times V_i$$
$$= 10 \times 0.01 = 0.1 \text{ V}$$

For second transistor, $V_o = A_V \times V_i = 20 \times 0.1 = 2$ V

$\therefore$ Net amplification is of 200 times.

11. (b,c,d) At $V_i = 1$V, $V_i \propto V_o$ and transistor is in active state and hence, it can be used as an amplifier.

At $V_i = 0.5$V, $V_o \approx$ maximum.

This is a switch ON condition.

At $V_i = 2.5$V, $V_o \approx 0$

This is a switch OFF condition.

12. (a,b,d) W_1 and W_3 are mutual induction coupling of W_2.

In an *L-C* oscillatory circuit, energy oscillates between inductor and capacitor.

13. (c,d) Input and output are out of phase always.

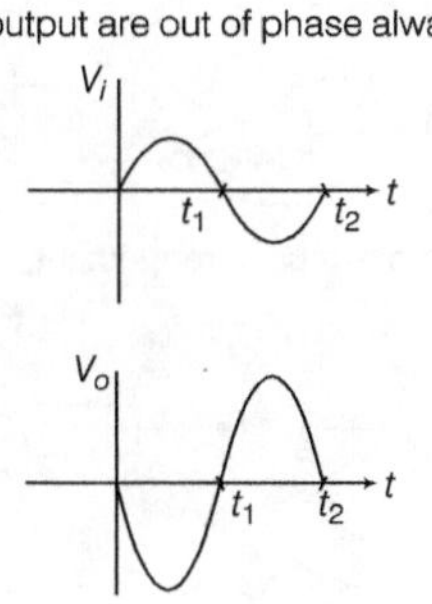

14. (b,d) $$\beta = \left.\frac{\Delta I_C}{\Delta I_B}\right|_{V_{CE} = 2\text{V}}$$

$\Rightarrow$ $$\beta = \frac{(9.5 - 2.5)\text{ mA}}{(60 - 10)\ \mu\text{A}} = \frac{7 \times 10^{-3}\text{A}}{50 \times 10^{-6}\text{ A}} = 140$$

Also, r_o = Output resistance

$$= \left.\frac{\Delta V_{CE}}{\Delta I_C}\right|_{I_B = 60\mu\text{A}}$$

$$= \frac{0.5}{10 \times 10^{-3}} = \frac{0.5}{10^{-2}} = 50\Omega$$

15. (b,d) Power is drained by transistor from the biasing batteries and is supplied to the output.

Voltage gain is low both at high and low frequencies.

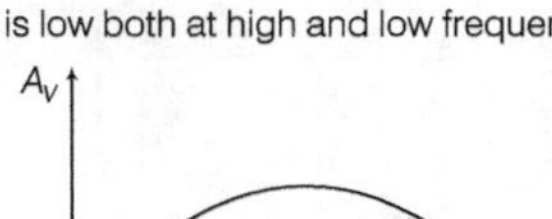

A_v is constant for mid-frequency range

16. (d) From output characteristics, at point Q,

$$V_{CE} = 8\text{V}, I_C = 4\text{mA}$$

As, $$V_{CE} = I_C R_C + v_{CE}$$

$\Rightarrow$ $$R_C = \frac{V_{CC} - V_{CE}}{I_C}$$
$$= \frac{16 - 8}{I_C}$$
$$= \frac{16 - 8}{4 \times 10^{-3}}$$
$$= 2\text{k}\Omega$$

As, $$V_{BB} = I_B R_B + V_{BE}$$
$$R_B = \frac{V_{BB} - V_{BE}}{I_B}$$
$$= \frac{16 - 0.7}{30 \times 10^{-6}}$$
$$= 510\ \Omega$$

$\therefore$ $$\frac{R_B}{R_C} = \frac{510}{2} = 225 : 1$$

17. (d) β = Current gain $= \frac{I_C}{I_B} = \frac{40 \times 10^{-3}}{30 \times 10^{-6}} = 133$

18. (c) Power gain $= A_P = \beta \times A_V$

$$= \beta^2 \times \left(\frac{R_C}{R_B}\right)$$
$$= (133)^2 \times \left(\frac{2 \times 10^3}{510 \times 10^3}\right) = 69$$

19. (a) Applying Kirchhoff's loop rule to base-emitter loop, we have

$$-I_B R_B - V_{BE} - I_E R_E + V_{EE} = 0$$

Also, $$I_C = \beta I_B$$

and $$I_C \approx I_E$$

$\Rightarrow$ $$I_B = \frac{I_C}{\beta} \approx \frac{I_E}{\beta}$$

So, from above equation we have,

$$V_{EE} = \frac{I_E}{\beta} \cdot R_B + I_E R_E + V_{BE}$$

$\Rightarrow$ $$I_C = I_E = \frac{V_{EE} - V_{BE}}{R_E + R_{B/\beta}}$$
$$= \frac{10 - 0.7}{\left(4.7 + \frac{47}{100}\right) \times 10^3}$$
$$= 1.8 \times 10^{-3}\text{ A} = 1.8 \text{ mA}$$

20. (c) From Kirchhoff's voltage rule on collector side, we have

$$V_{CC} - I_C R_C - V_{CE} - I_E R_E + V_{EE} = 0$$

$\Rightarrow$ $$V_{CE} = V_{CC} + V_{CE} - I_C (R_C + R_E)$$
$$= 10 + 10 - 1.8\ [(1 + 4.7) \times 10^{-3}]$$
$$= 9.74\text{V}$$

21. (d) Statement I is incorrect as a transformer uses part of input power so, it cannot be used where input is quite less.

Statement II is correct.

22. (c) Statement I is correct but Statement II is incorrect as a positive feed back to an oscillating system is possible, when there is resonance. Driving frequency is equal to systems natural frequency.

DPP-4 Logic Gates

1. (i) An algebra is a statement of rules for manipulating members of a given set.

Boolean algebra uses only two numbers 0 and 1 and few rules which can be used to find the alternative ways to define or express a logical function.

(ii) Rules for Boolean algebra are given in following table. They consist of definition of AND, OR and NOT operations and few theorems.

OR (+)	$0+A=A$ $1+A=1$ $A+A=A$ $A+\overline{A}=1$
AND (·)	$0\cdot A=0$ $1\cdot A=A$ $A\cdot A=0$ $A\cdot \overline{A}=0$
NOT (–)	$\overline{\overline{A}}=A$
Commutation	$A+B=B+A$ $A\cdot B=B\cdot A$
Association	$A+(B+C)=(A+B)+C$ $A\cdot(B\cdot C)=(A\cdot B)\cdot C$
Distribution	$A\cdot(B+C)=A\cdot B+A\cdot C$ $A+(B\cdot C)=(A+B)\cdot(A+C)$
Absorption	$A+(A\cdot B)=A$ $A\cdot(A+B)=A$
de-Morgan's Laws	$\overline{A+B}=\overline{A}\cdot\overline{B}$ $\overline{A\cdot B}=\overline{A}+\overline{B}$

(iii) Boolean algebra can be used to define/express logical functions as

(a) AND

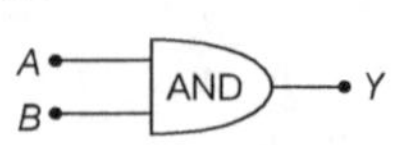

A	B	Y
0	0	0
0	1	0
1	0	0
1	1	1

Clearly (refer to basic law table)

$$A\cdot B \equiv Y$$

(b) OR

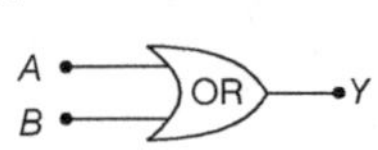

A	B	Y
0	0	0
0	1	1
1	0	1
1	1	1

$\Rightarrow$ $Y \equiv A + B$

(c) NOT

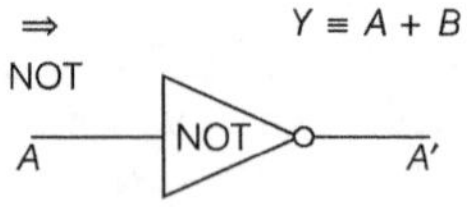

A	$A' \equiv Y$
0	1
1	0

$$Y \equiv A'$$

(iv) Boolean expression can be used to define/manipulate and design a logical circuit.

For example we have an exclusive OR gate, the XOR gate.

Here, logic is that the output will be high, when either input is high but not when both inputs are high.

Symbol XOR

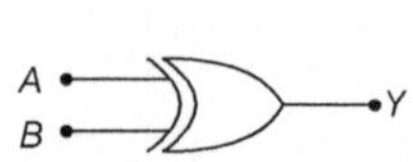

Truth Table

A	B	Y
0	0	0
0	1	1
1	0	1
1	1	0

In algebraic terms, XOR is

$$A \oplus B = A\cdot\overline{B} = B\cdot\overline{A}$$

Now, lets manipulate XOR gate

$$A \oplus B = (A\cdot\overline{B} + B\cdot\overline{A})$$
$$= A\cdot\overline{B} + B\cdot\overline{A} + A\cdot\overline{A} + B\cdot\overline{B}$$

(We uses the fact that $A\cdot\overline{A} = 0$ and $0 + A = A$; $A\cdot\overline{A} + A\cdot\overline{B} \equiv 0 + A\cdot\overline{B} \equiv A\cdot\overline{B}$)

$$= A\cdot(\overline{A}+\overline{B}) + B\cdot(\overline{A}+\overline{B}) \quad \text{(we make use of distribution)}$$
$$= A\cdot(\overline{A\cdot B}) + B\cdot(\overline{A\cdot B}) \quad \text{(we use de-Morgan's Laws)}$$
$$= (A+B)\cdot(\overline{A\cdot B})$$

So, we have 2-distinct XOR gates.

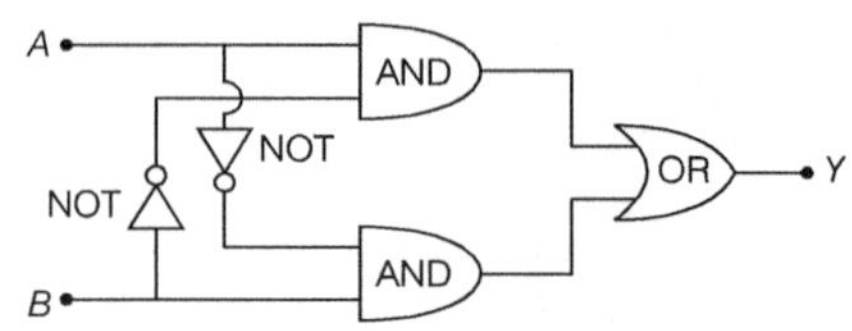

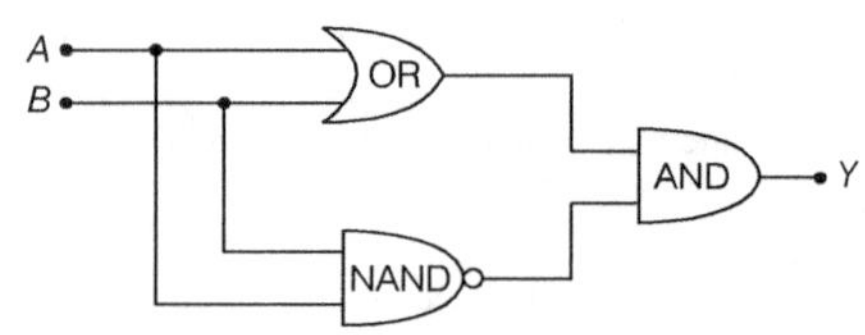

You, can make many more arrangement, if you like.

2. (c) Truth table for given gate is

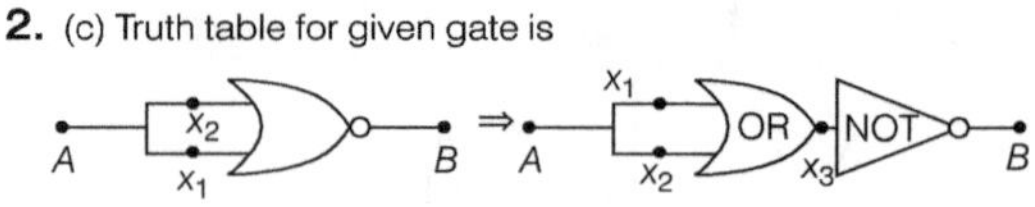

A	x_1	x_2	x_3	B
0	0	0	0	1
1	1	1	1	0

$\Rightarrow$

A	B
0	1
1	0

From truth table it is obvious that given gate is a "NOT" gate.

3. (b) Given logic circuit is

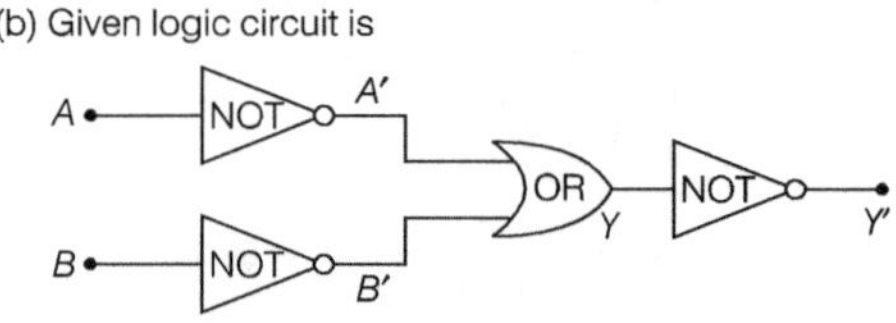

Let we draw truth table for this,

A	B	A′	B′	Y	Y′
0	0	1	1	1	0
0	1	1	0	1	0
1	0	0	1	1	0
1	1	0	0	0	1

Clearly, output resembles output of an AND gate.

4. (b) Two NOR gates in series are

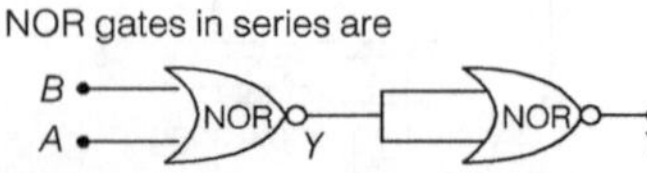

This is equivalent to

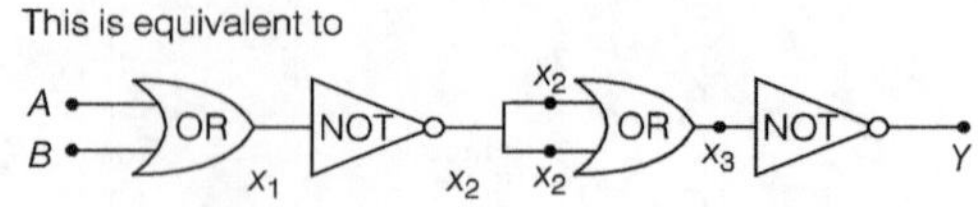

Truth table for above is

A	B	x_1	x_2	x_3	Y
0	0	0	1	1	0
1	0	1	0	0	1
0	1	1	0	0	1
1	1	1	0	0	1

Two NOR gates in series are equivalent to an OR gate.

5. (a) Truth table for given logic circuit is.

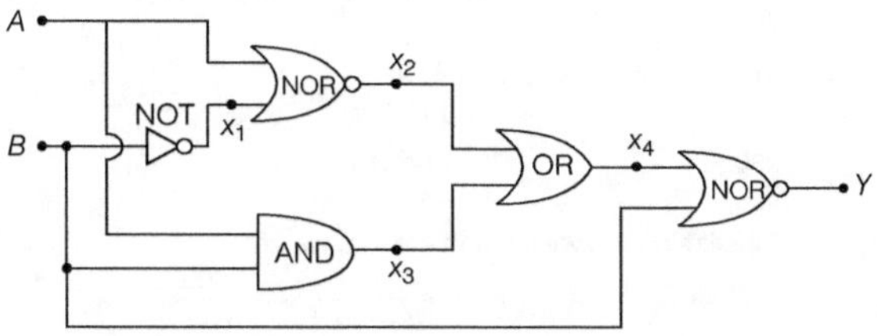

A	B	x_1	x_2	x_3	x_4	Y
0	0	1	0	0	0	0
1	0	1	1	0	1	0
0	1	0	0	0	0	0
1	1	0	0	1	1	1

Clearly above circuit is equivalent to AND gate.

6. (d) Inputs and outputs are as follows:

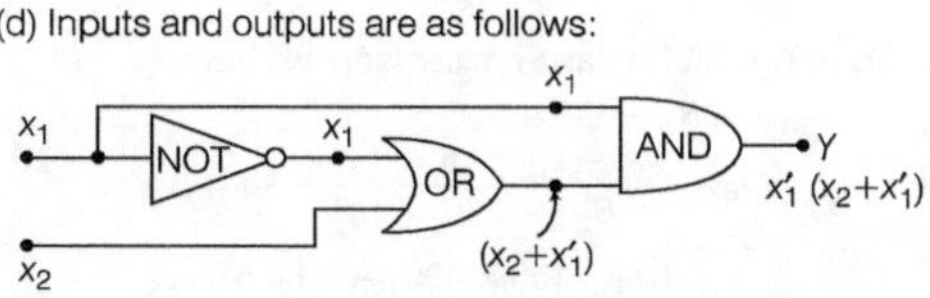

Clearly, $Y \equiv A \cdot (\bar{A} + B)$

7. (b) A logic gate works on a binary function. So, input must have 2-defined levels (0 and 1) which is a characteristics of a digital signal.

8. (c) 0 symbolises a particular group or a single lower value.

1 symbolises a particular group or a single higher value.

9. (b) Given circuit is

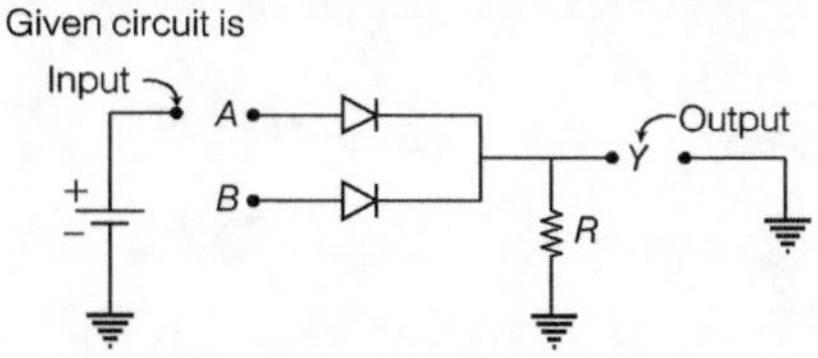

When supply is connected with either of diodes, as diodes are in forward bias, we get a higher voltage value in the gap (Y).

If supply is connected to both of diodes, we still get a higher voltage in gap Y. So, given circuit is an OR gate. Resistance R keeps potential of output so, given low.

10. (c) A NOT gate is a transistor. All other can be made by using diodes (p-n junction).

11. (d) From an OR gate an output is produced, when either of or both of inputs are at higher potential value. So, output waveform is

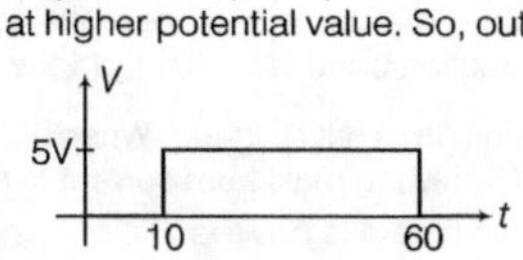

$\Rightarrow$ $V = 0$ from $t = [0, 10]$ second

and $V = 5\text{V}$ from $t = [10, 60]$ second

Note *Check inclusion of time properly.*

12. (c) Waveforms of inputs are,

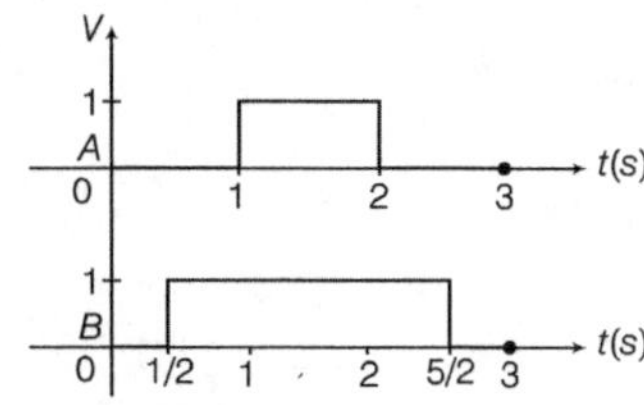

AND gates gives an output when both inputs are at higher voltage level and hence, output is

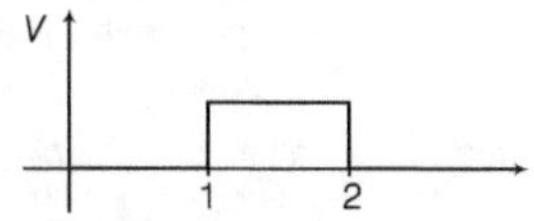

When this is fed to a NOT gate, output will be

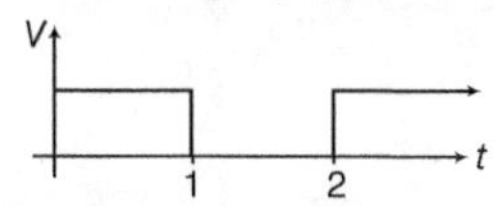

13. (b,c) NOR and NAND are known as universal gates as any of logical operations can be carried by a combination of only NOR or only NAND gates.

14. (a,b,c) Small scale integration means logic gates upto 10. Medium scale integration is for logic gates upto 100 large scale integration composses of upto 1000 logic gates.

For VLSI-very large scale integration, number of logic gates must be greater than 1000. There is no upper limit for that. So, option (d) is incorrect.

15. (a,b,c,d) In a negative logic lower value is 1 and higher value is 0. In a positive logic lower value is 0 and higher value is1.

16. (c,d) Logic gates may have more than 2-inputs but a logic gate performs a Boolean function so, it can not have more than one output.

17. (c) Output of NAND gate is

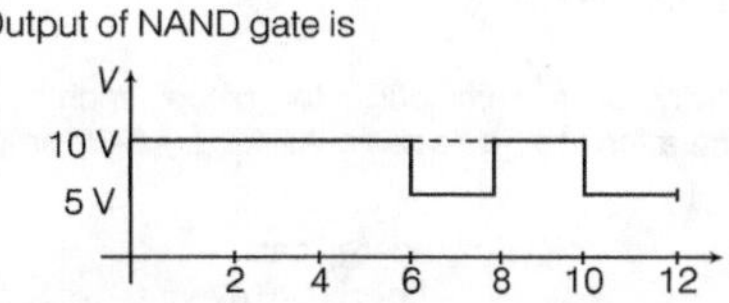

We can write this output in positive logic as 111010.

But in negative logic it becomes to 000101.

18. (b) Output for a NOR gates is,

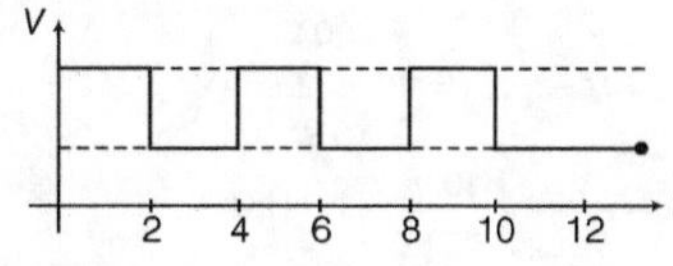

$\therefore$ It is 101010

19. (a) Any gate can be made using NAND or NOR gate. Confirm this by drawing truth tables.

Hence (i) $\rightarrow$ (p), (ii) $\rightarrow$ (r), (iii) $\rightarrow$ (q),(iv) $\rightarrow$ (p), (v) $\rightarrow$ (q), (vi) $\rightarrow$ (r).

20. (a) As, output of a function is unique, both Statement I and Statement II are correct and Statement II is a correct reason for Statement I.

21. (a) Any Boolean operation can be performed using either a combination of NOR gates or NAND gates. Both Statement I and Statement II are correct and Statement II is a correct explanation of Statement I.

Revisal Problems (JEE Main)

1. (b) In pure Si,

$$n_e = 7 \times 10^{15}$$

After doping, donor immunity $N_D = \dfrac{5 \times 10^{22}}{10^7}$

$$= 5 \times 10^5$$

Also, $n_e = \dfrac{N_D}{2} = \dfrac{5 \times 10^{15}}{2}$

$$= 2.5 \times 10^{15}$$

So, number of electrons in sample after doping

$$= n_e' = 2.5 \times 10^{15} + 5 \times 10^{15}$$

$$= 7.5 \times 10^{15}$$

Change in electron concentration due to doping.

$$= \frac{n_e(\text{final}) - n_e(\text{initial})}{n_e(\text{initial})}$$

$$= \frac{2.5 \times 10^{15}}{5 \times 10^{15}} = \frac{1}{2}$$

2. (c) Rectifier effeciency

$$\eta = \frac{P(\text{out put})}{P(\text{in put})} = \frac{I_{dc}^2 R_L}{I_{rms}^2 R_L}$$

As diodes are ideal, we are not considering forwared resistance of p-n junctions.

Now, for half wave rectification,

$$I_{rms} = \frac{I_0}{2}, I_{dc} = \frac{I_0}{\pi}$$

$$\eta_{\text{half wave}} = \frac{2^2}{\pi^2} = \frac{4}{\pi^2} \approx 0.4$$

For full wave rectification,

$$I_{rms} = \frac{I_0}{\sqrt{2}}, I_{dc} = \frac{2I_0}{\pi}$$

$$\eta_{\text{half wave}} = \frac{4}{\pi^2} \times 2 = \frac{8}{\pi^2} \approx 0.8$$

So, effeciency ration $= \dfrac{1}{2} = 0.5.$

3. (c) Resitivity of a semiconductor piece decreases due to photogeneration of e^--hole pairs. As, $\lambda_{max} = 2480$ nm, (from graph)

$$E_g = \text{energy gap}$$

$$= \frac{hc}{\lambda} = \frac{1240 \text{ eV-nm}}{2480 \text{ nm}} = 0.5 \text{ eV}$$

4. (c) $R_{L(\min)} = \dfrac{RV_z}{V_i - V_z} = \dfrac{10^3 \times 10}{50 - 10} = \dfrac{10^3}{40}$

$$= 250\ \Omega$$

Also, $I_{L(\min)} = I_R - I_{z(\max)} = 40 - 32$ (mA)

$$= 8 \text{ mA}$$

So, $R_{L(\max)} = \dfrac{V_z}{I_{L(\min)}} = \dfrac{10 \text{ V}}{8 \text{ mA}} = 1.25 \text{ k}\Omega$

5. (b) No (X) current goes in lower branch (due to reverse biased of diode). Now, total circuit resistance

$$= 50 + 150 + 100$$

$$= 300\ \Omega$$

So, circuit current, $I = \dfrac{V}{R} = \dfrac{6}{300} = \dfrac{1}{50}$

$$= 0.02 \text{ A}$$

6. (d) Antimony (Sb) is pentavalent it gives a n-type semiconductor, when Si is doped with it.

7. (a) Carbon is least conductive and Germenium is most conductive.

8. (a) For input (V_i) less than 0.7 diode is closed and

$\therefore \qquad V_i = V_o$

For $V_i \geq 0.7$ V, diode conducts and hence, output voltage equals to the diode's voltage (0.7 V). The rest of input voltage drops across the resistor R.

9. (c) When switch is turned ON, base is positive and transistor is in On condition. So, LED is ON. Before the switch is OFF, LED is not glowing.

10. (a) According to Kirchhoff's Law,

$$I = I_Z + I_L$$

Also, $\qquad V_{in} = (IR_s) + V_Z$

Also, $\qquad I_Z = 5I_L$

Substituting the values and solving, we get

$$R_s = \frac{4}{24 \times 10^{-3}}\ \Omega = \frac{1000}{6}\ \Omega$$

$$= 166.67\ \Omega$$

11. (c) $I_C = \dfrac{1\text{V}}{1 \text{ k}\Omega} = 1 \text{ mA}$

$$\beta = \frac{I_C}{I_B} \Rightarrow I_B = \frac{I_C}{\beta}$$

$$\Rightarrow \qquad I_B = \frac{1}{45} \text{ mA} = 0.022 \text{ mA}$$

12. (b) By using KVL, in base-emitter loop, we have

$$V_{BB} - I_B R_B - V_{BE} = 0$$

$$\Rightarrow \qquad I_B = \frac{V_{BB} - V_{BE}}{R_B} = \frac{5 - 0.7}{10 \times 10^3} = 430\ \mu\text{A}$$

$$I_C = \beta I_B = 150 \times 430\ \mu\text{A} = 64.5 \text{ mA}$$

$$V_{CE} = V_{CC} - I_C R_C = 10 - 64.5 \times 100 \times 10^{-3}$$

$$= 10 - 6.45 = 3.55 \text{ V}$$

As, $\qquad V_{CE} = V_{CE} + V_{BC}$

$\Rightarrow \qquad V_{CB} = V_{CE} - V_{BE}$

$$= 3.55 - 0.7$$

$$= 2.85 \text{ V}$$

13. (a)

$$I_B = \frac{V_{BB} - V_{BE}}{R_B}$$

$$= \frac{5 - 0.7}{1 \times 10^3} = 43 \times 10^{-3} \text{ A}$$

$$I_C = \beta I_B = 200 \times 4.3$$

$$= 860 \text{ mA}$$

$$V_{CE} = V_{CC} = I_C R_C$$

$$= 5 - I_C \times 0 = 5 \text{ V}$$

$\therefore \qquad$ Power dissipated $= V_{CE} \times I_C$

$$= 5 \times 860 \times 10^{-3}$$

$$= 4300 \times 10^{-3} \text{W}$$

$$= 4.3 \text{ W}$$

14. (c) As a n-p-n transistor is formed over same crystal and both E and C are of n-type so, conduction and valence band have same positions in them.

Also, base is p-type so, due to potential barrier of depletion layers, its conduction band is of slightly higher energy level.

15. (d) Given circuit is a NOT gate. When A is made positive, transistor is ON drawing maximum current to collector.

So, $V_3 = 0$ for the time A is positive.

16. (a) When $x_1 = x_3 = 0$ and $x_2 = 1$, inputs to AND gate are 1 and 1 so, $y = 0$ due to inversion of NOT gate.

17. (c) Output of AND gate is $(\overline{x_1} \cdot \overline{x_2})$ which is combined with $(x_1 \cdot x_2)$ by an OR gate.

$\therefore \qquad Y \equiv (x_1 \cdot x_2) + (\overline{x_1} \cdot \overline{x_2})$.

18. (a) For Si,

Energy gap, $(E_g) = 1.1$ eV

For Ge,

Energy gap, $(E_g) = 0.67$ eV

For Si diode, Threshold voltage = 0.7 V

For Ge diode, Threshold voltage = 0.3 V

Hence, (i) $\rightarrow$ (p), (ii) $\rightarrow$ (q), (iii) $\rightarrow$ (r), (iv) $\rightarrow$ (s).

19. (c) Statement I is correct but Statement II is incorrect.
A transformer cannot increase the frequency of AC.

20. (b) Both Statement I and Statement II are correct but Statement II is not an explanation of Statement I.
I_C is independent of V_{CE} because number of charge carriers crossing FB base emitter junction, ratio of doping % of E, B and C temperature and biasing potential of base determines number of charge carriers reaching collector which constitutes collector current.

21. (d) Statement I is incorrect as an NAND gate can be obtained by using NOR gates. Statement II is correct.

JEE Main & AIEEE Archive

1. (a) For forward bias, p-side must be a higher potential than n-side. So,

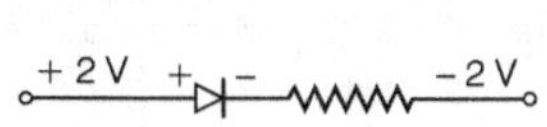

is forward biased.

2. (a) To increase frequency of light emitted from LED (Light Emitting Diode) potential barrier to diode is increased. So for same value of current, higher value of voltage is required for higher frequency.

3. (a)

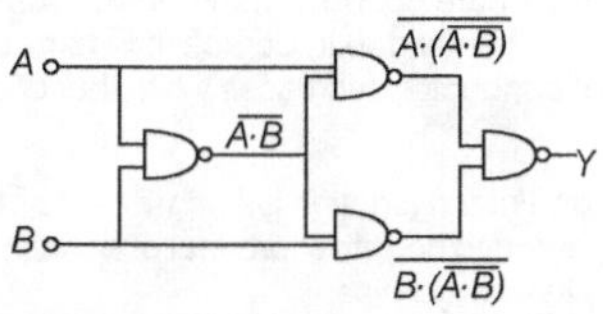

$$Y = \overline{\overline{\{A \cdot \overline{(A \cdot B)}\}} \cdot \overline{\{B \cdot \overline{(A \cdot B)}\}}} \quad \text{[using de-Morgan's rules]}$$

$$= \overline{(\overline{A} + A \cdot B)} \cdot \overline{(\overline{B} + A \cdot B)} \quad \text{[break the line, change the sign]}$$

$$= \overline{(\overline{A} + A \cdot B)} + \overline{(\overline{B} + A \cdot B)}$$

$$= A \cdot \overline{(A \cdot B)} + B \cdot \overline{(A \cdot B)}$$

$$= A \cdot (\overline{A} + \overline{B}) + B \cdot (\overline{A} + \overline{B})$$

$$= A \cdot \overline{B} + B \cdot \overline{A} \qquad [\because A \cdot \overline{A} = B \cdot \overline{B} = 0]$$

A	B	$\overline{A}$	$\overline{B}$	$A \cdot \overline{B}$	$B \cdot \overline{A}$	Y
0	0	1	1	0	0	0
0	1	1	0	0	1	1
1	0	0	1	1	0	1
1	1	0	0	0	0	0

It is the truth table of XOR gate.

4. (c)

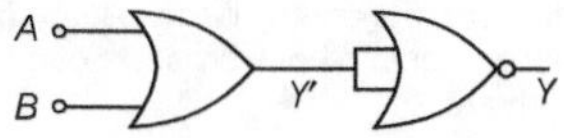

$$Y' = A + B$$

and $\qquad Y = \overline{Y'} = \overline{A + B}$

i.e. output of a NOR gate.

5. (a) Truth table for given combination is

A	B	X
0	0	0
0	1	1
1	0	1
1	1	1

This comes out to be truth table of OR gate.

6. (c) Given figure is half-wave rectifier.
[since, diode allows the signal to pass during positive half only i.e. forward bias]
So, graph (c) represents half-wave a rectifier.

7. (a) Truth Table

A	B	Y
0	0	0
0	1	0
1	0	0
1	1	1

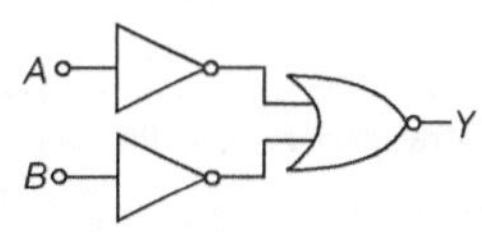

8. (d) If we give the following inputs to A and B, then corresponding output is shown in truth table.

A	B	C
0	0	0
0	1	1
1	0	1
1	1	1

The above table is similar to OR gate.

9. (d) During negative cycle, diode will not allow the signal to pass.

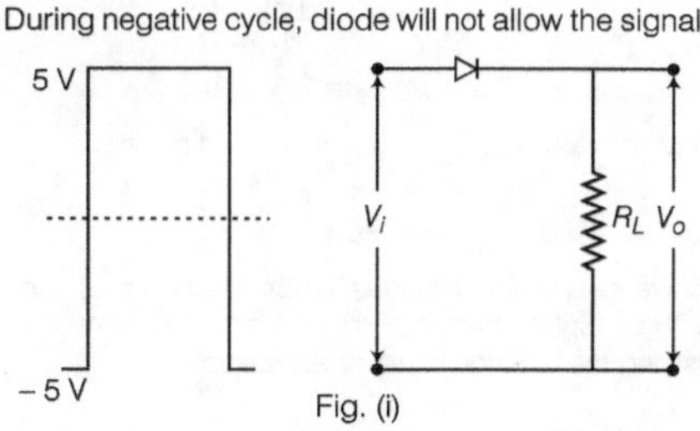

Fig. (i)

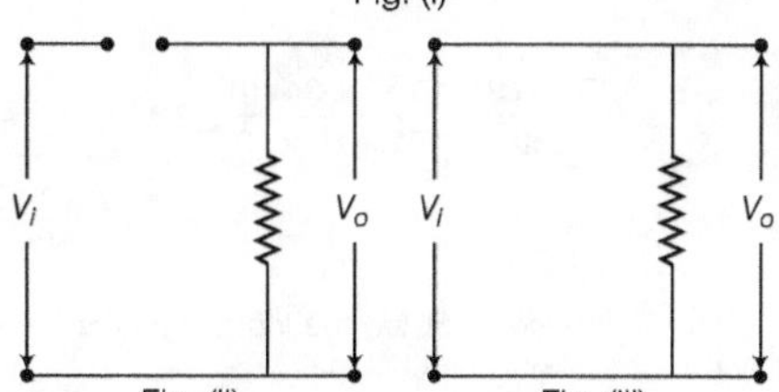

Fig. (ii) Fig. (iii)

For $V_i < 0$, the diode is reverse biased and hence offer infinite resistance, so circuit would be like as shown in Fig. (ii) and $V_o = 0$.

For $V_i > 0$, the diode is forward biased and circuit would be as shown in Fig. (iii) and $V_o = V_i$.

10. (d) The number of free electrons for conduction is significant only in Si and Ge but small in C, as C is an impurity.

11. (a) $\beta = I_C / I_B$

and $I_E = I_C + I_B$

$\therefore$ $\beta = \dfrac{I_C}{I_E - I_C} = \dfrac{5.488}{5.60 - 5.488} = 49$

12. (c) If lattice constant of semiconductor is decreased, then E_c and E_v decrease but E_g increases.

13. (b) Covalent bonding.

[they are simply semiconductors having 4th group]

14. (c) $I = n_e A v_d$

$\therefore$ $\dfrac{I_e}{I_h} = \dfrac{n_e \times (v_d)_e}{n_h \times (v_d)_h}$

Here, $\dfrac{n_e}{n_h} = \dfrac{7}{5}, \dfrac{I_e}{I_h} = \dfrac{7}{4}$

$\therefore$ $\dfrac{7}{4} = \dfrac{7}{5} \times \dfrac{(v_d)_e}{(v_d)_h}$

or $\dfrac{(v_d)_e}{(v_d)_h} = \dfrac{5}{7} \times \dfrac{7}{4} = \dfrac{5}{4}$

15. (b) In the given circuit, diode D_1 is reverse biased while D_2 is forward biased, so the circuit can be redrawn as

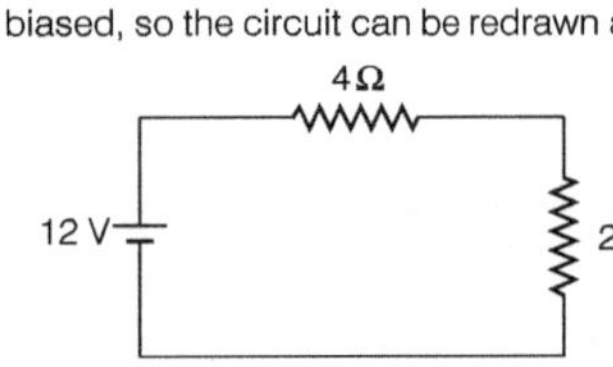

[$\because$ for ideal diodes, reverse biased means open and forward biased means short]

Apply KVL to get current following through the circuit

$$-12 + 4I + 2I = 0$$

or $I = \dfrac{12}{6}$

$= 2\text{ A}$

16. (d) For forward bias, p is positive and n is negative.

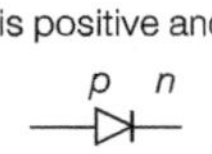

For reverse biasing of an ideal diode, the potential of n-side should be higher than potential of p-side. Only option (d) is satisfying the criterion for reverse biasing.

17. (c) $E_g = h\nu = \dfrac{hc}{\lambda}$

$= \left(\dfrac{6.63 \times 10^{-34} \times 3 \times 10^8}{2480 \times 10^{-9} \times 1.6 \times 10^{-19}}\right) \text{eV}$

[$\because 1\text{ eV} = 1.6 \times 10^{-19}$ J]

$= 0.5$ eV

18. (c) In common base amplifier, the input signal is amplified but remains in phase with output signal.

19. (c) Given, $f = 50$ Hz

$\therefore$ $T = \dfrac{1}{50}$ second

For full wave rectifier,

$$T_1 = \frac{T}{2} = \frac{1}{100}$$

$\therefore$ $f_1 = 100$ Hz $\left(\because f = \dfrac{1}{T}\right)$

20. (d) When forward bias is applied on n-p-n transistor, then it works as an amplifier. In forward biased n-p-n transistor, electrons move from emitter to base and holes move from base to emitter.

21. (d) For a transistor amplifier in common emitter configuration, current gain

$$A_i = -\frac{h_{fe}}{1 + h_{oe} R_L}$$

where, h_{fe} and h_{oe} are hybrid parameters of a transistor.

$\therefore$ $A_i = -\dfrac{50}{1 + 25 \times 10^{-6} \times 1 \times 10^3} = -48.78$

22. (c) We know that resistance of conductor is directly proportional to temperature (i.e. $R \propto \Delta t$), while resistance of semiconductor is inversely proportional to temperature $\left(\text{i.e. } R \propto \dfrac{1}{\Delta t}\right)$.

Therefore, it is clear that resistance of conductor decreases with decrease in temperature or *vice-versa*, while in case of semiconductor, resistance increases with decrease in temperature or *vice-versa*.

Since, copper is pure conductor and germanium is a semiconductor, hence due to decrease in temperature, resistance of conductor decreases while that of semiconductor increases.

23. (c) When p-end of p-n junction is connected to positive terminal of battery and n-end to negative terminal of battery, then p-n junction is said to be in forward bias.

In forward bias, the more numbers of electrons go from n-region to p-region and more numbers of holes go from p-region to n-region.

Therefore, major current due to both types of carriers takes place through the junction, causing more recombination of electron-hole pairs, thus causing reduction in height of depletion region and barrier potential.

24. (b) Metal has number of free electrons while semiconductor has not at room temperature.

The difference in the variation of resistance with temperature in metal and semiconductor is caused due to difference in the variation of the number of charge carriers with temperature.

25. (a) Due to the reverse biasing, the width of depletion region increases and current flowing through the diode is almost zero. In this case, electric field is almost zero at the middle of the depletion region.

26. (c) At absolute zero, the energy band is large (> 3 eV) and hence Si acts as insulator or in other words, we can say that there is no free charge carriers.

27. (a) Because emitter has largest size and maximum number of charge carriers.

28. (c) Metals have minimum energy gap and insulators have maximum energy gap and semi-conductors have energy gap lying between insulators and metals.

29. (c) If we increase the temperature, the specific resistance (resistivity) of conductor increases while that of semiconductor decreases.

5. Communication System

DPP-1 Propogation of Electromagnetic Waves

1. (b) Speed of electromagnetic wave in a medium is

$$v = \frac{1}{\sqrt{\mu\varepsilon}} = \frac{1}{\sqrt{\varepsilon_0\mu_0\varepsilon_r\mu_r}} = \frac{c}{\sqrt{\varepsilon_r\mu_r}}.$$

But for most of the transparent medium χ_m is very less than 1 so, $\mu_r = 1 + \chi_m \approx 1$

$\therefore v = \frac{c}{\sqrt{\varepsilon_r}}$ (speed is mostly affected by electrical properties of medium.)

$$\Rightarrow \quad v = \frac{3 \times 10^8 \text{ ms}^{-1}}{\sqrt{9}} = 1 \times 10^8 \text{ ms}^{-1}$$

2. (b) Given, arrangement is a crude model of a half-wave dipole antenna

Length of each rod $= \frac{1}{4} \times$ wavelength of radiation produced

So, $\quad \lambda = 4 \times l = 4 \times 0.5 \text{ m} = 2\text{m}$

So, $\quad \text{frequency} = f = \frac{c}{\lambda} = \frac{3 \times 10^8 \text{ms}^{-1}}{2\text{m}}$

$$= 1.5 \times 10^8 \text{ s}^{-1} = 1.5 \times 10^8 \text{ Hz}$$

3. (a) A ground wave when glides over the surface of the earth, (over water in this case), horizontal component of electric field in contact with earth's (water) surface is short circuitted.

As a result wave soon becomes vertically polarised and so a vertically oriented antenna is used for reception.

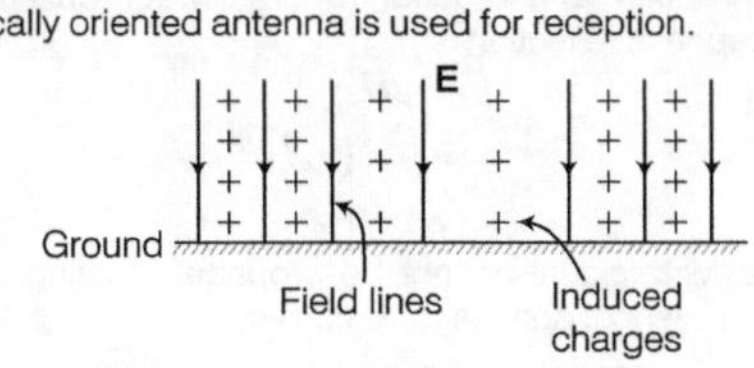

4. (d) High attenuation (loss of energy of strength of signal) which increases with frequency makes ground wave propagation unsuitable for higher frequencies and long ranges.

5. (c) Signal from our cell phone travels through roof tops, plants, vehicles etc. to the nearest service providers tower by ground wave propagation mode.

6. (b) Ground wave mode is preferred for 500 kHz to 1500 kHz frequency range.

7. (c) Using $f \cdot \lambda = c$,

$$\lambda = \frac{3 \times 10^8 \text{ ms}^{-1}}{3 \times 10^6 \text{ Hz}} = 10^2 = 100 \text{ kms}$$

So, only in two wavelengths signal can be reaching the space.

8. (d) F_2 layer is topmost layer of ionosphere so, skip distance will be largest for F_2 layer.

9. (a) E layer of ionosphere extends from 90 km to 120 km. So, its thickness is $= x_1 - x_2 = 120 - 90 \approx 30$ km.

10. (a) AM radio transmission from one state to another or one continent to another employs ionospheric reflection mode. But AM radio broadcast at a place uses ground wave.

11. (a,c,d) Fading of signal occurs when signal is partially absorbed (or lost partially due to refraction). This occurs when electron density is changing at a very rapid rate, i.e. in morning or evening hours or during sun storms. A solar eclipse occurs very speedily. Thus, it has not much effect.

12. (b,c,d) A geostationary satellite can be parked only at those places where gravitational pull is exactly equal to centripetal force. With $T = 24$ h, this distance is around 36000 kms above the earth's surface in equatorial plane.

As the sun is exactly above the equator at 21st March and 22nd September, noise is excessively high on these dates.

13. (a,b,c,d) AM radio broadcast uses ground wave. Signal to your mobile phone reaches *via* line of sight communication. Signal may be direct or diffracted or reflected.

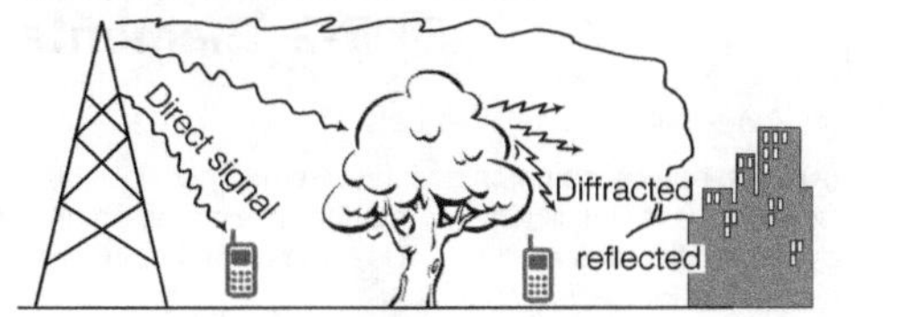

14. (a,b,c,d) Characteristic impedance of free space is

$$z_0 = \sqrt{\frac{\mu_0}{\varepsilon_0}} = 120\pi = 377\ \Omega$$

Also, wave number

$$k = \frac{2\pi}{\lambda} = \frac{\varepsilon_0}{c}$$

$$c = 3 \times 10^8 \text{ms}^{-1}$$

$$\text{Flux density} = W = |\mathbf{S}| = |\mathbf{E} \times \mathbf{B}|$$

$$= c\varepsilon_0|\mathbf{E}|^2 = \frac{1}{z_0}|\mathbf{E}|^2$$

$$= c\mu_0|\mathbf{B}|^2 = z_0|\mathbf{B}|^2$$

So, we have

$$E_{\max} = \sqrt{2z_0 \times W} = 275\ \mu\text{Vm}^{-1}$$

$$\therefore \quad E_{\text{rms}} = \text{Effective} = \frac{E_{\max}}{\sqrt{2}}$$

$$= 196\ \mu\text{Vm}^{-1}$$

and $\quad B_{\text{rms}} = 0.52\ \mu\text{Am}^{-1}$

15. (a,b,c) Distance upto which a tower of height h covers $d = \sqrt{2Rh}$ (where, R = radius of earth)

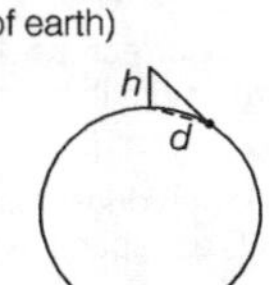

$$\Rightarrow \quad d = \sqrt{2 \times 6.4 \times 10^6 \times h}, h$$

$$= \left[3.57\sqrt{h \times 10^6}\right]$$

$$= 3.57 \times 10^3 \times \sqrt{h}$$

$$= 3.57\sqrt{h} \text{ km}$$

A range of a tower is usually measured in square kilometmetres, a more practical formula for d is

$$d = 3.57\sqrt{h} \text{ km}$$

where, h is in metres. Last option is incorrect as a repeater may transmit at a frequency different from received frequency.

16. (d) Ground wave mode is generally used for low frequencies upto 300 kHz. Sky wave mode is for moderate frequencies which can be reflected by F layers of ionosphere, upto 30 MHz. For $f > 30$ MHz, space wave mode is used.

Hence, (i) → (r), (ii) → (r,q), (iii) → (q), (iv) → (p).

17. (5) For F_1 layer,

$$5 \times 10^6 = 9\,(N_{max})^{1/2}$$

For F_2 layer,

$$8 \times 10^6 = 9\,(N_{max})^{1/2}$$

$$\therefore \qquad \text{Ratio} = \left(\frac{5}{8}\right)^2 = \frac{5^2}{64}$$

18. (d) Statement I is incorrect but Statement II is correct.

Frequency value of TV signals is very high to be reflected from sky (Ionosphere).

19. (d) Statement I is incorrect but Statement II is correct.

At low altitudes, due to high pressure, rate of ionisation is less then the rate of recombination. At high altitudes due to low pressure and high energy falling rate, ionisation is more than recombination. So higher layers have more ions and electrons and hence, more conductive.

20. (c) Statement II is incorrect. Short wave means a high frequency which must be produced for a large distance transmission using a suitable mode.

DPP-2 Elements of Communication System

1. (c) X is communication channel.

2. (a) In an analog communication system, input signal is similar to (analogous to) intelligent information. The signal which is relayed by transmitter after modulation is either a continuous waveform or it may be a digital system.

3. (c) In a digital system input is converted into a coded binary system, then it is modulated and relayed.

4. (d) For more uses bandwidth per user may be as low as possible.

5. (a) Frequency of signal now a days is increased, this reduces need for a long receiving antenna.

6. (d) Size (length) of antenna is around $\frac{1}{4}$ of wavelength of signal transmitted.

As, $\qquad \lambda = \frac{c}{f} = \frac{3 \times 10^8}{20 \times 100} = 1.5 \times 10^4\,\text{m} = 15\,\text{km}.$

7. (c) A microphone is a transducer it converts one type of signal (sound) into another type of signal (electrical).

8. (d) A modem is a modulator and de-modulator. Its function is to increase frequency (by mixing noise to signal). An increased frequency corresponds to a lower bandwidth.

9. (b) Only in case of an analog system a signal has only one value at an instant of time.

10. (d) A GPS system has 24 satellites to locate exact position and speed of a user.

11. (a,b,c) Option (d) is incorrect as no prism is required, signal created is already monochromatic.

12. (a,b,d) Option (c) is incorrect as frequency for an optical communication system is more than or equal to frequency of light 10^{14} Hz.

13. (b,c,d) Troposphere is lower ionosphere so, option (b) is correct, option (a) is incorrect as space wave is used for satellite communication also.

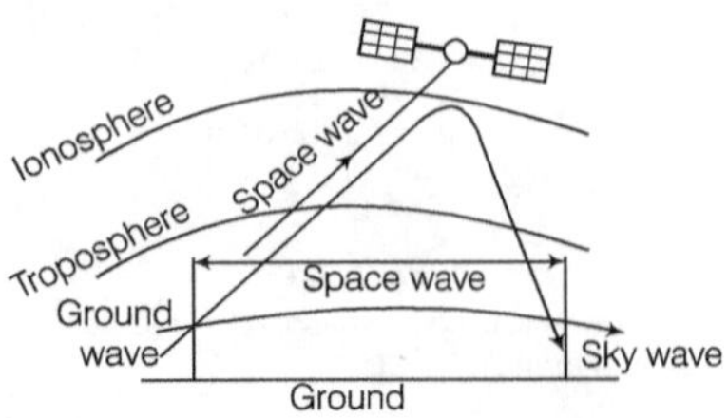

14. (a,b,c) A is source which creates intelligent information. So, an intelligent signal created by A have only one value at an instant of time.

B is transmitter it modifies the base band signal.

C is transmission channel.

D is demodulator and receiver.

15. (a,b,c) Maximum distance covered by using space wave communication by transmitter

$$= \sqrt{2rh}$$
$$= \sqrt{2 \times 6.4 \times 10^6 \times 300}$$
$$= 62\text{ km}$$

So, both option (a) and option (b) are correct. Critical frequency for ionospheric reflection is

$$f_c = 9(N_{max})^{1/2}$$
$$= 9 \times (10^{12})^{1/2}$$
$$= 9\text{ MHz}$$

So, 100 MHz signal cannot be propagated using ionospheric reflection. Hence, options (c) is correct.

16. (c) We have to fulfill 3-condition.

(i) Low power requirement

(ii) Efficient system

(iii) Economical system

Ionospheric reflection mode is most suitable under given conditions.

17. (a) Very high frequencies are not suitable for ionospheric reflection.

18. (b) E layer will be suitable as skip distance is around 1000 kms and this clears option (a) of solution of Q. No. 17. f_c for E layer is 27 MHz.

19. (a) Both Statement I and Statement II are correct and Statement II explains Statement I.

Maximum frequency that can be reflected off the ionosphere

$$f_c = 9(N_{max})^{1/2}$$

Any frequency higher than this goes in space without reflection.

20. (d) Statement I is incorrect but Statement II is correct.

An optical communication system is one which is not jammed because 2-light waves which are not travelling in same direction along same line does not produces interference.

DPP-3 **Modulation : AM and FM**

1. (d) Modulation process superimposes or translates a high frequency carrier wave such that base band variation is read as change in any of property like amplitude, frequency or phase of carrier.

2. (c) Modulating signal (waveform) is

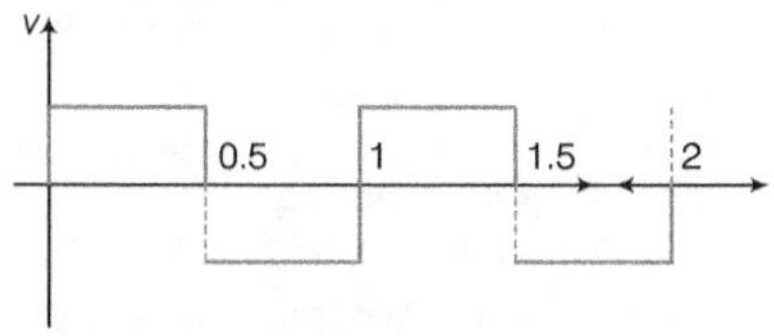

So, modulated carrier is as shown in option (c).

Also, note that carrier's frequency is

$$2\pi f = 8\pi$$

$$\Rightarrow \quad f = 4 \text{ Hz or } T = \frac{1}{4} = 0.25 \text{ s}$$

3. (d) As value of modulating signal is added or subtracted from the peak value of carrier, the instantaneous value of (top or bottom) envelope is

$$v_1 = V_c + v_m = V_c + V_m \sin \omega_m t$$

$$= V_c + V_m \sin 2\pi f_m t$$

So, modulated carrier will be

$$v_{AM} = (V_c + V_m \sin 2\pi f_m t) \cdot \sin (2\pi f_c t)$$

$$= V_c \sin 2\pi f_c t + V_m \cdot \sin 2\pi f_m t \cdot \sin 2\pi f_c t$$

4. (a) Carrier wave frequency $= \frac{1}{T} = \frac{1}{0.25} = 4$ Hz

$$\therefore \quad \omega_c = 2\pi f_c = 8\pi \text{ rads}^{-1}$$

So, carrier wave is

$$v_c = V_c \sin 8\pi t$$

In ideal modulation condition, modulation index = 1

$$\Rightarrow \quad V_m = V_c$$

$$\therefore \quad v_c = 2 \sin 8\pi t$$

Now, AM signal will be

$$v_{AM} = V_c \sin 2\pi f_c t + (V_m \sin 2\pi f_m t) \cdot \sin 2\pi f_c t$$

$$= 2(\sin 8\pi t) + 2(\sin 2\pi t)(\sin 8\pi t)$$

5. (b) Modulation index

$$m = \frac{V_{max} - V_{min}}{V_{max} + V_{min}}$$

$$= \frac{\frac{V_{max}\text{ (peak to peak)}}{2} - \frac{V_{min}\text{ (peak to peak)}}{2}}{\frac{V_{max}\text{ (peak to peak)}}{2} + \frac{V_{min}\text{ (peak to peak)}}{2}}$$

$$= \frac{\frac{12}{2} - \frac{4}{2}}{\frac{12}{2} + \frac{4}{2}} = \frac{4}{8} = \frac{1}{2} = 0.5$$

6. (d) $v_{AM} = V_c \sin 2\pi f_c t + \frac{V_m}{2} \cos (2\pi t (f_c - f_m))$

$$- \frac{V_m}{2} \cos [2\pi t (f_c + f_m)]$$

Here, $f_c - f_m$ = LSB frequency

$= 300000 + 400 = 300.4$ kHz

and $f_c + f_m$ = USB frequency

$= 300000 - 400 = 299.6$ kHz

Also, $V_c = 2, V_m = 2$ as $m = 1$

So, substituting values we have

$$v_{AM} = 2 \sin (6 \times 10^5 \pi t) + \cos (600.8 \times 10^3 \pi t)$$

$$- \cos (599.2 \times 10^3 \pi t)$$

7. (d) V_c or $A_c = 9$ V

$$\frac{A_m}{A_c} = m = \text{modulation index} = 0.75$$

$$\Rightarrow \quad A_m = 0.75 \times A_c$$

$$= \frac{3}{4} \times 9 = 6.75 \text{ V}$$

Also, $f_{LSB} = f_c - f_m = 300000 - 400$

$= 299600$ Hz

$= 299.6$ kHz

and $f_{USB} = f_c + f_m = 300.4$ kHz

8. (d)

$$v_{AM} = V_c \sin 2\pi f_c t + \frac{V_m}{2} [\cos (2\pi t (f_c - f_m)) - \cos (2\pi t (f_c + f_m))]$$

$$\text{So, } (v_{AM})_{rms} = \frac{V_c}{\sqrt{2}} \cdot \sin 2\pi f_c t + \frac{V_m}{2\sqrt{2}}$$

$$[\cos 2\pi t (f_c - f_m) - \cos 2\pi t (f_c + f_m)]$$

$$\text{Power} = \frac{V^2}{R}$$

$$\text{So,} \quad P_{AM} = \frac{(V_c / \sqrt{2})^2}{R} + 2 \cdot \frac{(V_m / (2\sqrt{2}))^2}{R}$$

$$\Rightarrow \quad \frac{\frac{V_c^2}{\sqrt{2}}}{R} + 2 \cdot \frac{\left(\frac{V_m}{2\sqrt{2}}\right)^2}{R}$$

$$\Rightarrow \quad \frac{\frac{V_c^2}{2}}{\frac{R}{1}} + 2 \cdot \frac{\frac{V_m^2}{4 \times 2}}{\frac{R}{1}}$$

$$\Rightarrow \quad \frac{V_c^2}{2R} + \frac{V_m^2}{4R}$$

9. (c) $P_T = P_c \left(1 + \frac{m^2}{2}\right)$

With 100% modulation, $m = 1$

$$\Rightarrow \quad P_T = P_c \times \frac{3}{2}$$

$$\Rightarrow \quad P_c = \text{carrier power} = \frac{2}{3} \times P_T$$

Also, $P_T = P_c + P_{LSB} + P_{USB}$

With $P_{LSB} = P_{USB}$

$$\therefore \quad P_T - P_c = P_{SB}$$

$$\Rightarrow \quad P_{SB} = P_T - \frac{2}{3} P_T$$

$$= \frac{1}{3} P_T = \frac{1}{3} \times \frac{3}{2} P_c$$

$$\text{So,} \quad \frac{P_{SB}}{P_c} = \frac{1}{2}$$

$\therefore$ Total side band power is $\frac{1}{2}$ part (50%) of carrier power.

10. (b) When percentage of modulation is less than optimum (100%), there is much less power in the side bands.

$$\text{As,} \quad P_T = P_c \left(1 + \frac{m^2}{2}\right).$$

11. (a) $P_c = I^2 R = (4.8)^2 \times 40 = 921.6$ W

12. (b) As $I_T = I_c \sqrt{1 + \frac{m^2}{2}}$

$$I_T = 4.8 \sqrt{1 + \frac{(0.9)^2}{2}} = 4.8 \sqrt{1.405} = 5.7 \text{ A}$$

$$P_T = I_T^2 R = (5.7)^2 \times 40 = 1295 \text{ W}$$

13. (d) $P_{SB} = P_T - P_c = 1295 - 921.6 = 373.4$ W

So, power in each side band is $= \frac{373.4}{2} = 186.7$ W

14. (c) In standard AM broadcast,

$$P_T = P_c + P_{LSB} + P_{USB}$$

Also, $$P_T = P_c\left(1 + \frac{m^2}{2}\right)$$

$\Rightarrow$ $$P_C = \frac{2}{3}P_T$$

and $$P_{LSB} = P_{USB}$$

So, $\frac{2}{3}$rd of total power (when $m = 1$) is in the carrier.

So, $\frac{2}{3}$rd total power is wasted.

So, % of wasted power $= \frac{\text{Wasted power}}{\text{Total power}} \times 100$

$$= \frac{2}{3} \times 100 = 66.67\%$$

15. (a) In an standard AM transmission

$$P_T = P_c\left(1 + \frac{m^2}{2}\right)$$

and with an 100% modulation,

$$P_T = P_c\left(\frac{3}{2}\right)$$

or $$P_c = \frac{2}{3}P_T$$

So, $\frac{1}{3}$rd of total power is lies in the side bands.

In a SSB AM transmission, LSB and carrier both are suppressed.

Clearly only $\frac{1}{2} \times \frac{1}{3}P_T$ is required in SSB transmission.

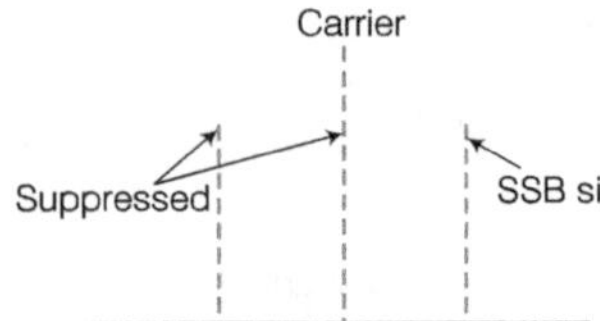

So, power required for standard AM = P_T

and power required for SSB AM = $\frac{P_T}{6}$,

$$\frac{P_{SSB\ AM}}{P_{AM}} = \frac{1}{6}.$$

16. (c) In frequency modulation bandwidth required is very large compared to other modulations (AM, phase, etc.).

This results in a higher frequency and power requirement is much higher (than AM, phase, ..., etc.).

17. (c) As amplitude of modulating signal × frequency of carrier

$\therefore$ When amplitude decreases, frequency must decrease.

18. (c) FM signals are not prone to man made noises, like sparks, weldings, etc. So, they are preferred for music transmission where quality is a primary requirement.

19. (a,b) The square law device produces an output

$$y(t) = BA_m \sin \omega_m t + BA_c \sin \omega_c t + \frac{CA_m^2}{2} + \frac{CA_c^2}{2} - \frac{CA_m^2}{2}\cos 2\omega_m t$$
$$- \frac{CA_c^2}{2}\cos 2\omega_c t + CA_m A_c \cos(\omega_c - \omega_m)t - CA_m A_c \cos(\omega_c - \omega_m t)$$

There is a DC term $\frac{C}{2}(A_m^2 + A_c^2)$ and sinusoids with frequencies $\omega_m, 2\omega_m, 2\omega_c, \omega_c, \omega_c - \omega_m$ and $\omega_c + \omega_m$.

This signal when goes through a band pass filter, it rejects low (DC and ω_m) and high ($2\omega_c$ and $2\omega_m$) and allows rest of the band (ω_c, $\omega_m + \omega_c$ and $\omega_m - \omega_c$) to pass through it.

20. (a,b) $v_{AM} = 5(1 + 0.6 \cos 6280t) \sin 211 \times 10^4 t$

Comparing with standard equation,

$$v_{AM} = A_c(1 + m \cos \omega_m t) \sin \omega_c t$$

We have,

A_c = carrier amplitude = 5 V

m = modulation index = 0.6

ω_m = signal frequency = 6280 rads^{-1} = 1 kHz

ω_c = carrier frequency = 211×10^4 rads^{-1} = 336 kHz

So, $$A_{max} = A_c + mA_c = 5 + 0.6 \times 5 = 8\text{V}$$

$$A_{min} = A_c - mA_c = 2\text{V}$$

Frequencies contained by AM

$$= f_c, f_c - f_s \text{ and } f_c + f_s$$
$$= 336, 335 \text{ and } 337 \text{ kHz}$$

Amplitudes of 3 components are

$$\frac{mA_c}{2} = \frac{0.6 \times 5}{2} = 1.5 \text{ V}$$

$$A_c = 5 \text{ and } \frac{mA_c}{2} = 1.5 \text{ V}$$

21. (a,b,c) Total power of AM is

$$P_T = \frac{(\text{rms voltage})^2}{R} = \frac{(0.707 \times V_{max})^2}{R}$$

Now, power delivered by carrier,

$$P_c = \frac{(0.707 \times 5)^2}{2000} = 6.25 \text{ MW}$$

Power delivered by lower side band component is

$$P_{LSB} = \frac{(0.707 \times 2.5)^2}{2000} = 1.562 \text{ MW}$$

Also, $P_{USB} = 1.562$ MW

Total power of AM = 6.25 + 1.562 + 1.562 = 9.374 MW

22. (d)

(i) Without any modulating signal, AM contains only carrier.

(ii) With a modulating signal of amplitude = $A_c = A$, modulated signal have an amplitude of

$$A_{AM} = A_c + A_m = A + A = 2A$$

(iii) With $A_c = A$ and $A_m = 0.5$ A amplitude of modulated wave is 1.5 A.

(iv) With $A_c = A$ and $A_m = 1.5$ A, $A_{AM} = 2.5$ A and signal is distorted.

Hence, (i) $\rightarrow$ (s), (ii) $\rightarrow$ (q), (iii) $\rightarrow$ (r), (iv) $\rightarrow$ (p)

Revisal Problems (JEE Main)

1. (b) A and B are far off to communicate *via* ground waves. A and B are not able to see each other. So, LOS or space wave is not possible.

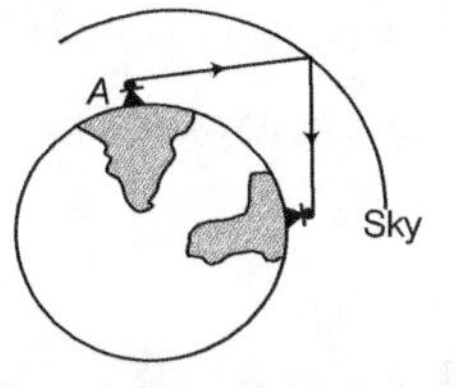

Hence, only possible mode is ionospheric reflection.

This is possible when frequency range is from 3 MHz to 30 MHz.

2. (d) UHF band is 50 MHz to 1000 MHz. So, space wave mode.

3. (d) Digital signals have only two discreate values (0 and 1).

4. (c) A digital signal has only two discreate values and only binary system is utilised for coding and decoding system.

5. (b) in LOS communication using space waves, the transmitter and receiver must see each other.

6. (c) Modulation index = $\mu = \dfrac{A_m}{A_c}$

$$\therefore \quad \mu = \frac{9V}{12V} = \frac{3}{4} = 0.75$$

7. (c) Statement I is incorrect, as multiple side bands or distortion appears only when μ (modulation index) is greater than 1. Modulation index is kept upto 1.

Ideal condition for AM is $V_m = V_c$ or $\mu = 1$, which gives 100% modulation. This results in greatest power output voltage at receiver with no distortion. All other Statements are correct.

8. (d) When antenna impedance is known, power transmitted

$$P_T = I_T^2 R$$

and $$I_T = I_C\sqrt{1 + \left(\frac{\mu^2}{2}\right)}$$

where μ = modulation index

I_C = unmodulated carrier current in load.

Also, power in each side band

$$P_{SB} = P_T - P_C$$

$$P_C = I_C^2 R = \text{carrier power}$$

Now, $$I_T = 4.8\sqrt{1 + \frac{(0.9)^2}{2}} = 5.7 \text{ A}$$

$$P_T = (5.7)^2 \times 40 = 1295 \text{ W}$$

$$P_C = (4.8)^2 \times 40 = 921.6 \text{ W}$$

$$P_{SB}(\text{total}) = P_T - P_C = 373.4 \text{ W}$$

$$P_{SB} \text{ (each side band)} = \frac{373.4}{2} = 186.7 \text{ W}$$

9. (a) A frequency domain

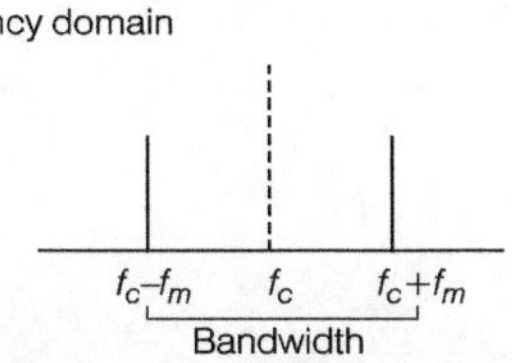

Display of DSB signal is shown in the above figure.

Clearly, bandwidth of signal = $(f_c + f_m) - (f_c - f_m)$

$$= 2f_m = 2 \times 5 = 10 \text{ kHz}$$

10. (c) For carrier cosine signal

$$v_c = V_c \cos(2\pi f_c t).$$

and modulating sine signal,

$$v_m = V_m \sin(2\pi f_m t)$$

We have a modulated wave,

$$v = (V_c + V_m \sin 2\pi f_m t) \cdot \cos(2\pi f_c t)$$

So, $V_c = 20\,V$ and $V_m = 4\,V$

$\therefore$ Modulation index, $\mu = \dfrac{A_m}{A_c} = \dfrac{V_m}{V_c}$

$$= \frac{4}{20} = \frac{1}{5} = 0.2$$

11. (b) 1600 kHz – Ground wave

5 MHz – Sky wave

60 MHz – Space wave

12. (d) $$P_T = P_C\left(1 + \frac{\mu^2}{2}\right)$$

$$25 = P_C\left(1 + \frac{(0.3)^2}{2}\right)$$

$$\Rightarrow \quad P_C = \frac{25}{1.045}$$

Percentage of power carried by carrier = $\dfrac{P_C}{P_T} \times 100 = 95.69\%$

13. (d) For both carrier and massage signal, amplitudes are same, $A_c = A_m = 2A$.

Hence, 100% modulation occurs. So, AM waves have displacements from 0 to $2A$.

14. (a) Number of channels possible

= Bandwidth of frequency range ÷ Bandwidth per channel.

$$= \frac{2 \times 75 \text{ MHz}}{200 \text{ kHz}} = \frac{2 \times 75 \times 10^6 \text{ Hz}}{200 \times 1000 \text{ Hz}}$$

= 750 channels

= 750 users

15. (b) A high frequency wave is bent lesser than a low frequency wave. So, $f_B > f_A$ or $\lambda_B < \lambda_A$.

16. (c) Modulation index, $\mu = \dfrac{A_{max} - A_{min}}{A_{max} + A_{min}}$

$$= \frac{V_{max} - V_{min}}{V_{max} + V_{min}}$$

$$= \frac{\frac{20}{2} - \frac{6}{2}}{\frac{20}{2} + \frac{6}{2}} = \frac{20 - 6}{20 + 6}$$

$$= \frac{14}{26} = 0.53$$

17. (d) At least four transducers are required.

To communicate a voice signal *via* optical link, first voice signal must be transformed into electrical, then in optical signal.

18. (d) Given circuit of a demodulator as $f << RC$

So, output does not contain any of carrier frequency.

So, output is the envelope of input.

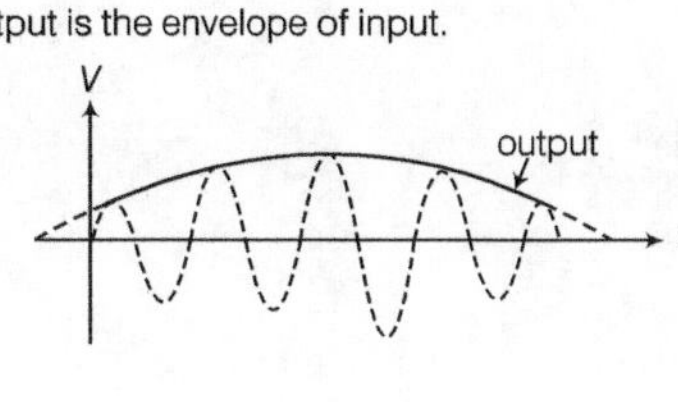

19. (d) Due to higher pressure, number of gas atoms is large and recombination rate of ions and electrons is greater than ionisation rate.

20. (d) Given circuit is a modulator. Output is a modulated wave

As $V_{out} = (V_c + V_m \sin 2\pi f_m t) \cdot \sin 2\pi f_c t$

So, $V_{out} = (V_c + V_m) \sin 2\pi f_c t$

$$= \begin{cases} 3\sin 8\pi t & t = 0 \text{ to } 0.5\text{ s} \\ \sin 8\pi t & t = 0.5 \text{ to } t = 1\text{ s} \end{cases}$$

21. (b) A low frequency signal may get reflected by D or E layer in day time due to sufficient ionisation. It does not reach F layer due to reflection from D or E layer. Atmospheric attenuation is large for low frequency signals.

22. (d) Frequency values (in s^{-1}) are AM 10^6, FM 10^8 microwaves 10^{10}, visible 10^{14}, X-rays 10^{18}, gamma 10^{21}. So, energy also follows same ascending order. $E = hf$ can be used to find the values.

Hence, (i) → (u), (ii) → (t), (iii) → (s), (iv) → (r), (v) → (q), (vi) → (p).

JEE Main & AIEEE Archive

1. (d)

(i) Infrared rays are used to treat muscular strain.

(ii) Radiowaves are used for broadcasting purposes.

(iii) X-rays are used to detect fracture of bones.

(iv) Ultraviolet rays are absorbed by ozone.

Hence, (i) → (p), (ii) → (q), (iii) → (r), (iv) → (s).

2. (b)

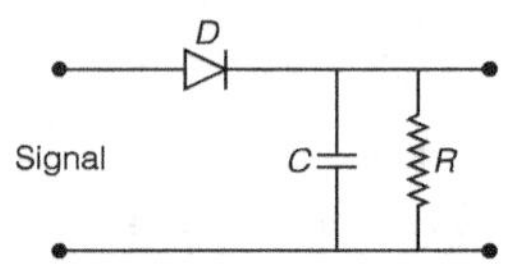

$$\tau = RC = 100 \times 10^3 \times 250 \times 10^{-12}\text{ s}$$
$$= 2.5 \times 10^7 \times 10^{-12}\text{ s}$$
$$= 2.5 \times 10^{-5}\text{ s}$$

The highest frequency which can be detected with tolerable distortion is

$$f = \frac{1}{2\pi m_a RC} \quad \text{[where, } m_a \text{ is modulation]}$$
$$= \frac{1}{2\pi \times 0.6 \times 2.5 \times 10^{-5}}\text{ Hz}$$
$$= \frac{100 \times 10^4}{25 \times 1.2\,\pi}\text{ Hz}$$
$$= \frac{4}{1.2\,\pi} \times 10^4\text{ Hz}$$
$$= 10.61\text{ kHz}$$

This condition is obtained by applying the condition that rate of decay of capacitor voltage must be equal or less than the rate of decay modulated signal voltage for proper detection of modulated signal.

3. (a) Range of radar on earth surface (optical distance, for space wave i.e. line of view).

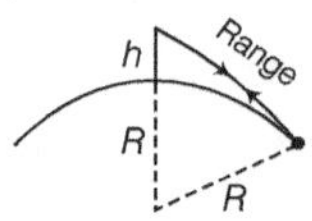

$$\therefore \quad \text{Range} = \sqrt{(R + h)^2 - R^2}$$
$$= \sqrt{2Rh + h^2} \simeq \sqrt{2Rh}$$
$$= \sqrt{2 \times 6400 \times \frac{1}{2}}\text{ km} = 80\text{ km}$$

4. (b) Modulation does not change time lag between transmission and reception.

5. (b) In radio communication, sky wave refers to the propagation of radio waves reflected or refracted back towards earth from the ionosphere.

Since, it is not limited by the curvature of the earth, sky wave propagation can be used to communicate beyond horizon. Ionosphere is a region of upper atmosphere and induces the thermosphere and parts of mesosphere and exosphere. It is distinguished because it is ionised by solar radiation. It plays an important part in atmospheric electricity.

6. (b) Some of the characteristics of an optical fibre are as follows :

(i) It works on the principle of total internal reflection.

(ii) It consists of core made up of glass/ silica /plastic with refractive index n_1, which is surrounded by a glass or plastic cladding with refractive index n_2, $(n_2 > n_1)$. The refractive index of cladding can be either changing abruptly or gradually changing (graded index fibre).

(iii) There is a very little transmission loss through optical fibres.

(iv) There is no interference from stray electric and magnetic fields to the signals through optical fibres.

Last 3 Years' Questions
JEE Main & Advanced
(2016-2018)

JEE Main

2016

1. Arrange the following electromagnetic radiations in the order of increasing energy. **[Communication System]**

A. Blue light B. Yellow light C. X-ray D. Radio wave

(a) D, B, A, C (b) A, B, D, C
(c) C, A, B, D (d) B, A, D, C

2. Radiation of wavelength λ, is incident on a photocell. The fastest emitted electron has speed v. If the wavelength is changed to $\frac{3\lambda}{4}$, the speed of the fastest emitted electron will be **[Atoms]**

(a) $> v\left(\frac{4}{3}\right)^{1/2}$ (b) $< v\left(\frac{4}{3}\right)^{1/2}$ (c) $= v\left(\frac{4}{3}\right)^{1/2}$ (d) $= v\left(\frac{3}{4}\right)^{1/2}$

3. Half-lives of two radioactive elements A and B are 20 min and 40 min, respectively. Initially, the samples have equal number of nuclei. After 80 min, the ratio of decayed numbers of A and B nuclei will be **[Nuclei]**

(a) 1 : 16 (b) 4 : 1
(c) 1 : 4 (d) 5 : 4

4. If a, b, c and d are inputs to a gate and x is its output, then, as per the following time graph, the gate is **[Electronic Devices]**

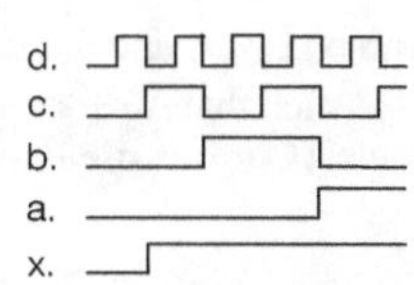

(a) NOT (b) AND
(c) OR (d) NAND

5. Choose the correct statement. **[Communication System]**

(a) In amplitude modulation, the amplitude of the high frequency carrier wave is made to vary in proportion to the amplitude of the audio signal
(b) In amplitude modulation, the frequency of the high frequency carrier wave is made to vary in proportion to the amplitude of the audio signal
(c) In frequency modulation, the amplitude of the high frequency carrier wave is made to vary in proportion to the amplitude of the audio signal
(d) In frequency modulation, the amplitude of the high frequency carrier wave is made to vary in proportion to the frequency of the audio signal

6. Identify the semiconductor devices whose characteristics are as given below, in the order (a),(b),(c),(d).

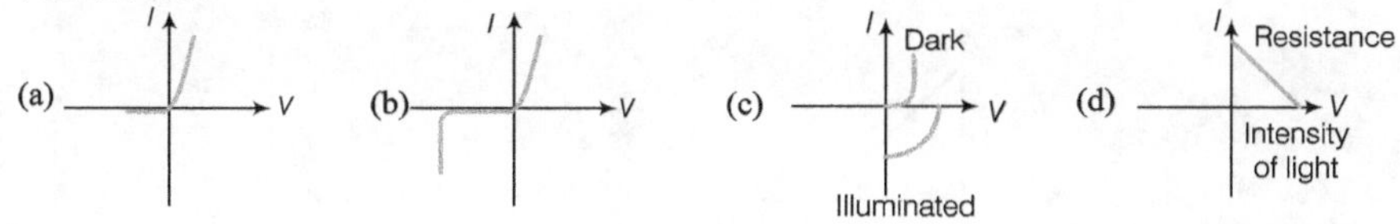

[Electronic Devices]

(a) Simple diode, Zener diode, Solar cell, Light dependent resistance
(b) Zener diode, Simple diode, Light dependent resistance, Solar cell
(c) Solar cell, Light dependent resistance, Zener diode, Simple diode
(d) Zener diode, Solar cell, Simple diode, Light dependent resistance

7. For a common-emitter configuration, if α and β have their usual meanings, the **incorrect** relationship between α and β is **[Electronic Devices]**

(a) $\frac{1}{\alpha} = \frac{1}{\beta} + 1$ (b) $\alpha = \frac{\beta}{1-\beta}$

(c) $\alpha = \frac{\beta}{1+\beta}$ (d) $\alpha = \frac{\beta^2}{1+\beta^2}$

2017

8. Some energy levels of a molecule are shown in the figure. The ratio of the wavelengths $r = \lambda_1 / \lambda_2$ is given by **[Atoms]**

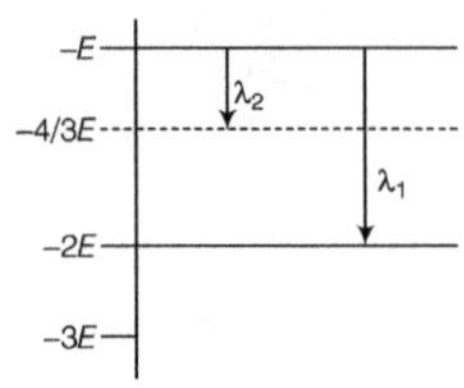

(a) $r = \frac{2}{3}$ (b) $r = \frac{3}{4}$

(c) $r = \frac{1}{3}$ (d) $r = \frac{4}{3}$

9. A radioactive nucleus A with a half-life T, decays into a nucleus B. At $t = 0$, there is no nucleus B. After sometime t, the ratio of the number of B to that of A is 0.3. Then, t is given by **[Nuclei]**

(a) $t = T \frac{\log 1.3}{\log_e 2}$ (b) $t = T \log 1.3$

(c) $t = \frac{T}{\log 1.3}$ (d) $t = \frac{T \log_e 2}{2 \log 1.3}$

10. In amplitude modulation, sinusoidal carrier frequency used is denoted by ω_c and the signal frequency is denoted by ω_m. The bandwidth $(\Delta\omega_m)$ of the signal is such that $\Delta\omega_m << \omega_c$. Which of the following frequencies is not contained in the modulated wave?**[Communication System]**

(a) ω_c (b) $\omega_m + \omega_c$
(c) $\omega_c - \omega_m$ (d) ω_m

11. In a common emitter amplifier circuit using an *n-p-n* transistor, the phase difference between the input and the output voltages will be **[Electronic Devices]**

(a) 90° (b) 135°
(c) 180° (d) 45°

12. An electron beam is accelerated by a potential difference V to hit a metallic target to produce X-rays. It produces continuous as well as characteristic X-rays. If $\lambda_{\min}$ is the smallest possible wavelength of X-rays in the spectrum, the variation of $\log \lambda_{\min}$ with $\log V$ is correctly represented in **[Atoms]**

(a) 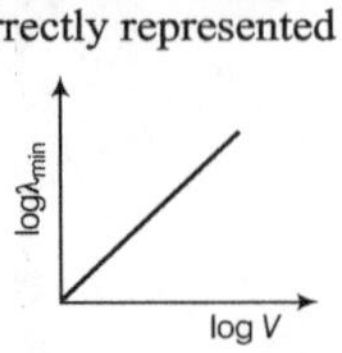

(b) 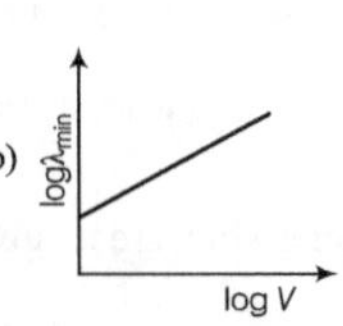

(c) 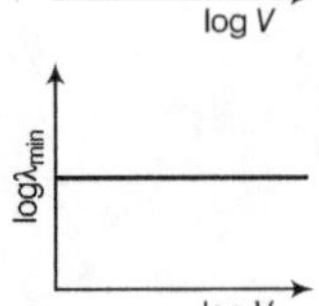

(d) 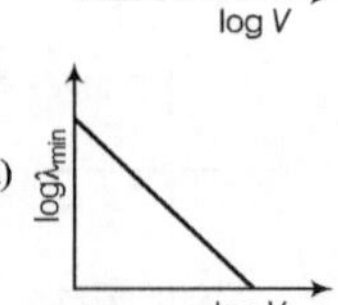

13. A particle A of mass m and initial velocity v collides with a particle B of mass $\frac{m}{2}$ which is at rest. The collision is head on, and elastic. The ratio of the de-Broglie wavelengths λ_A to λ_B after the collision is **[Dual Nature of Matter and Radiation]**

(a) $\frac{\lambda_A}{\lambda_B} = 2$ (b) $\frac{\lambda_A}{\lambda_B} = \frac{2}{3}$

(c) $\frac{\lambda_A}{\lambda_B} = \frac{1}{2}$ (d) $\frac{\lambda_A}{\lambda_B} = \frac{1}{3}$

2018

14. An electron from various excited states of hydrogen atom emit radiation to come to the ground state. Let λ_n, λ_g be the de-Broglie wavelength of the electron in the nth state and the ground state, respectively. Let Λ_n be the wavelength of the emitted photon in the transition from the nth state to the ground state. For large n, (A, B are constants) **[Dual Nature of Matter and Radiation]**

(a) $\Lambda_n \approx A + \frac{B}{\lambda_n^2}$ (b) $\Lambda_n \approx A + B\lambda_n^2$

(c) $\Lambda_n^2 \approx A + B\lambda_n^2$ (d) $\Lambda_n^2 \approx \lambda$

15. If the series limit frequency of the Lyman series is ν_L, then the series limit frequency of the Pfund series is **[Atoms]**

(a) $25\ \nu_L$ (b) $16\ \nu_L$

(c) $\frac{\nu_L}{16}$ (d) $\frac{\nu_L}{25}$

16. The reading of the ammeter for a silicon diode in the given circuit is **[Electronic Devices]**

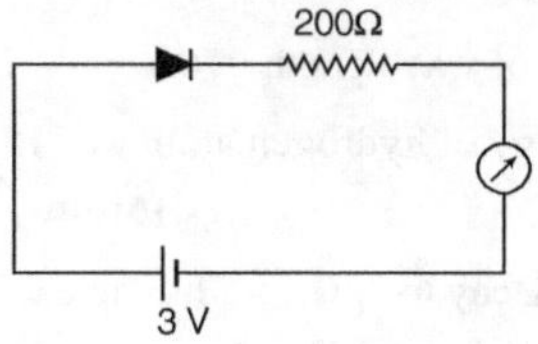

(a) 0 (b) 15 mA

(c) 11.5 mA (d) 13.5 mA

17. A telephonic communication service is working at carrier frequency of 10 GHz. Only 10% of it is utilised for transmission. How many telephonic channels can be transmitted simultaneously, if each channel requires a bandwidth of 5 kHz? **[Communication System]**

(a) 2×10^3 (b) 2×10^4

(c) 2×10^5 (d) 2×10^6

18. An EM wave from air enters a medium. The electric fields are $\mathbf{E}_1 = E_{01}\hat{\mathbf{x}}\cos\left[2\pi\nu\left(\frac{z}{c} - t\right)\right]$ in air and $\mathbf{E}_2 = E_{02}\hat{\mathbf{x}}\cos[k(2z - ct)]$ in medium, where the wave number k and frequency ν refer to their values in air. The medium is non-magnetic.

If ε_{r_1} and ε_{r_2} refer to relative permittivities of air and medium respectively, which of the following options is correct? **[Communication System]**

(a) $\frac{\varepsilon_{r_1}}{\varepsilon_{r_2}} = 4$ (b) $\frac{\varepsilon_{r_1}}{\varepsilon_{r_2}} = 2$

(c) $\frac{\varepsilon_{r_1}}{\varepsilon_{r_2}} = \frac{1}{4}$ (d) $\frac{\varepsilon_{r_1}}{\varepsilon_{r_2}} = \frac{1}{2}$

JEE Advanced

2016

1. In a historical experiment to determine Planck's constant, a metal surface was irradiated with light of different wavelengths. The emitted photoelectron energies were measured by applying a stopping potential. The relevant data for the wavelength (λ) of incident light and the corresponding stopping potential (V_0) are given below:

λ (μm)	V_0 (Volt)
0.3	2.0
0.4	1.0
0.5	0.4

Given that $c = 3\times10^8\ \text{ms}^{-1}$ and $e = 1.6\times10^{-19}$ C, Planck's constant (in units of J-s) found from such an experiment is) **[Dual Nature of Matter and Radiation, Type 1]**

(a) 6.0×10^{-34}
(b) 6.4×10^{-34}
(c) 6.6×10^{-34}
(d) 6.8×10^{-34}

2. Highly excited states for hydrogen-like atoms (also called Rydberg states) with nuclear charge Ze are defined by their principle quantum number n, where $n >> 1$. Which of the following statement(s) is (are) true? **[Atoms, Type 2]**

(a) Relative change in the radii of two consecutive orbitals does not depend on Z
(b) Relative change in the radii of two consecutive orbitals varies as $1/n$
(c) Relative change in the energy of two consecutive orbitals varies as $1/n^3$
(d) Relative change in the angular momenta of two consecutive orbitals varies as $1/n$

3. A hydrogen atom in its ground state is irradiated by light of wavelength 970Å.

Taking $hc/e = 1.237\times10^{-6}$ eVm and the ground state energy of hydrogen atom as -13.6 eV, the number of lines present in the emission spectrum is **[Atoms, Type 3]**

4. The isotopes ${}^{12}_{5}\text{B}$ having a mass 12.014 u undergoes β-decay to ${}^{12}_{6}\text{C}$. ${}^{12}_{6}\text{C}$ has an excited state of the nucleus (${}^{12}_{6}\text{C}^*$) at 4.041 MeV above its ground state. If ${}^{12}_{5}\text{B}$ decays to ${}^{12}_{5}\text{C}^*$, the maximum kinetic energy of the β-particle in units of MeV is ($1\text{u} = 931.5\ \text{MeV}/c^2$, where c is the speed of light in vacuum) **[Nuclei, Type 3]**

5. An accident in a nuclear laboratory resulted in deposition of a certain amount of radioactive material of half-life 18 days inside the laboratory. Tests revealed that the radiation was 64 times more than the permissible level required for safe operation of the laboratory. What is the minimum number of days after which the laboratory can be considered safe for use? **[Nuclei, Type 2]**

(a) 64
(b) 90
(c) 108
(d) 120

6. The electrostatic energy of Z protons uniformly distributed throughout a spherical nucleus of radius R is given by

$$E = \frac{3}{5}\frac{Z(Z-1)e^2}{4\pi\varepsilon_0 R}$$

The measured masses of the neutron, ${}^{1}_{1}\text{H}$, ${}^{15}_{7}\text{N}$ and ${}^{15}_{8}\text{O}$ are 1.008665u, 1.007825u, 15.000109u and 15.003065 u, respectively. Given that the radii of both the ${}^{15}_{7}\text{N}$ and ${}^{15}_{8}\text{O}$ nuclei are same, $1\ \text{u} = 931.5\ \text{MeV}/c^2$ (c is the speed of light) and $e^2/(4\pi\varepsilon_0) = 1.44$ MeV fm. Assuming that the difference between the binding energies of ${}^{15}_{7}\text{N}$ and ${}^{15}_{8}\text{O}$ is purely due to the electrostatic energy, the radius of either of the nuclei is ($1\text{fm} = 10^{-15}$ m) **[Nuclei, Type 2]**

(a) 2.85 fm
(b) 3.03 fm
(c) 3.42 fm
(d) 3.80 fm

Note *Type-1 Only one correct option, Type-2 More then one correct option, Type-3 Single digit Answer, Type-4 Numerical value based*

7. Light of wavelength λ_{ph} falls on a cathode plate inside a vacuum tube as shown in the figure. The work function of the cathode surface is ϕ and the anode is a wire mesh of conducting material kept at a distance d from the cathode. A potential difference V is maintained between the electrodes. If the minimum de-Broglie wavelength of the electrons passing through the anode is λ_e, which of the following statements(s) is (are) true?

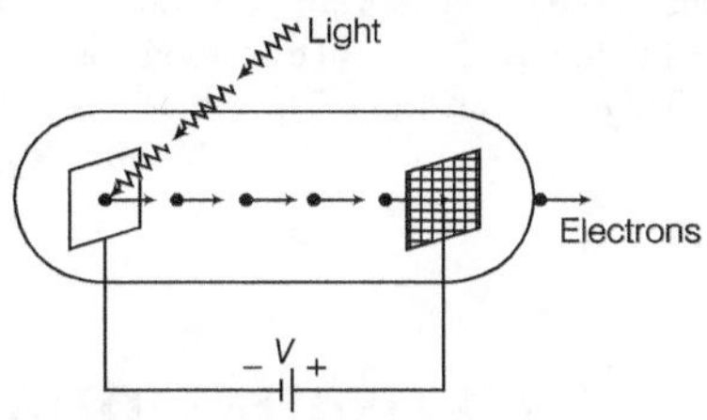

[Atoms, Type 2]

(a) λ_e increases at the same rate as λ_{ph} for $\lambda_{ph} < hc/\phi$

(b) λ_e is approximately halved, if d is doubled

(c) λ_e decreases with increase in ϕ and $0\lambda_{ph}$

(d) For large potential difference ($V >> \phi/e$), λ_e is approximately halved if V is made four times

2017

8. An electron in a hydrogen atom undergoes a transition from an orbit with quantum number n_i to another with quantum number n_f. v_i and v_f are respectively the initial and final potential energies of the electron. If $v_i / v_f = 6.25$, then the smallest possible n_f is **[Atoms, Type 3]**

9. ^{131}I is an isotope of Iodine that β decays to an isotope of Xenon with a half-life of 8 days. A small amount of a serum labelled with ^{131}I is injected into the blood of a person. The activity of the amount of ^{131}I injected was 2.4×10^5 Becquerel (Bq). It is known that the injected serum will get distributed uniformly in the blood stream in less than half an hour. After 11.5 h, 2.5 ml of blood is drawn from the person's body, and gives an activity of 115 Bq. The total volume of blood in the person's body, in litres is approximately (you may use $e^2 \approx 1 + x$ for $|x| << 1$ and $\ln 2 \approx 0.7$). **[Nuclei, Type 3]**

10. A photoelectric material having work-function ϕ_0 is illuminated with light of wavelength λ $\left(\lambda < \frac{hc}{\phi_0}\right)$. The fastest photoelectron has a de-Broglie wavelength λ_d. A change in wavelength of the incident light by $\Delta\lambda$ results in a change $\Delta\lambda_d$ in λ_d. Then, the ratio $\frac{\Delta\lambda_d}{\Delta\lambda}$ is proportional to **[Dual Nature of Matter and Radiation, Type 1]**

(a) $\frac{\lambda_d^2}{\lambda^2}$ (b) $\frac{\lambda_d}{\lambda}$

(c) $\frac{\lambda_d^3}{\lambda}$ (d) $\frac{\lambda_d^3}{\lambda^2}$

2018

11. In a radioactive decay chain, $^{232}_{90}Th$ nucleus decays to $^{212}_{82}Pb$ nucleus. Let N_α and N_β be the number of α and β- particles respectively, emitted in this decay process. Which of the following statements is (are) true? **[Nuclei, Type 2]**

(a) $N_\alpha = 5$ (b) $N_\alpha = 6$

(c) $N_\beta = 2$ (d) $N_\beta = 4$

12. In a photoelectric experiment, a parallel beam of monochromatic light with power of 200 W is incident on a perfectly absorbing cathode of work function 6.25 eV. The frequency of light is just above the threshold frequency, so that the photoelectrons are emitted with negligible kinetic energy. Assume that the photoelectron emission efficiency is 100%. A potential difference of 500 V is applied between the cathode and the anode. All the emitted electrons

are incident normally on the anode and are absorbed. The anode experiences a force $F = n \times 10^{-4}$ N due to the impact of the electrons. The value of n is .... (Take mass of the electron, $m_e = 9\times10^{-31}$ kg and eV = 1.6×10^{-19} J)

[Dual Nature of Matter and Radiation, Type 4]

13. Consider a hydrogen-like ionised atom with atomic number Z with a single electron. In the emission spectrum of this atom, the photon emitted in the $n = 2$ to $n = 1$ transition has energy 74.8 eV higher than the photon emitted in the $n = 3$ to $n = 2$ transition. The ionisation energy of the hydrogen atom is 13.6 eV. The value of Z is .... **[Atoms, Type 4]**

Answer *with* Explanations

JEE Main

1. *(a)* Theoretical question. Therefore, no solution is required.

2. *(a)* According to the law of conservation of energy, i.e. Energy of a photon $(h\nu)$ = Work function (ϕ) + Kinetic energy of the photoelectron $\left(\frac{1}{2}mv^2_{max}\right)$

According to Einstein's photoelectric emission of light

i.e. $$E = (\text{KE})_{max} + \phi$$

As, $$\frac{hc}{\lambda} = (\text{KE})_{max} + \phi$$

If the wavelength of radiation is changed to $\frac{3\lambda}{4}$, then

$$\Rightarrow \quad \frac{4}{3}\frac{hc}{\lambda} = \left(\frac{4}{3}(\text{KE})_{max} + \frac{\phi}{3}\right) + \phi$$

$(\text{KE})_{max.}$ for fastest emitted electron $= \frac{1}{2}mv'^2 + \phi$

$$\Rightarrow \quad \frac{1}{2}mv'^2 = \frac{4}{3}\left(\frac{1}{2}mv^2\right) + \frac{\phi}{3}$$

i.e $$v' > v\left(\frac{4}{3}\right)^{1/2}$$

3. *(d)* **A :** Numbers left :

$$N \rightarrow \frac{N}{2} \rightarrow \frac{N}{4} \rightarrow \frac{N}{8} \rightarrow \frac{N}{16}$$

∴ Number decayed,

$$N_A = N - \frac{N}{16} = \frac{15}{16}N$$

B : Numbers left : $N \rightarrow \frac{N}{2} \rightarrow \frac{N}{4}$

∴ Numbers decayed,

$$N_B = N - \frac{N}{4} = \frac{3}{4}N$$

Ratio : $$\frac{N_A}{N_B} = \frac{(15/16)N}{(3/4)N} = \frac{5}{4}$$

4. *(c)* Output of OR gate is 0 when all inputs are 0 and output is 1 when atleast one of the input is 1.

Observing output *x* It is 0 when all inputs are 0 and it is 1 when atleast one of the input is 1.

∴ The gate is OR.

Alternative Method

OR Gate

A	B	C	D	X
0	0	0	0	0
0	0	0	1	1
0	0	1	0	1
0	0	1	1	1
0	1	0	0	1
0	1	0	1	1
0	1	1	0	1
0	1	1	1	1
1	0	0	0	1

5. *(a)* In amplitude modulation, $\mu = \frac{A_m}{A_c}$

6. *(a)* Theoretical question. Therefore, no solution is required.

7. *(b, d)* $$I_b + I_c = I_e$$

$$\Rightarrow \quad \frac{I_b}{I_c} + 1 = \frac{I_e}{I_c}$$

$$\Rightarrow \quad \frac{1}{\beta} + 1 = \frac{1}{\alpha}$$

$$\Rightarrow \quad \frac{1}{\alpha} = \frac{1+\beta}{\beta}$$

$$\Rightarrow \quad \alpha = \frac{\beta}{1+\beta}$$

8. *(c)* We have, $\lambda = \frac{hc}{\Delta E}$

So, ratio of wave lengths

$$\frac{\lambda_1}{\lambda_2} = \frac{hc/\Delta E_1}{hc/\Delta E_2} = \frac{\Delta E_2}{\Delta E_1}$$

$$= \frac{\left(\frac{4}{3}E - E\right)}{2E - E} = \frac{1}{3}$$

9. *(a)* Decay scheme is ,

Given, $\frac{N_B}{N_A} = 0.3 = \frac{3}{10}$

$\Rightarrow \quad \frac{N_B}{N_A} = \frac{30}{100}$

So, $\quad N_0 = 100 + 30 = 130$ atoms

By using $\quad N = N_0 e^{-\lambda t}$

We have, $\quad 100 = 130 e^{-\lambda t}$

$\Rightarrow \quad \frac{1}{1.3} = e^{-\lambda t}$

$\Rightarrow \quad \log 1.3 = \lambda t$

If T is half-life, then $\lambda = \frac{\log_e 2}{T}$

$\Rightarrow \quad \log 1.3 = \frac{\log_e 2}{T} \cdot t$

$\therefore \quad t = \frac{T \cdot \log (1.3)}{\log_e 2}$

10. *(d)* Frequency spectrum of modulated wave is

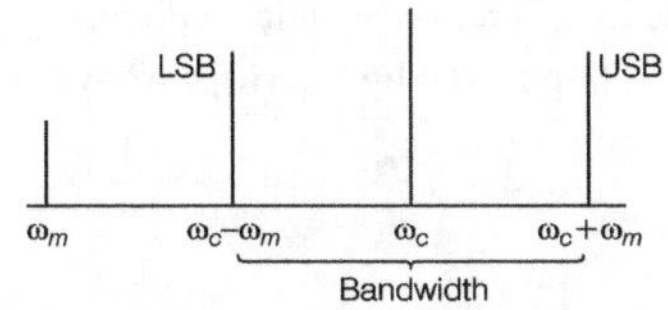

Clearly, ω_m is not included in the spectrum.

11. *(c)*

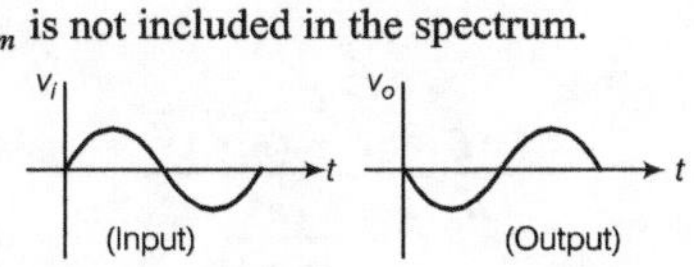

In a CE *npn* transistor amplifier output is 180° out of phase with input.

12. *(d)* $\lambda_{\min} = \frac{hc}{eV}$

$\log (\lambda_{\min}) = \log \left(\frac{hc}{e}\right) - \log V \quad y = c - mx$

So, the required graph is given in option (d).

13. *(a)* For elastic collision,

$$p_{\text{before collision}} = p_{\text{after collision}}.$$

$$mv = mv_A + \frac{m}{2} v_B$$

$$2v = 2v_A + v_B \qquad \text{...(i)}$$

Now, coefficient of restitution,

$$e = \frac{v_B - v_A}{u_A - v_B}$$

Here, $u_B = 0$ (Particle at rest) and for elastic collision $e = 1$

$\therefore \quad 1 = \frac{v_B - v_A}{v}$

$\Rightarrow \quad v = v_B - v_A \qquad \text{...(ii)}$

From Eq. (i) and Eq. (ii)

$$v_A = \frac{v}{3}$$

and $\quad v_B = \frac{4v}{3}$

Hence, $\quad \frac{\lambda_A}{\lambda_B} = \frac{\left(\frac{h}{mV_A}\right)}{\frac{h}{\frac{m}{2}.V_B}}$

$$= \frac{V_B}{2V_A} = \frac{4/3}{2/3} = 2$$

14. *(a)* If wavelength of emitted photon in de-excitation is Λ_n;

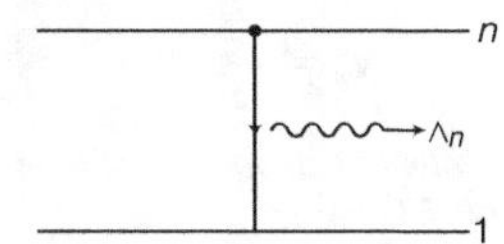

Then, $\quad \frac{hc}{\Lambda_n} = E_n - E_g$

$$\frac{hc}{\Lambda_n} = \frac{p_n^2}{2m} - \frac{p_g^2}{2m}, \qquad \left[\because E = \frac{p^2}{2m}\right]$$

As energies are negative, we get

$$\frac{hc}{\Lambda_n} = \frac{p_g^2}{2m} - \frac{p_n^2}{2m}$$

$$\Rightarrow \quad = \frac{p_g^2}{2m}\left(1 - \left(\frac{p_n}{p_g}\right)^2\right)$$

$$= \frac{h^2}{2m\lambda_g^2}\left(1 - \frac{\lambda_g^2}{\lambda_n^2}\right) \qquad \left[\because p \propto \lambda^{-1}, p = \frac{h}{\lambda}\right]$$

$$\Rightarrow \quad \Lambda_n = \frac{2m\lambda_g^2 c}{h}\left(1 - \frac{\lambda_g^2}{\lambda_n^2}\right)^{-1}$$

$$\Rightarrow \quad \Lambda_n = \frac{2m\lambda_g^2 c}{h}\left(1 + \frac{\lambda_g^2}{\lambda_n^2}\right) \qquad [\because (1 - x)^{-n} = 1 + nx]$$

$$\Rightarrow \quad \Lambda_n \simeq A + \frac{B}{\lambda_n^2}$$

where, $\quad A = \left[\frac{2mc\lambda_g^2}{h}\right]$

and $\quad B = \left[\frac{2mc\lambda_g^4}{h}\right]$ are constants.

15. *(d)* Series limit occurs in the transition $n_2 = \infty$ to $n_1 = 1$ in Lyman series and $n_2 = \infty$ to $n_1 = 5$ in *p*fund series.

For Lyman series,

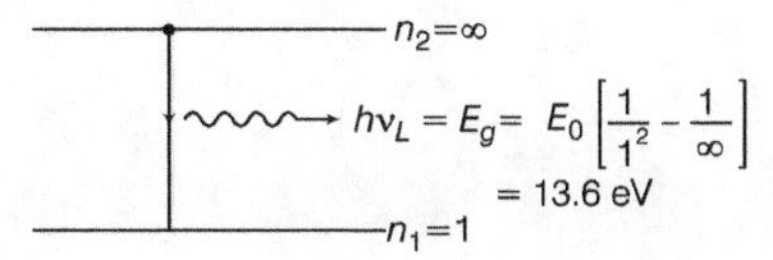

$$h\nu_L = 13.6 \qquad \text{...(i)}$$

In *pfund* series

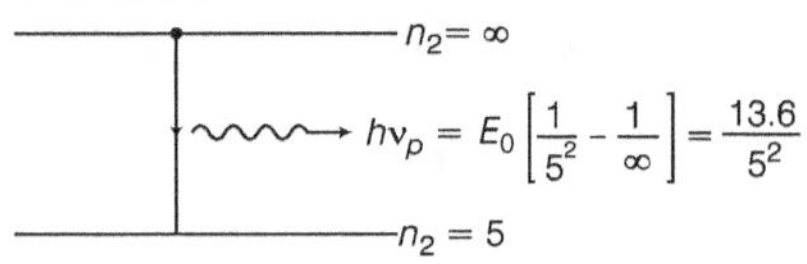

$$h\nu_p = \frac{13.6}{5^2} \quad \ldots\text{(ii)}$$

From Eqs. (i) and (ii), we get

$$25h\nu_p = h\nu_L$$

$$\therefore \quad \nu_p = \frac{\nu_L}{25}$$

16. *(c)* Key Idea *Potential drop in a silicon diode in forward bias is around 0.7 V.*

In given circuit, potential drop across 200 Ω resistor is

$$I = \frac{\Delta V_{\text{net}}}{R} = \frac{3 - 0.7}{200}$$

$$\Rightarrow \quad I = 0.0115\text{A}$$

$$\Rightarrow \quad I = 11.5 \text{ mA}$$

17. *(c)* Only 10% of 10 GHz is utilised for transmission.

∴ Band available for transmission

$$= \frac{10}{100} \times 10 \times 10^9 \text{ Hz} = 10^9 \text{ Hz}$$

Now, if there are n channels each using 5 kHz then,

$n \times 5 \times 10^3 = 10^9$

$$\Rightarrow \quad n = 2 \times 10^5$$

18. *(c)* Key Idea *Speed of progressive wave is given by,* $v = \frac{\omega}{k}$

As electric field in air is,

$$E_1 = E_{01} \textbf{ x} \cos\left(\frac{2\pi\nu z}{c} - 2\pi\nu t\right)$$

$$\therefore \quad \text{Speed in air} = \frac{2\pi\nu}{\left(\frac{2\pi\nu}{c}\right)} = c$$

$$\text{Also,} \quad c = \frac{1}{\sqrt{\mu_0 \varepsilon_{r_1} \varepsilon_0}} \quad \ldots\text{(i)}$$

In medium, $E_2 = E_{02}\textbf{x} \cos(2kz - kct)$

∴ Speed in medium $= \frac{kc}{2k} = \frac{c}{2}$

$$\text{Also,} \quad \frac{c}{2} = \frac{1}{\sqrt{\mu_0 \varepsilon_{r_2} \varepsilon_0}} \quad \ldots\text{(ii)}$$

As medium is non-magnetic medium, $\mu_{\text{medium}} = \mu_{\text{air}}$

On dividing Eq. (i) by Eq. (ii), we have

$$2 = \sqrt{\frac{\varepsilon_{r_2}}{\varepsilon_{r_1}}} \Rightarrow \frac{\varepsilon_{r_1}}{\varepsilon_{r_2}} = \frac{1}{4}$$

JEE Advanced

1. *(b)* $\frac{hc}{\lambda} - \phi = eV_0$ $\quad$ (ϕ = work function)

$$\frac{hc}{0.3 \times 10^{-6}} - \phi = 2e \quad \ldots\text{(i)}$$

$$\frac{hc}{0.4 \times 10^{-6}} - \phi = 1e \quad \ldots\text{(ii)}$$

Subtracting Eq. (ii) from Eq. (i)

$$hc\left(\frac{1}{0.3} - \frac{1}{0.4}\right)10^6 = e$$

$$hc\left(\frac{0.1}{0.12} \times 10^6\right) = e$$

$$h = 0.64 \times 10^{-33}$$

$$= 6.4 \times 10^{-34} \text{ J-s}$$

2. *(a,b,d)* As radius $r \propto \frac{n^2}{z}$

$$\Rightarrow \quad \frac{\Delta r}{r} = \frac{\left(\frac{n+1}{z}\right)^2 - \left(\frac{n}{z}\right)^2}{\left(\frac{n}{z}\right)^2}$$

$$= \frac{2n+1}{n^2} \approx \frac{2}{n} \propto \frac{1}{n}$$

Energy $\quad E \propto \frac{z^2}{n^2}$

$$\Rightarrow \quad \frac{\Delta E}{E} = \frac{\frac{z^2}{n^2} - \frac{z^2}{(n+1)^2}}{\frac{z^2}{(n+1)^2}}$$

$$= \frac{(n+1)^2 - n^2}{n^2 \cdot (n+1)^2} \cdot (n+1)^2$$

$$\Rightarrow \quad \frac{\Delta E}{E} = \frac{2n+1}{n^2} \simeq \frac{2n}{n^2} \propto \frac{1}{n}$$

Angular momentum, $L = \frac{nh}{2\pi}$

$$\Rightarrow \quad \frac{\Delta L}{L} = \frac{\frac{(n+1)h}{2\pi} - \frac{nh}{2\pi}}{\frac{nh}{2\pi}} = \frac{1}{n} \propto \frac{1}{n}$$

3. **(6)** Energy of incident light (in eV)

$$E = \frac{12375}{970} = 12.7\text{eV}$$

After excitation, let the electron jumps to nth state, then

$$\frac{-13.6}{n^2} = -13.6 + 12.7$$

Solving this equation, we get

$$n = 4$$

∴ Total number of lines in emission spectrum,

$$= \frac{n(n-1)}{2} = \frac{4(4-1)}{2} = 6$$

4. **(9)** ${}^{12}_{5}B \longrightarrow {}^{12}_{6}C + {}^{0}_{-1}e + \bar{\nu}$

Mass of ${}^{12}_{6}C = 12.000$ u (by definition of 1 amu)

Q-value of reaction, $Q = (M_B - M_C) \times c^2$

$$= (12.014 - 12.000) \times 931.5$$
$$= 13.041 \text{ MeV}$$

4.041MeV of energy is taken by ${}^{12}_{6}C^*$

$\Rightarrow$ Maximum KE of β-particle is $(13.041 - 4.041) = 9$ MeV

5. *(c)* Using the relation

$$R = R_0\left(\frac{1}{2}\right)^n$$

Here, R is activity of radioactive substance, R_0 initial activity and n is number of half lives.

$$1 = 64\left(\frac{1}{2}\right)^n$$

Solving we get, $n = 6$

Now, $t = n(t_{1/2})$

$$= 6(18 \text{ days})$$
$$= 108 \text{ days}$$

6. *(c)* Electrostatic energy

$$= \text{Binding energy of N} - \text{Binding energy of O}$$
$$= [(7M_H + 8M_n - M_N)] - [(8M_H + 7M_n - M_O)] \times c^2$$
$$= [-M_H + M_n + M_O - M_N]c^2$$
$$= [-1.007825 + 1.008665 + 15.003065 - 15.000109] \times 931.5$$
$$= +3.5359 \text{ MeV}$$

$$\Delta E = \frac{3}{5} \times \frac{1.44 \times 8 \times 7}{R} - \frac{3}{5} \times \frac{1.44 \times 7 \times 6}{R}$$
$$= 3.5359 \text{ MeV}$$
$$R = \frac{3 \times 1.44 \times 14}{5 \times 3.5359} = 3.42 \text{ fm}$$

7. *(d)* $K_{max} = \dfrac{hc}{\lambda_{ph}} - \phi$

Kinetic energy of electron reaching the anode will be

$$K = \frac{hc}{\lambda_{ph}} - \phi + eV$$

Now, $\lambda_e = \dfrac{h}{\sqrt{2mK}}$

$$= \frac{h}{\sqrt{2m\left(\dfrac{hc}{\lambda_{ph}} - \phi + eV\right)}}$$

If $eV >> \phi$, $\lambda_e = \dfrac{h}{\sqrt{2m\left(\dfrac{hc}{\lambda_{ph}} + eV\right)}}$

If $V_f = 4V_i$, $(\lambda_e)_f \simeq \dfrac{(\lambda_e)_i}{2}$

8. *(5)* Potential energy of hydrogen atom ($Z = 1$) in nth orbit (in eV)

$$\text{P, E} = -\frac{27.2}{n^2}$$

$$\frac{v_f}{v_i} = \frac{-\dfrac{27.2}{n_f^2}}{-\dfrac{27.2}{n_i^2}} = \frac{1}{6.25}$$

$$6.25 = \frac{n_f^2}{n_i^2}$$

$$\frac{n_f}{n_i} = 2.5 = \frac{5}{2}$$

Hence the answer is 5.

9. *(5)* $I^{131} \xrightarrow[T_{1/2} = 8 \text{ Days}]{} Xe^{131} + \beta$

$$A_0 = 2.4 \times 10^5 \text{ Bq} = \lambda N_0$$

Let the volume is V,

$$t = 0$$
$$A_0 = \lambda N_0$$
$$t = 11.5 \text{ h } A = \lambda N$$
$$115 = \lambda\left(\frac{N}{V} \times 2.5\right)$$
$$115 = \frac{\lambda}{V} \times 2.5 \times (N_0 e^{-\lambda t})$$
$$115 = \frac{(N_0\lambda)}{V} \times (2.5) \times e^{-\frac{\ln 2}{8 \text{ day}}(11.5 \text{ h})}$$
$$115 = \frac{(2.4 \times 10^5)}{V} \times (2.5) \times e^{-1/24}$$
$$V = \frac{2.4 \times 10^5}{115} \times 2.5\left[1 - \frac{1}{24}\right]$$
$$= \frac{2.4 \times 10^5}{115} \times 2.5\left[\frac{23}{24}\right]$$
$$= \frac{10^5 \times 23 \times 25}{115 \times 10^2}$$
$$= 5 \times 10^3 \text{ ml} = 5 \text{ L}$$

10. *(d)* According to photoelectric effect equation

$$KE_{max} = \frac{hc}{\lambda} - \phi_0$$

$$\frac{p^2}{2m} = \frac{hc}{\lambda} - \phi_0 \qquad [KE = p^2/2m]$$

$$\frac{(h/\lambda_d)^2}{2m} = \frac{hc}{\lambda} - \phi_0 \qquad [p = h/\lambda]$$

Assuming small changes, differentiating both sides,

$$\frac{h^2}{2m}\left(-\frac{2d\lambda_d}{\lambda_d^3}\right) = -\frac{hc}{\lambda^2}d\lambda$$

$$\frac{d\lambda_d}{d\lambda} \propto \frac{\lambda_d^3}{\lambda^2}$$

11. *(a, c)* ${}^{232}_{90}Th$ is converting into ${}^{212}_{82}Pb$.

Change in mass number $(A) = 20$

$\therefore$ Number of α-particle emitted $= \frac{20}{4} = 5$

Due to 5 α-particles, Z will change by 10 units.
Since, given change is 8, therefore number of β-particles emitted is 2.

12. *(24)* $\therefore$ Power $= nhf$

(where, n = number of photons incident per second)

Since, KE = 0, hf = work-funcition W

$$200 = nW = n[6.25 \times 1.6 \times 10^{-19}]$$

$$\Rightarrow \quad n = \frac{200}{1.6 \times 10^{-19} \times 6.25}$$

As photon is just above threshold frequency KE_{max} is zero and they are accelerated by potential difference of 500 V.

$$\therefore \quad KE_f = q\Delta V$$

$$\frac{P^2}{2m} = q\Delta V$$

$$\Rightarrow \quad P = \sqrt{2mq\Delta V}$$

Since, efficiency is 100%, number of electrons emitted per second = number of photons incident per second.
As, photon is completely absorbed, force exerted

$$= n(mV) = nP = n\sqrt{2mq\Delta V}$$

$$= \frac{200}{6.25 \times 1.6 \times 10^{-19}} \times \sqrt{2(9 \times 10^{-31}) \times 1.6 \times 10^{-19} \times 500}$$

$$= 24$$

13. *(3)* $\Delta E_{2-1} = 13.6 \times Z^2 \left[1 - \frac{1}{4}\right]$

$$= 13.6 \times Z^2 \left[\frac{3}{4}\right]$$

$$\Delta E_{3-2} = 13.6 \times Z^2 \left[\frac{1}{4} - \frac{1}{9}\right]$$

$$= 13.6 \times Z^2 \left[\frac{5}{36}\right]$$

$$\therefore \quad \Delta E_2 = \Delta E_{3-2} + 74.8$$

$$13.6 \times Z^2 \left[\frac{3}{4}\right] = 13.6 \times Z^2 \left[\frac{5}{36}\right] + 74.8$$

$$13.6 \times Z^2 \left[\frac{3}{4} - \frac{5}{36}\right] = 74.8$$

$$Z^2 = 9$$

$$\therefore \quad Z = 3$$

www.ingramcontent.com/pod-product-compliance
Lightning Source LLC
Chambersburg PA
CBHW080211170726
48282CB00030B/1080
* 9 7 8 9 3 1 3 1 9 3 3 9 5 *